Acoustic Communication in Birds

Volume 2

COMMUNICATION AND BEHAVIOR

AN INTERDISCIPLINARY SERIES

Under the Editorship of ***Duane M. Rumbaugh,***
Georgia State University and Yerkes Regional Primate Research Center of Emory University

DUANE M. RUMBAUGH (ED.), LANGUAGE LEARNING BY A CHIMPANZEE: THE LANA PROJECT, 1977

ROBERTA L. HALL AND HENRY S. SHARP (EDS.), WOLF AND MAN: EVOLUTION IN PARALLEL, 1978

HORST D. STEKLIS AND MICHAEL J. RALEIGH (EDS.), NEUROBIOLOGY OF SOCIAL COMMUNICATION IN PRIMATES: AN EVOLUTIONARY PERSPECTIVE, 1979

P. CHARLES-DOMINIQUE, H. M. COOPER, A. HLADIK, C. M. HLADIK, E. PAGES, G. F. PARIENTE, A. PETTER-ROUSSEAUX, J. J. PETTER, AND A. SCHILLING (EDS.), NOCTURNAL MALAGASY PRIMATES: ECOLOGY, PHYSIOLOGY, AND BEHAVIOR, 1980

JAMES L. FOBES AND JAMES E. KING (EDS.), PRIMATE BEHAVIOR, 1982

DONALD E. KROODSMA AND EDWARD H. MILLER (EDS.), ACOUSTIC COMMUNICATION IN BIRDS, VOLUME 1: PRODUCTION, PERCEPTION, AND DESIGN FEATURES OF SOUNDS, 1982. VOLUME 2: SONG LEARNING AND ITS CONSEQUENCES, 1982

Acoustic Communication in Birds

Volume 2

Song Learning and Its Consequences

Edited by

DONALD E. KROODSMA
Department of Zoology
University of Massachusetts
Amherst, Massachusetts

EDWARD H. MILLER
Vertebrate Zoology Division
British Columbia Provincial Museum
Victoria, British Columbia, Canada

Taxonomic Editor
HENRI OUELLET
Vertebrate Zoology Division
Museum of Natural Sciences
National Museums Canada
Ottawa, Ontario, Canada

1982

ACADEMIC PRESS
A Subsidiary of Harcourt Brace Jovanovich, Publishers

New York London
Paris San Diego San Francisco São Paulo Sydney Tokyo Toronto

ACADEMIC PRESS, INC.
111 Fifth Avenue, New York, New York 10003

United Kingdom Edition published by
ACADEMIC PRESS, INC. (LONDON) LTD.
24/28 Oval Road, London NW1 7DX

Library of Congress Cataloging in Publication Data
Main entry under title:

Acoustic communication in birds.

(Communication and behavior)
Includes bibliographical references and index.
1. Bird-song. 2. Animal communication. I. Kroodsma, Donald E. II. Miller, Edward H. III. Series.
QL698.5.A26 598.2'59 82-6736
ISBN 0-12-426802-1 (v. 2) AACR2

PRINTED IN THE UNITED STATES OF AMERICA

82 83 84 85 9 8 7 6 5 4 3 2 1

The contributors dedicate these volumes to Peter Marler. Peter's research and writings have placed the study of animal communication in the fore of modern evolutionary and ecological thought in ethology, and his impact is apparent from the sheer number and diversity of citations he receives in these two volumes. Research into almost any facet of bird acoustics puts one into contact with his work, on topics such as learning, geographic variation, species-specificity, "design features," individual variation, and grading. In addition, Peter has touched many workers individually, as graduate students, postdoctoral researchers, and peers, whereby his influence upon research and thinking on animal communication has carried yet further.

We are sure that this treatise reflects Peter Marler's influence upon the field of avian acoustics, and hope that it also reflects our esteem and respect for him as a scientist, teacher, and colleague.

Contents

6 Microgeographic and Macrogeographic Variation in Acquired Vocalizations of Birds

PAUL C. MUNDINGER

7 Genetic Population Structure and Vocal Dialects in *Zonotrichia* (Emberizidae)

MYRON CHARLES BAKER

8 Individual Recognition by Sound in Birds

J. BRUCE FALLS

9 Conceptual Issues in the Study of Communication

COLIN G. BEER

Contributors

Numbers in parentheses indicate the pages on which the authors' contributions begin.

Myron Charles Baker (209), Department of Zoology and Entomology, Colorado State University, Fort Collins, Colorado 80523

Jeffrey R. Baylis (51, 311), Department of Zoology, University of Wisconsin, Madison, Wisconsin 53706

Colin G. Beer (279), Institute of Animal Behavior, Rutgers University, Newark, New Jersey 07102

J. Bruce Falls (237), Department of Zoology, University of Toronto, Toronto, Ontario M5S 1A1, Canada

Susan M. Farabaugh (85), Department of Zoology, University of Maryland, College Park, Maryland 20742, and Department of Zoological Research, National Zoological Park, Smithsonian Institution, Washington, D.C. 20008

Donald E. Kroodsma (1, 125, 311), Department of Zoology, University of Massachusetts, Amherst, Massachusetts 01003

Peter R. Marler (25), Rockefeller University Field Research Center, Millbrook, New York 12545

Paul C. Mundinger (147), Department of Biology, Queens College of the City University of New York, Flushing, New York 11367

Susan Peters (25), Rockefeller University Field Research Center, Millbrook, New York 12545

Foreword

It is entirely appropriate that these volumes of collected essays on aspects of the study of bird vocalizations should be dedicated to Peter Marler. While the scientific study of bird vocalizations was initiated by W. H. Thorpe, its subsequent expansion and elaboration was due primarily to Peter Marler.

W. H. Thorpe was initially an entomologist. Finding that some aspects of insect behavior were more labile than had formerly been supposed, he came to feel that the most pressing problems in the study of animal behavior concerned the interface between "instinct," as it was then called, and learning. As an amateur ornithologist, he realized that birds provided exceptionally suitable material for this work. Birds have a repertoire of relatively stereotyped movement patterns and yet at the same time exhibit marked learning ability. Accordingly, in 1950 he set up "An Ornithological Field Station" (now called the Sub Department of Animal Behaviour) at Madingley, Cambridge. Knowing that Chaffinch (*Fringilla coelebs*) song was subject to at least some individual variation and some flexibility (e.g., Poulsen, 1951), he decided to study the ontogeny of Chaffinch vocalization, and initiated a program of hand-rearing Chaffinches and studying their song development after varying exposure to the species' song (Thorpe, 1963).

Peter Marler was the first graduate student at Madingley. When he came, he already had a Ph.D. in botany and a tenured job. It must have been a decision of considerable courage to give up that job in order to take a Ph.D. in a subject that really interested him. His Ph.D. was a field study of the Chaffinch: he wanted to work in the field and the choice of the Chaffinch as subject meshed well with Thorpe's own work. At that time I was studying the courtship of captive Chaffinches and I am sure that Peter will remember the long discussions, and sometimes disputes, that we had over whether this or that posture should be called the lopsided wings-drooped posture or something a bit snappier—issues that seemed terribly important to us then!

Peter stayed on as a postdoctoral worker and took on the song learning work. His training as an all-round biologist stood him in good stead, and at that time he became interested not only in the ontogenetic problem that Thorpe was studying—how does Chaffinch song develop—but also in the other three questions with which ethologists are concerned—those of causation, function, and evolution. When he subsequently moved to Berkeley, he took the song problem with him. Changing his British speech for what seemed to some of us a rather extreme American accent, and the Chaffinch for White-crowned Sparrows (*Zonotrichia leucophrys*) and juncos, he started his well-known studies of ontogeny. At around that time the study of bird song really took off and it can now be regarded as having produced material of crucial importance for all four major questions of ethology.

Taking the ontogenetic question first, three fundamentally important principles have been established largely or entirely through work on bird song. The principle of sensitive periods in development, though apparent from earlier work on imprinting and other phenomena, owes a great deal to studies of bird song. The fact that what an animal learns is constrained in part by its species, though again an idea coming also from other sources of evidence, was established most firmly in the 1950s and early 1960s by the work on bird song. And third, the view that the elaboration and perfection of the song pattern depend on comparison between the vocal output and a previously established template casts a new light on many aspects of ontogeny. In all of these issues Peter Marler and his colleagues played a leading role. Understanding the processes of song development was facilitated by comparisons between studies of different species (e.g., Immelmann, 1969; Konishi, 1964; Nicolai, 1956), and Marler himself used a comparative approach to good effect, relating plasticity to the ecology of the species (Marler, 1967).

Marler's field studies of the Chaffinch had inevitably alerted him to problems concerned with the causation of bird song, and his thesis contains a great deal of observational material on the factors determining when a Chaffinch sings. He soon became interested also in the patterning of bird song, and was one of the first people to study the detailed sequencing of different song types in an individual's vocal output (e.g., Isaac and Marler, 1963). However, a more detailed study of the neural mechanisms underlying bird song arose from one of the findings of the studies of song ontogeny. Domestic Fowl (*Gallus domesticus*) and Ringed Turtle-Doves (*Streptopelia risoria*) were found to be capable of developing all the normal species' vocal signals, even though deafening took place soon after hatching. On the other hand, Konishi and Nottebohm (e.g., 1969) found that, in a number of species of songbirds, deafening before full song had developed resulted in consistent abnormalities in their song and often in a regression to a rather amorphous type of vocal output. However, birds that had already learned to sing could continue singing after deafening. Since it was known that the control of human speech is influenced by the speaker's perception of himself speaking, the finding that deaf birds could continue to sing normally was surprising. One possibility was that the deafened bird was using feedback from the muscles. Nottebohm therefore investigated the effects of severing the hypoglossal nerves in the syrinx. This has led to the discovery of laterality in the motor control of bird vocalizations, and what is more of a partially reversible laterality. Furthermore, it has led to a detailed investigation of the brain mechanisms underlying bird song, and has provided us with quite new data on the role of hormones in the development of neural mechanisms (e.g., Nottebohm, 1980).

Marler's training as an all-round naturalist led him early on to ask functional questions. He was concerned with the functions of different calls in the Chaffinch's repertoire. What information did each carry? This led him into an attempt to categorize the nature of the information carried in animal signals (Marler, 1961). While not everyone will agree with the view that natural selection always acts to enhance the effectiveness of signals in transmitting information about the signaler, Marler's formulations greatly facilitated the study of communication.

He was also concerned with the diversity of avian vocalizations. Noticing that the alarm calls of different species tended to resemble each other, while the songs were very different, he speculated about the selective factors controlling the form of avian vocalizations—selective factors that included the optimal degree of audibility in the habitat in question as well as the particular response to be elicited in the responding individual (e.g., Marler, 1955). In addition, in comparing the ontogeny of different avian species he was forced to ask why learning plays a much greater part in some species than others, why the sensitive period occurs at different ages in different species, why learning is constrained in one way in some species and in another way in others, and so on. We do not yet know the answers to all these questions, but it was Marler's pioneering work in the early 1950s that posed them.

Finally, and inevitably, functional questions led to questions about the course of evolution of avian vocalizations. Peter Marler was concerned with this problem at a very early stage: as a student, he went on an expedition to the Canary Islands and became concerned with the differences between the songs of the species found there and their mainland counterparts (Marler and Boatman, 1951). His work on bird ontogeny led him into questions of the relations between song and call notes, and the functional questions he raised were inevitably linked with evolutionary questions about how song evolved.

While Peter Marler's work on avian vocalizations has led to progress in answering all four of the major questions in which ethologists are interested, that is not all. Perhaps influenced by Thorpe, who was a pioneer in emphasizing the importance of perceptual processes at a time when most ethologists were thinking in relatively mechanistic terms and were not yet sensitive to his suggestions, Peter Marler became interested at an early stage in the relations between the physical structure of bird vocalization and their quality as perceived by the recipient (e.g., Marler, 1969). And while outside the scope of this volume, Marler's more recent work on primate vocalizations, and his emphasis on common features between avian communication and human language (Marler, 1970) has done much to stimulate research.

I would suggest that the study of bird song is the example *par excellence* of the ethological approach. It involves the study of a naturally occurring pattern of behavior against a background of the natural history of the species concerned, but employs an experimental methodology. It involves questions about the causation, ontogeny, function, and evolution of the pattern in question, questions that are at the same time independent and interfertile. It is probably true to say that the study of bird song has done as much for the advancement of ethology as the study of any other specific aspect of behavior. In this, Peter Marler has played a major role.

Robert A. Hinde

REFERENCES

Immelmann, K. (1969). Song development in the Zebra Finch and other estrilidid finches. *In* "Bird Vocalizations. Their Relation to Current Problems in Biology and Psychology" (R. A. Hinde, ed.), pp. 61–74. Cambridge Univ. Press, London and New York.

Isaac, D., and Marler, P. (1963). Ordering of sequences of singing behaviour of Mistle Thrushes in relation to timing. *Anim. Behav.* **11,** 179–188.

Konishi, M. (1963). The role of auditory feedback in the vocal behavior of the Domestic Fowl. *Z. Tierpsychol.* **20,** 349–367.

Konishi, M. (1964). Effects of deafening on song development in two species of juncos. *Condor* **66,** 85–102.

Konishi, M., and Nottebohm, F. (1969). Experimental studies in the ontogeny of avian vocalizations. *In* "Bird Vocalizations. Their Relation to Current Problems in Biology and Psychology" (R. A. Hinde, ed.), pp. 29–48. Cambridge Univ. Press, London and New York.

Marler, P. (1955). Characteristics of some animal calls. *Nature (London)* **176,** 6.

Marler, P. (1961). The logical analysis of animal communication. *J. Theoret. Biol.* **1,** 295–317.

Marler, P. (1967). Comparative study of song development in sparrows. *Proc. 14th Int. Ornith. Cong. Oxford,* pp. 213–244, Blackwell, Oxford.

Marler, P. (1969). Tonal quality of bird sounds. *In* "Bird Vocalizations. Their Relation to Current Problems in Biology and Psychology" (R. A. Hinde, ed.), pp. 5–18. Cambridge Univ. Press, London and New York.

Marler, P. (1970). Birdsong and speech development: could there be parallels? *Am. Sci.* **58** (6), 669–673.

Marler, P., and Boatman, D. J. (1951). Observations on the birds of Pico, Azores. *Ibis* **93,** 90–99.

Nicolai, J. (1956). Zur Biologie und Ethologie des Gimpels (*Pyrrhula pyrrhula* L.). *Z. Tierpsychol.* **13,** 93–132.

Nottebohm, F. (1980). Brain pathways for vocal learning in birds. A review of the first 10 years. *Prog. in Psychobiol. and Physiol. Psychol.* **9,** 85–124.

Poulsen, H. (1951). Inheritance and learning in the song of the Chaffinch (*Fringilla coelebs*). *Behaviour* **3,** 216–228.

Thorpe, W. H. (1963). "Learning and Instinct in Animals." Methuen, London.

Preface

We began corresponding about co-editing a book on bird sounds in 1978, excited by the enormous increase in evolutionary understanding and interpretation of communication systems since Robert Hinde's edited volume, *Bird Vocalizations,* appeared in 1969. The rapid mushrooming and splintering of ideas and observations on animal communication were daunting, but we nevertheless shared the belief that a representative and useful collection of writings on evolution and ecology of bird acoustics could be assembled. Our original intent was to compile both taxonomic and conceptual reviews, but the meager knowledge of acoustic signals of many important avian taxa made the first of these impossible. Consequently, we solicited contributions from active researchers in bird acoustics, for chosen areas of evolutionary and behavioral ecology, and sought to complement them with reviews of sound recording techniques, and sound production, reception, and processing.

Volume 1 begins with several of these background chapters. The first discusses sound recording, makes certain recommendations, and points out common errors. The others outline some of the complex events and processes between sound production and behavioral response to sound. The remaining chapters stand apart from the first ones, a gap that accurately reflects our poor understanding of the processes which ultimately link individual physiological responses to population-genetical changes, and to larger-scale evolutionary trends. Bridging this gap will require diverse research endeavors and theorizing, some of which are touched on in Chapters 5–9 of Volume 1 and Chapters 1–9 of Volume 2.

The chapters are varied. They range from lengthy, well balanced, detailed syntheses of subjects such as coding of species-specificity (Becker), individuality (Falls), and environmental acoustics (Wiley and Richards), etc., to briefer more partisan explications of ideas and observations on dialects (Baker), subsong (Marler and Peters), sexual selection (Catchpole), "motivation–structural rules" (Morton), etc. Others present novel views and reviews on doggedly troublesome concepts such as duetting (Farabaugh), vocal mimicry (Baylis), and geographic variation (Mundinger). We have contributed chapters on ontogeny, the evolution of complex vocalizations, and character shift, and Beer wrote a general chapter on conceptual issues relevant to bird acoustics. Despite the inevitable deficiencies and uneven style and coverage in a volume of this sort, we feel that many important, current topics are refreshingly well reviewed, and that these volumes will serve as a stimulating and valuable reference for students of communication and bioacoustics. We thank the authors for their care and labor; Henri Ouellet, for assuming the unrewarding role of taxonomic editor; and Robert Hinde, for writing the Foreword.

Donald E. Kroodsma
Edward H. Miller

Note on Taxonomy

The editors, acknowledging the current unstable state of avian taxonomy and nomenclature, invited me to read the manuscript in order to ensure uniformity in taxonomic and nomenclatural usage. This proved to be particularly appropriate because none of the contributors is a taxonomist and because the sources used in this work are very diverse and cover an extensive time span.

Scientific names were nearly all standardized here after the *Reference List of the Birds of the World* (Morony *et al.*, 1975); this reference is readily available, even to the nonornithologists, includes nearly all the species currently recognized, and is the most up-to-date reference of its kind. Domestic or laboratory birds have usually been referred to their wild counterpart with the exception of "the chicken," which has been designated here as "Domestic Fowl" and "*Gallus domesticus*" because this work addresses itself to a broader audience than ornithologists. However, the Canary remains *Serinus canaria* and the Turkey *Meleagris gallopavo,* but the quail has become the Common Quail *Coturnix coturnix* and the Ring Dove the Ringed Turtle-Dove *Streptopelia risoria*. Otherwise Morony *et al.* (1975) have been followed closely, although I have been inclined in several instances to diverge from it.

For English names a variety of sources has been used. The A.O.U. Checklist (1957) and its supplements (1973, 1976) were used as the standard reference for all Holarctic and Neotropical species appearing in the checklist. I have referred to Voous (1973, 1977) for Palearctic species, to Hall and Moreau (1970) and Snow (1978) for Africa, to Ridgeley (1976) for Central America, and to Meyer de Schauensee (1970) for South America. In the few instances where Asiatic and Australian species were mentioned, I have referred to a variety of sources and had to use my best judgment.

Henri Ouellet

REFERENCES

American Ornithologists' Union. (1957). "Check-list of North American Birds," 5th ed. Lord Baltimore Press, Baltimore.

American Ornithologists' Union. (1973). Thirty-second supplement to the American Ornithologists' Union check-list of North American birds. *Auk* **90,** 411–419.

American Ornithologists' Union. (1976). Thirty-third supplement to the American Ornithologists' Union check-list of North American birds. *Auk* **93,** 875–879.

Hall, B. P., and Moreau, R. E. (1970). "An Atlas of Speciation in African Passerine Birds." Trustees of the British Museum (Natural History), London.

Meyer de Schauensee, R. (1970). "A Guide to the Birds of South America." Livingston, Wynnewood, Pennsylvania.

Morony, J. J., Jr., Bock, W. J., and Farrand, J., Jr. (1975). "Reference List of the Birds of the World." American Museum of Natural History, New York.

Ridgeley, R. S. (1976). "A Guide to the Birds of Panama." Princeton Univ. Press, Princeton, New Jersey.

Snow, D. W. (1978). "An Atlas of Speciation in African Non-Passerine Birds." Trustees of the British Museum (Natural History), London.

Voous, K. H. (1973, 1977). List of Recent Holarctic bird species. *Ibis* **115,** 612–638; **119,** 223–250, 376–406.

Introduction

Natural historians study four major aspects of animal behavior: phylogeny, causation, ontogeny, and adaptation (Hailman, 1977). These approaches have imparted the evolutionary flavor to ethology since its inception, so that ethology has come to be viewed as a truly biological discipline. Contributions to this self-conception have been diverse and numerous, from Darwin's writings on expressive behavior, to current mathematical renderings of optimal foraging. With the exception of phylogeny, the main concerns of ethologists have likewise been central to research in avian acoustics, as is evident from these volumes. In the following comments, we discuss some issues in the study of bird sounds, with particular reference to evolutionary considerations.

I. PHYLOGENY

The promise of ethology to systematics and phylogenetics remains largely unfulfilled, despite the long-lived comparative tradition in ethology (Lanyon, 1969; Mundinger, 1979). The tradition arose and is sustained mainly by studies on stereotyped motor patterns in visual displays of birds and fish (e.g., Lorenz, 1941; van Tets, 1965). It has proved less successful in research on taxa like mammals, in which senses like olfaction are so important, and in which movements and postures are less readily classified and quantified (e.g., Eisenberg, 1967; Golani, 1976). Comparative studies on bird sounds have not led to significant systematic advances for very different reasons, related to sound variability. There are several reasons for this. First, sounds can be emitted easily and cheaply, and thus can often assume numerous forms with little selective penalty (see comments below on nonadaptive evolution). In addition, sounds used over short distances often encode minor variations in the emitter's state, and need not identify the signaling species, so are often highly graded. Variability in sounds is increased further by vocal learning in many or most species, and by short- and long-term neuroendocrine effects in individuals. Finally, the physical simplicity of sound, and the extent to which anatomical and neuromuscular changes affect sound properties, can result in rapid divergence between homologous sounds of related taxa. The high variability consequent on these factors is daunting to taxonomic and phylogenetic research, and makes quantitative analyses very difficult. All in all, bird sounds make difficult material for systematics because of their numerous variations resulting from emitters' developmental, physical, genetic, and physiological characteristics.

Taxa (or call types) in which vocal learning is slight and in which species-specific characteristics are unimportant (as in short-range calls) offer the best material for systematics (Mundinger, 1979). Thus, contact calls are remarkably conservative in Anatidae, and their detailed study could resolve certain affinities

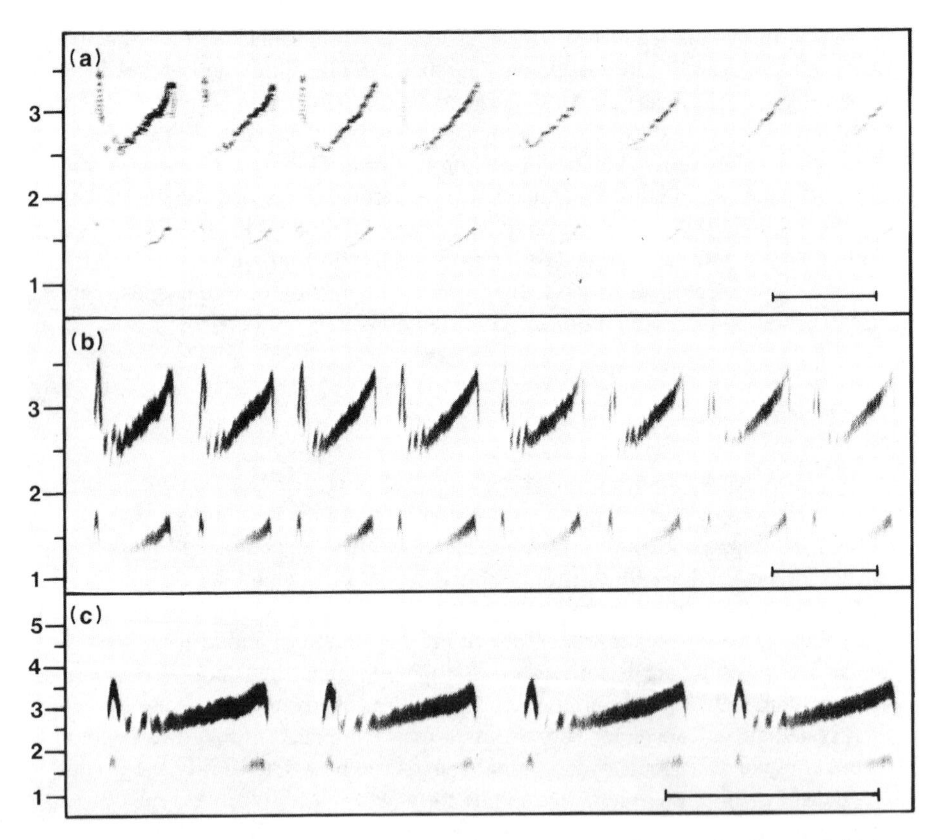

Fig. 1. Sound spectrograms of aerial display calls by a male Least Sandpiper (*Calidris minutilla*), showing no rhythmic FM and lack of pulsing. Part of a long rhythmic sequence is illustrated. Panel c corresponds to the first four calls of panels a and b, which are identical except for analyzing filter bandwidth (45 Hz for a, 300 Hz for b and c). The frequency scales are in kiloHertz; the time marker in the bottom right corner of each panel is 250 msec.

(Thielcke, 1970). Other call characteristics which should be evolutionarily conservative are those which conform with "motivation–structural rules" (see below).

The utility of acoustic characteristics in systematics depends on the recognition of shared derived homologies (Eldredge and Cracraft, 1980). Only through study of their distribution in taxa of interest can monophyletic groups be defined. For example, consider the Calidridinae, a subfamily of Scolopacidae which includes 24 closely related arctic and subarctic species. Unpaired males in these species employ simple and conspicuous aerial displays to simultaneously attract mates and repel competing males. Certain species, like Dunlin (*Calidris alpina*) and Baird's Sandpiper (*Calidris bairdii*), emit long series of rhythmically repeated buzzy (pulsed) calls. Others, like the Least Sandpiper (*Calidris minutilla*) and Stilt Sandpiper (*Micropalama himantopus*) emit remarkably similar series of calls which are presumably homologous, but the calls are never pulsed, and contain rhythmic frequency modulation at most (Figs. 1 and 2). The nonpulsed form is also present in many related, non-calidridine taxa, including Willets (*Catoptrophorus semipalmatus*) and curlews (*Numenius* species), so is presumably ancestral within the Calidridinae (Miller, 1982). Based on this characteristic then, *C. alpina* and *C. bairdii* share a common ancestor with one another more recently than they do with *C. minutilla* or *Micropalama*. By such reasoning, and using various characteristics, it is possible to construct a testable phylogeny. To date, only Mundinger (1979) has employed a cladistic approach in studying affinities of bird species based on acoustic characteristics.

II. CAUSATION

Several chapters in these two volumes deal with what Bates (1960) has termed "skin-in" biology. They concern sound production, neural control of song, and auditory perception (Chapters 2, 3, and 4 of Vol. 1). The latter area has received most attention from evolutionists, because of the prospect of detecting close adaptation of hearing abilities to species-specific sound characteristics. Such adaptation is not restricted to invertebrates and lower vertebrates, but is also known for certain mammals (Suga, 1978). Evidence for similar specificity in birds is weak, and in general species-typical frequency spectra correspond only approximately to hearing curves, while the latter differ little across species.

Neurophysiology lends itself less easily to evolutionary interpretation, because linkages of cause and effect are so difficult to trace. Some evidence points to unsuspectedly close adaptation even now, however. For example, male Long-billed Marsh Wrens (*Cistothorus palustris*) in nature learn about 50 and 150 songs in New York and San Francisco populations, respectively, and correlated

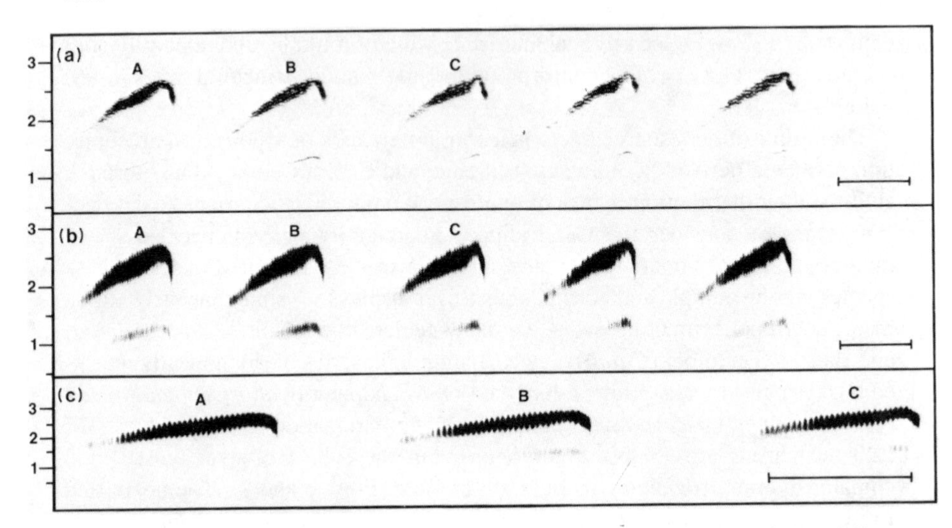

Fig. 2. Sound spectrograms of aerial display calls by a male Stilt Sandpiper (*Micropalama himantopus*), showing rhythmic FM and lack of pulsing. Part of a long rhythmic sequence is illustrated. Panel c corresponds to the first three calls of panels a and b, which are identical except for analyzing filter bandwidth (45 Hz for a, 300 Hz for b and c). The frequency scales are in kiloHertz; the time marker in the bottom right corner of each panel is 500 msec.

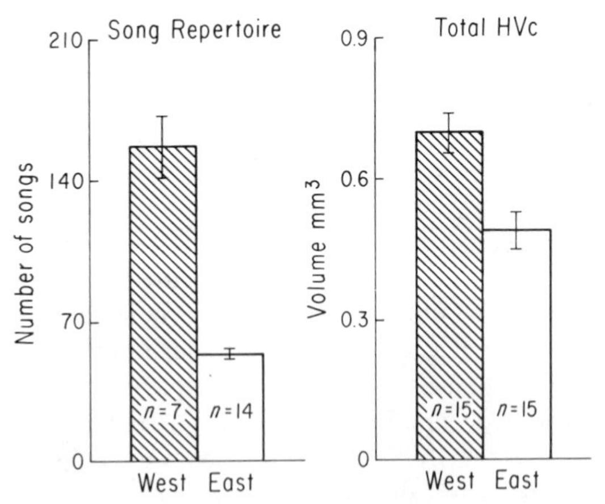

Fig. 3. Song repertoire size and total HVc (left plus right hemispheres—see Vol. 1, Chapter 3, for further discussion of these brain nuclei) for Long-billed Marsh Wrens from San Francisco, California (West) and Hudson River, New York (East) populations. Song repertoire sizes and the brain space devoted to singing are larger in Western populations (R. Canady, D. E. Kroodsma, F. Nottebohm, unpublished data).

with this 3:1 ratio in repertoire size is an approximate 3:2 ratio in the volume of several song control nuclei in the forebrain (Fig. 3). Laboratory experiments indicate that San Francisco males can learn more songs than New York males, a difference which is probably related to higher population density and overall intensity of competition for territories and mates in western populations (R. Canady, D. E. Kroodsma, and F. Nottebohm, unpublished data).

III. ONTOGENY

Most aspects of the ontogeny of acoustic behavior in birds are poorly known. Research has emphasized song learning in oscines, and little effort has been put into other groups (but see Andrew, 1969; Cosens, 1981; Wilkinson and Huxley, 1978). One reason for this bias is the relatively stereotyped, discrete nature of most oscine song, and the variable, highly graded nature of sounds in most other taxa (e.g., Huxley and Wilkinson, 1977; Mace, 1981; Mairy, 1979a,b). Few trends of broad evolutionary significance have emerged from studies of ontogeny yet, despite the documentation of widespread characteristics. For example, most young songbirds need models for normal song acquisition (see the Appendix of Vol. 2), and are maximally sensitive to appropriate models only during discrete time periods at a certain age. But why is song learning so significant for some sound types and not for others? Why is it so pronounced in oscines, and not in other groups? What roles have phylogeny and adaptation to local conditions played in the origin and maintenance of open developmental programs of song acquisition?

Vocal ontogenies have undoubtedly coevolved with other life history strategies, yet searches for correlations have yielded few generalizations. Further comparative work is needed, not only among closely related species, but also among populations of the same species and among individuals of the same population. The timing of sensitive periods for song learning may vary between migratory and nonmigratory populations, or between populations at different latitudes. Juveniles within a population may face different social and physical environments, and such environmental factors may influence the timing of vocal learning. Thus, Long-billed Marsh Wrens hatching early in the season experience longer days and hear more adult song than those hatching later in the season; birds hatching later can learn songs better their next spring, indicating that these environmental factors do have an important impact on vocal development (Kroodsma and Pickert, 1981). Much research on environmental influences and neuroendocrine bases of song development is needed before the evolutionary significance of different vocal ontogenies can be appreciated.

IV. ADAPTATION

When bird sounds are viewed as phenotypic components, their assessment as adaptations is sure to follow. There has been much recent study of adaptiveness of sound structure related to transmission in natural environments. This work has resulted in important findings which lead to prediction of optimal sound characteristics. For example, if maximal transmission distance is important, sounds should have characteristics which are resistant to attenuation and degradation. This prediction has been supported for long-distance signals even within species (Gish and Morton, 1981; Wasserman, 1979). More comprehensive studies suggest some problems with the assumption, though. Thus, Lemon *et al.* (1981) found that frequency characteristics of songs of 19 species of Parulidae were significantly related to singing height, in a manner contrary to prediction (Fig. 4). They comment (p. 1174) that perhaps "warblers are not trying to maximize the distance their songs carry but rather are optimizing the distance, sometimes greater, sometimes lesser, so as to communicate mainly with those individuals who are most significant to them biologically" (see also Krebs and Davies, 1981). Their suggestion can be generalized for different kinds of sounds, and not just long-distance ones: natural selection should promote the selective transmission of sound characteristics to intended receivers under average conditions; facultative responses may improve transmission, for example, by singing more

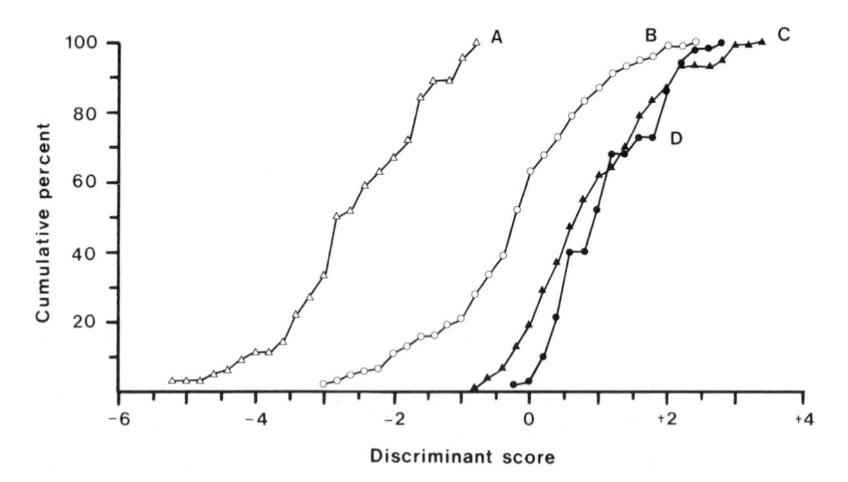

Fig. 4. Cumulative frequency plots of discriminant scores (from single function discriminant analysis) for songs of groups of warbler species (Parulidae) defined by singing height. For example, group A consists of four species with the highest mean singing height, group B contains six species, group C four species, and group D five species with lowest mean singing height. [After Fig. 1 of Lemon *et al.* (1981), using data provided by R. E. Lemon (*in litt.*).]

loudly or persistently when a receiver is known to be distant, or under noisy conditions, etc. Few situations must exist where maximal and optimal transmission distances are the same. In any case, since most displays have multiple functions, average adaptations and optimal solutions are probably the rule (Nuechterlein, 1981).

Another possibility to consider is that many or most spatio-temporal changes in sounds have no selective or adaptive significance (e.g., Wiens, 1982). P. J. B. Slater and colleagues have extensively documented factors influencing song features in the Chaffinch (*Fringilla coelebs*), and point out that immigration, learning errors, nonlearning of rare song types, and other factors can all lead to nonadaptive evolutionary change (e.g., Ince *et al.*, 1980; Slater and Ince, 1979). Various possible explanations for the origin of dialects in Japanese monkeys (*Macaca fuscata*) have been reviewed by Green (1975). He points out that several causes may contribute to dialectical differentiation, some concerned with the initial establishment of differences, some with their spread within local populations, and some with their maintenance there. Thus, "a behavioral founder-effect . . . with socially facilitated mimicry, enhanced by cultural propagation with feeding as a reinforcer" could account for his observations (Green, 1975, p. 308). The founder event is probably never totally random, but rather "the rise of the tonal theme at the three [study] sites indicates that the broad acoustic structure is circumscribed by the genetic constraints linking tonal sounds to affinitive circumstances" (p. 309); fine differences are probably under less stringent control. Comparable comments can be applied to birds.

Predictions about the adaptiveness of sound characteristics depend crucially upon the paradigm within which one works. A model of "honest" nonmanipulative communication predicts a straightforward match between optimal transmission distance and transmissibility of sound characteristics; other models yield different predictions (Dawkins and Krebs, 1978; Gish and Morton, 1981; Krebs and Davies, 1981; Richards and Wiley, 1980). Thus, if it is advantageous to inform a neighbor of one's distance from him, a song should exhibit predictable differential attenuation or degradation of its characteristics. If it is advantageous to conceal this information from a neighbor, characteristics should show unpredictable patterns of attenuation and degradation and should have highly transmissible features.

The selective value to communicating derives from the average effect induced or maintained in receivers. The evaluation of "optimality" must extend beyond the physical features of signals before and after average transmission; it must also include the relationship of signal structure to physiological/motivational states induced in receivers (see Collias, 1960; Eisenberg, 1981; Morton, 1977). General principles that apply to this realm are not clearly documented or understood, but it seems likely that the evolutionary potential of sound structure is limited for neurophysiological reasons, and other organismal properties are probably linked

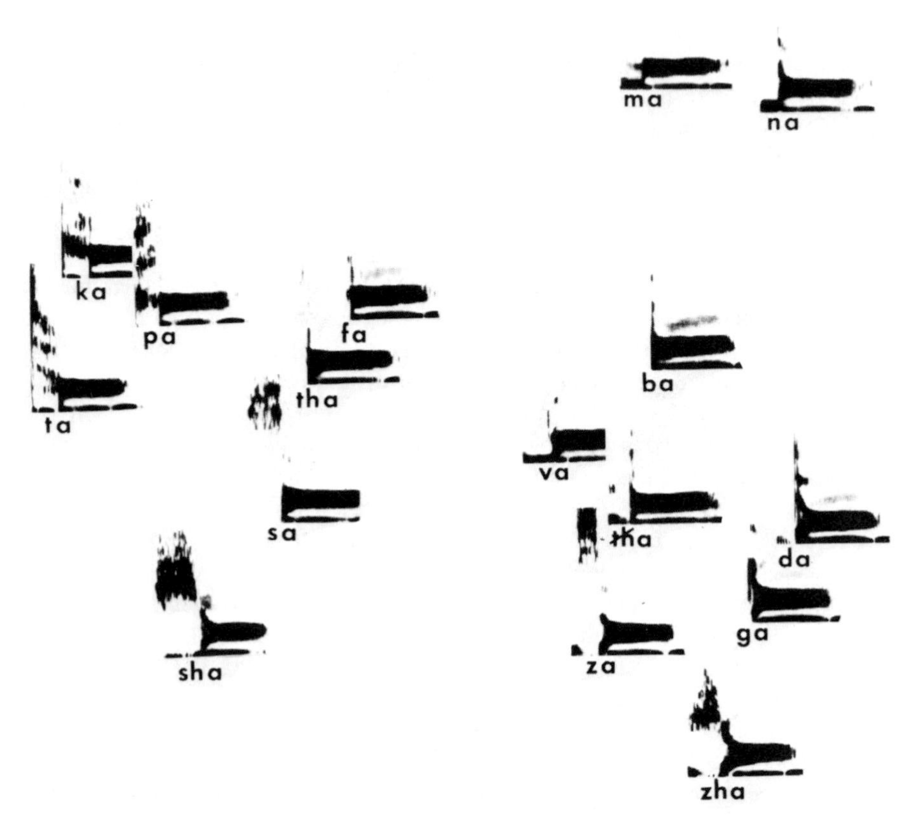

Fig. 5. Two-dimensional multidimensional scaling solution for 16 syllables differing in the initial consonant. The proximity of the syllables in the diagram reflects how easily they were confused with one another by subjects: large distances indicate low confusability, and small distances indicate high confusability. Morphological characteristics of the sounds suggest some structural and perceptual congruence. Thus, ka, pa, and ta are voiceless stops; fa, tha, sa, and sha are voiceless fricatives, etc. [From Shepard (1980, Fig. 2); reproduced with permission of the American Association for the Advancement of Science, and R. N. Shepard.]

to sound characteristics, too, but for yet unknown reasons. For example, duration plus three frequency measures of alarm calls in antelope ground squirrels (*Ammospermophilus*) are highly correlated with measures of the rostrum, and two of these are further correlated with habitat (Bolles, 1980).

The physical structure of sounds is under intense study to reveal adaptations to the physical environment, and such investigations will increasingly extend to the physiological realm. Despite these important trends, and the critical need for good detailed descriptions, it must be noted that the significance of different sounds to receivers can be most directly appraised without any reference to sounds' physical attributes. Thus, by pairwise testing of different sounds' confusability and similarity to subjects, the proximity of sound signals to one another in subjects' perceptual space can be estimated (Fig. 5; Shepard, 1980). Congruence between physical and perceptual proximity can be investigated subsequently.

V. CONCLUDING COMMENTS

There is a growing trend for research on bird acoustics to become more rigorous in scope, design, execution, and analysis. This is desirable and healthy, but purely descriptive studies and documentary audio recording also need to be sharply increased. Displays of most extant species have not even been taped or filmed, much less studied in detail. There is an urgent need for extensive audio recording and filming of species which are declining in abundance or range, before they become extinct. This need is acute for faunas in many areas of the world, especially those of the vast but rapidly disappearing tropical forests (Stuessy and Thomson, 1981). Without such material, we will be forever unable to know, appreciate, or explore the remarkable and diverse acoustic behavior of birds. Marshall (1978, p. 1) poignantly expresses this view in his monograph on small Asian night birds:

> Natural forests are hard to find and difficult to reach, being mostly contracted to high altitudes. They are disappearing at an appalling rate, invariably by clear-felling, to be replaced by wretched crops for a couple of seasons, then abandoned to desolate *Imperator, Eupatorium*, or other bushes. My study devolved into a crash program just to hear and tape-record the owls before they become extinct. Frills such as play-back experiments gave way to anguished efforts at identification.

REFERENCES

Andrew, R. J. (1969). The effects of testosterone on avian vocalizations. *In* "Bird Vocalizations. Their Relation to Current Problems in Biology and Psychology" (R. A. Hinde, ed.), pp. 97–130. Cambridge Univ. Press, London and New York.

Bates, M. (1960). "The Forest and the Sea; a Look at the Economy of Nature and the Ecology of Man." Random House, New York.

Bolles, K. (1980). "Variation and Alarm Call Evolution in Antelope Squirrels, *Ammospermophilus* (Rodentia: Sciuridae)." Ph.D. thesis, Univ. of California, Los Angeles, California.

Collias, N. E. (1960). An ecological and functional classification of animal sounds. *In* "Animal Sounds and Communication" (W. E. Lanyon and W. N. Tavolga, eds.), pp. 368–391. Amer. Inst. Biol. Sci., Washington, D.C.

Cosens, S. E. (1981). Development of vocalizations in the American Coot. *Can. J. Zool.* **59,** 1921–1928.

Dawkins, R., and Krebs, J. R. (1978). Animal signals: information or manipulation? *In* "Behavioural Ecology. An Evolutionary Approach" (J. R. Krebs and N. B. Davies, eds.), pp. 282–309. Sinauer Assoc., Sunderland, Massachusetts.

Eisenberg, J. F. (1967). A comparative study in rodent ethology with emphasis on evolution of social behavior, I. *Proc. U.S. Nat. Mus.* **122,** 1–51.

Eisenberg, J. F. (1981). "The Mammalian Radiations. An Analysis of Trends in Evolution, Adaptation, and Behavior." Univ. of Chicago Press, Chicago, Illinois.

Eldredge, N., and Cracraft, J. (1980). "Phylogenetic Patterns and the Evolutionary Process." Columbia Univ. Press, New York.

Gish, S. L., and Morton, E. S. (1981). Structural adaptations to local habitat acoustics in Carolina Wren songs. *Z. Tierpsychol.* **56,** 74–84.

Golani, I. (1976). Homeostatic motor processes in mammalian interactions: a choreography of display. *In* "Perspectives in Ethology" (P. P. G. Bateson and P. Klopfer, eds.), Vol. 2, pp. 69–134. Plenum, New York.

Green, S. (1975). Dialects in Japanese monkeys: vocal learning and cultural transmission of locale-specific vocal behavior? *Z. Tierpsychol.* **38,** 304–314.

Hailman, J. P. (1977). "Optical Signals, Animal Communication and Light." Indiana Univ. Press, Bloomington, Indiana.

Huxley, C. R., and Wilkinson, R. (1977). Vocalizations of the Aldabra White-throated Rail *Dryolimnas cuvieri aldabranus. Proc. R. Soc. London Ser. B* **197,** 315–331.

Ince, S. A., Slater, P. J. B., and Weismann, C. (1980). Changes with time in the songs of a population of Chaffinches. *Condor* **82,** 285–290.

Krebs, J. R., and Davies, N. B. (1981). "An Introduction to Behavioural Ecology." Blackwell, Oxford.

Kroodsma, D. E., and Pickert, R. (1980). Environmentally dependent sensitive periods for avian vocal learning. *Nature (London)* **288,** 477–479.

Lanyon, W. E. (1969). Vocal characters and avian systematics. *In* "Bird Vocalizations. Their Relation to Current Problems in Biology and Psychology" (R. A. Hinde, ed.), pp. 291–310. Cambridge Univ. Press, London, England.

Lemon, R. E., Struger, J., Lechowicz, M. J., and Norman, R. F. (1981). Song features and singing heights of American warblers: maximization or optimization of distance? *J. Acoust. Soc. Am.* **69,** 1169–1176.

Lorenz, K. (1941). Vergleichende Bewegungsstudien an Anatinen. *J. Orn.* **89** (*Suppl.*), 194–294.

Mace, T. R. (1981). "Causation, Function, and Variation of the Vocalizations of the Northern Jacana, *Jacana spinosa*." Ph.D. thesis, Univ. of Montana, Missoula, Montana.

Mairy, F. (1979a). Le roucoulement de la tourterelle rieuse domestique, *Streptopelia risoria* (L.) I. Variation morphologique de sa structure acoustique. *Bull. Soc. Roy. Sci. Liège* **9–10,** 355–377.

Mairy, F. (1979b). Le roucoulement de la tourterelle rieuse domestique, *Streptopelia risoria* (L.) II. Aspects causaux et sémantiques de la variation de sa morphologie acoustique. *Bull. Soc. Roy. Sci. Liège* **9–10,** 378–390.

Marshall, J. T. (1978). Systematics of smaller Asian night birds based on voice. *Amer. Orn. Union, Orn. Monogr.* **25,** 1–58.

Morton, E. S. (1977). On the occurrence and significance of motivation-structural rules in some bird and mammal sounds. *Amer. Natur.* **111,** 855–869.

Miller, E. H. (1982). Aerial displays of calidridine sandpipers, their structure and systematic significance. Unpublished data.

Mundinger, P. (1979). Call learning in the Carduelinae: ethological and systematic considerations. *Syst. Zool.* **28,** 270–283.

Nuechterlein, G. L. (1981). Variations and multiple functions of the advertising display of Western Grebes. *Behaviour* **76,** 289–317.

Richards, D. G., and Wiley, R. H. (1980). Reverberations and amplitude fluctuations in the propagation of sound in a forest: implications for animal communication. *Amer. Natur.* **115,** 381–399.

Shepard, R. N. (1980). Multidimensional scaling, tree-fitting, and clustering. *Science* **210,** 390–398.

Slater, P. J. B., and Ince, S. A. (1979). Cultural evolution in Chaffinch song. *Behaviour* **71,** 146–166.

Stuessy, T. F., and Thomson, K. S. (eds.) (1981). "Trends, Priorities and Needs in Systematic Biology." Report to the Systematic Biology Program, National Science Foundation. Association of Systematics Collections, Lawrence, Kansas.

Suga, N. (1978). Specialization of the auditory system for reception and processing of species-specific sounds. *Fed. Proc., Fed. Am. Soc. Exp. Biol.* **37,** 2342–2354.

Thielcke, G. (1970). Die sozialen Funktionen der Vogelstimmen. *Vogelwarte* **25,** 204–229.

Van Tets, G. F. (1965). A comparative study of some social communication patterns in the Pelecaniformes. *Amer. Orn. Union, Orn. Monogr.* **2,** 1–88.

Wasserman, F. E. (1979). The relationship between habitat and song in the White-throated Sparrow. *Condor* **81,** 424–426.

Wiens, J. A. (1982). Song pattern variation in the Sage Sparrow (*Amphispiza belli*): dialects or epiphenomena? *Auk* **99,** 208–229.

Wilkinson, R., and Huxley, C. R. (1978). Vocalizations of chicks and juveniles and the development of adult calls in the Aldabra White-throated Rail *Dryolimnas cuvieri aldabranus* (Aves: Rallidae). *J. Zool.,* **186,** 487–505.

1

Learning and the Ontogeny of Sound Signals in Birds

DONALD E. KROODSMA

I. INTRODUCTION

Vocal learning is the ability to use auditory information (including feedback) to modify or enhance vocal development. Humans represent the epitome of vocal learning, but it is by no means unique to humans. It is undoubtedly the basis for the elaborate songs of the humpback whale (*Megaptera novaeangliae;* Payne and Payne, in press). Among other mammals, though, including the nonhuman primates, evidence for vocal learning is sparse (see Green, 1975). On the other hand, in birds, and especially in songbirds, vocal learning is documented almost routinely with each new species that is closely studied (e.g., Nottebohm, 1972). Vocal learning appears to have reached a pinnacle in humans and in (song) birds, and the parallels in ontogeny, nervous control, and population consequences of vocal learning between humans and birds are intriguing. Thus vocal learning is most pronounced early in life: children and young birds babble (or "subsing");

ACOUSTIC COMMUNICATION IN BIRDS
VOLUME 2

ISBN 0-12-426802-1

hearing of both others and self is essential for normal vocal development; predispositions for learning conspecific sounds may exist; neural lateralization of vocal control is pronounced; and offspring may perpetuate local dialects by settling near where they were born or hatched (Marler, 1970a; Petrinovich, 1972).

Vocal learning in birds is intimately involved in almost all aspects of acoustic communication, and it has profound consequences for population structure and species integrity (see Chapters 7 and 8, Volume 1). We must understand comparative aspects of vocal learning to account for the evolution of complex sounds and dialects (see Chapters 5, 6, and 7 of this volume), duetting (see Chapter 4, this volume), and mimicry (see Chapter 3, this volume). In this chapter I examine the nature, extent, and evolution of vocal learning in birds. After discussing several well-documented examples in which vocal development appears to proceed independently of audition (and therefore independently of vocal learning, as defined), I indicate what is minimally adequate evidence for implicating vocal learning and review a world survey provided in Appendix, pp. 311–337. In the last sections, I discuss aspects of selective vocal learning and the timing of vocal learning, and then comment on selective forces that may have promoted the evolution of vocal learning in birds (for other reviews of vocal development in birds, see Konishi and Nottebohm, 1969; Marler and Mundinger, 1972; Slater, 1983).

II. VOCAL DEVELOPMENT THAT IS INDEPENDENT OF AUDITION

It is difficult to determine whether a given vocalization develops normally without access to an external model *or* without auditory feedback. In the first case, acoustic isolation during development must be insured. Deafening achieves complete auditory isolation (e.g., Nottebohm and Nottebohm, 1971), but degrees of acoustic isolation from external models can be achieved with intact birds which are either hand-raised or cross-fostered from an early age (preferably from the egg; Lanyon, 1979). Furthermore, it must be demonstrated that vocalizations which develop are identical to those characteristic of the species under natural conditions. By far the best approach is to assess the response of wild conspecifics to experimental and natural sounds. A second but less satisfactory approach is to compare the details of the experimental and wild-type vocalizations spectrographically; however, a sonagram is very crude, and relevant features of the vocalizations may not even be represented on sonagrams (e.g., see Hall-Craggs, 1979). Still another approach assumes that humans hear bird sounds roughly as birds do, and involves a panel of experienced listeners rating songs as natural or unnatural (e.g., Kroodsma, 1977).

There are few studies verifying the existence of vocalizations which develop independently of auditory information. In various songbird species, wild birds have in all cases failed to respond in typical fashion to songs of conspecifics which were isolated during vocal development (e.g., Thielcke, 1973; Shiovitz, 1975; Ewert, 1979; Lanyon, 1979; see also Chapter 2 of this volume). Among non-songbirds, Lade and Thorpe (1964, p. 366) cross-fostered several individuals from five dove species and observed that "rearing of one species by another failed to change the song in any way," and Nottebohm and Nottebohm (1971, p. 313) found that deafened Ringed Turtle-Dove (*Steptopelia risoria*) squabs "develop vocalizations that are normal in structure and temporal delivery." Hybrid doves often exhibit vocal characteristics which are intermediate between the parent species (Lade and Thorpe, 1964), and deafened Ringed Turtle-Doves emit vocalizations more or less normally throughout the nesting cycle (Nottebohm and Nottebohm, 1971). Thus, in these dove species it seems fair to invoke an "endogenous" mechanism for call acquisition.

The ontogeny of vocalization in the Domestic Fowl (*Gallus domesticus*) is similar to that in the doves. Calls of normally reared, intact birds show much individual variation and grading, and Konishi (1963) found that birds deafened at an early age developed vocalizations similar to those of intact birds. Further spectrographic evidence and analyses might have been warranted in both the dove and fowl studies, but the data certainly contrast sharply with results from comparable studies of oscine song (e.g., Konishi, 1965, and Section III).

While many species of songbirds learn songs, some of their simpler vocalizations may not require external models. For example, Güttinger and Nicolai (1973) cross-fostered several species of estrildids and noted that songs often developed abnormally, whereas simple calls seemed to be structurally normal. Likewise, the Marsh Tit (*Parus palustris*) has two types of songs; the simpler one appears to develop normally in acoustically isolated individuals, but the more complex songs require external models for normal development (Becker, 1978; see also Chapter 2 of this volume).

Some observations of vocalizations which vary little over the range of a species also suggest a largely "closed" developmental program. For example, the relatively simple songs of several *Myiarchus* species scarcely vary throughout their geographic ranges (Lanyon, 1978); vocal learning coupled with typical dispersal patterns would be expected to generate some spatial variation (see Chapters 6 and 7 of this volume). In addition, body size and fundamental frequency of these relatively simple *Myiarchus* songs covary, suggesting further that development of vocalizations in these species may be rigidly controlled.

Data on song ontogeny of two *Empidonax* species support these inferences for tyrannids. The Alder and Willow flycatchers (*Empidonax alnorum* and *E. traillii*, respectively) are nearly indistinguishable morphologically, and speciation has undoubtedly occurred very recently (Stein, 1963). Both species have limited song repertoires; during the dawn chorus, the Alder sings "fee-bee-o" and the Willow sings "creet" and two forms of "fitz-bew" (Fig. 1) (Stein, 1963). Each

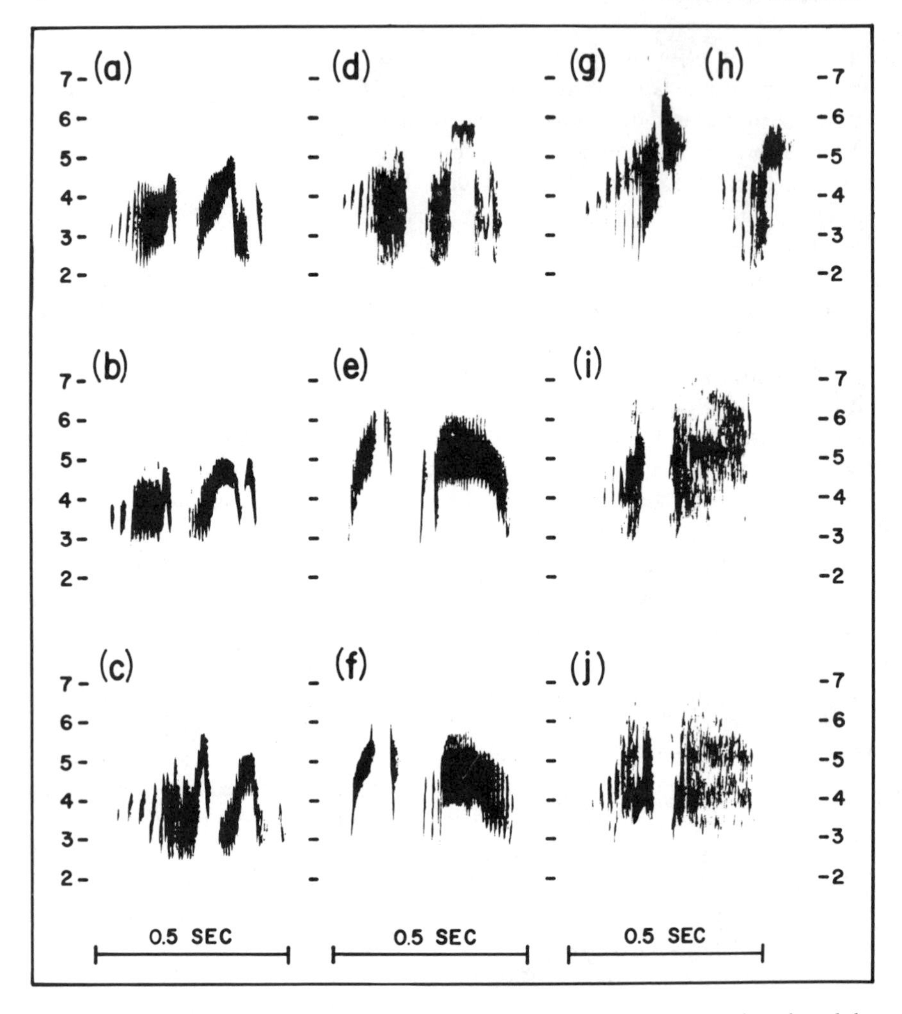

Fig. 1. Sonagrams of fledgling Willow and Alder flycatchers compared to the adult songs. (a) Adult male Alder Flycatcher song "fee-bee-o." (b) Fledgling Alder Flycatcher, at approximately 15 days of age, in nature. (c,d) Fledgling Alder Flycatchers, at 19 days of age, in captivity; these two birds heard only Willow Flycatcher vocalizations from 7 days of age. (e,f,g) Adult male Willow Flycatcher "high fitz-bew," "fitz-bew," and "creet," respectively (see Stein, 1963). (h) Fledgling Willow Flycatcher incipient "creet" (?) at 18 days of age, in captivity. (i,j) Fledgling Willow Flycatchers incipient "fitz-bew" (?) at 18 and 26 days, respectively, from two different individuals in the laboratory. Fledgling Willow Flycatchers in captivity heard only the Alder "fee-bee-o" from 10 days of age. The ordinate is KHz.

species may, in addition, use a variety of calls during encounters. If these two were typical songbirds, vocal learning might begin at 15–20 days, perhaps simultaneously with the very first traces of subsong. Subsong and plastic song would remain highly amorphous, however, until the next spring, when the adult song(s) would crystallize (see Chapter 2, this volume). In contrast, Alder Flycatchers in nature fledge at 12 days and at once produce an accurate version of the adult ''fee-bee-o''; even if tutored with the Willow Flycatcher songs from day 7 in the laboratory, Alder fledglings still utter a typical ''fee-bee-o.'' Tonal quality differs from that of the adult version, but the three-noted structure is clearly evident (see Fig. 1).

The vocal development of Willow Flycatchers appears slightly different. Fledglings first utter what may be an incipient ''creet,'' followed several days later in some birds (males only?) by a two-noted call which could be the future ''fitz-bew,'' though it certainly is less clear than the corresponding ''fee-bee-o'' in Alders of the same age. The Willow Flycatcher adult song repertoire contains three different call or song types as opposed to only one for the Alder Flycatcher, and it is interesting that vocal ontogenies may also differ.

It is quite likely that no vocal learning occurs in the Alder and probably the Willow Flycatcher, yet given the subtle differences in development, it would be dangerous to generalize to the hundreds of other tyrannids. The Tyrannidae is the third largest passerine family (Storer, 1971), and studies of vocal ontogenies for other flycatchers and other non-oscines as well would be very informative (see Nottebohm, 1972).

III. WORLD SURVEY OF VOCAL LEARNING IN BIRDS

A. Types of Evidence Implicating Vocal Imitation

The possibility of a fortuitous resemblance between a model sound and the ''copy'' is highest with simpler models and versatile songsters, and adequate controls are necessary to document vocal learning unequivocally. Several forms of evidence, some direct and some indirect, can implicate vocal learning as playing a role in vocal development. I discuss the evidence here, and in Appendix are listed bird species for which each type of evidence has been provided.

1. Vocal Imitation of (a) Conspecific Sounds, (b) Heterospecific Avian Sounds, or (c) Nonavian Sounds (Including Human Voice), under Controlled Laboratory Conditions

In many studies, very young birds (e.g., nestling songbirds) are removed from the nest and ''tutored'' with vocalizations characteristic of their own or another species. Similarly, adult birds can be taken into captivity and exposed to vocalizations different from those in their home populations. If the subject accurately reproduces these new sounds, then vocal imitation has occurred (see Figs. in Chapter 2, this volume).

2. *Interspecific Vocal Imitation (Mimicry) by Free-Living Birds*

Mimicry is a second source of evidence for vocal imitation. Here, again, it must be demonstrated that convergence on a vocal pattern could not have arisen by chance alone: as Armstrong (1963, p. 73) warns, ''very loquacious birds are apt to utter calls fortuitously resembling those of other species just as a silly person who talks incessantly will occasionally say something sensible.'' Thus, a relatively simple vocalization uttered by a single individual is unsatisfactory evidence for mimicry, even though sonagrams of model and proposed copy may appear identical (e.g., Borror, 1968). However, when many relatively complex vocalizations from one or more individuals appear very similar to the sounds of other species, vocal mimicry is likely; consider here, for example, the extensive spectrographic evidence for two House Wrens (*Troglodytes aedon*) mimicking numerous complex songs of Bewick's Wrens (*Thryomanes bewickii;* Kroodsma, 1973).

3. *Intraspecific Vocal Imitation among Free-Living Birds*

Interspecific vocal imitation and limited dispersal from the site of imitation (not necessarily the site of hatching) often lead to the phenomenon of song dialects or song sharing among neighboring males (see Chapters 6 and 7, this volume). Vocal characteristics are quite homogenous within a locality but nearby localities may have different song characteristics even though they are within the dispersal distance of young birds. When birds disperse to a new area and acquire typical song characteristics at the new location, vocal learning has occurred (e.g., Kroodsma, 1974; Jenkins, 1977). Extensive sharing of complex vocalizations among birds at a single location is also strongly suggestive of vocal learning (e.g., see Heckenlively, 1970; Lein, 1978). However, ''complex'' is a relative (and nebulous) term, and the extent to which vocal learning is required to produce homogenous yet complex vocal behaviors among neighboring males is unknown (Nottebohm, 1972).

4. *Abnormal Vocalizations*

Abnormal vocalizations develop in acoustical isolation if an external model is needed. It is possible that isolation can produce physiological or psychological abnormalities which in turn affect vocal development, and recommended controls involve tutoring with sounds of the same or different species. Ultimate isolation from sound experience occurs if a subject is deafened, for then a bird can neither imitate an external song nor match its own sounds to memorized sounds.

B. Literature Survey of Vocal Learning in Birds

Anecdotal and often conflicting evidence of interspecific vocal mimicry abounds in the literature (see Chapter 3, this volume). For example, Scott (1902) and Grinnell *et al.* (1930) extol the mimicking abilities of the Yellow-breasted Chat (*Icteria virens*), but A. A. Saunders believed that such resemblances between proposed models and copies were merely fortuitous (Bent, 1953). Some of these discrepancies in opinions may reflect real observed differences. Because they are isolated from conspecific sounds, occasional individuals of a given species may mimic extensively (e.g., Lanyon, 1957). Also, geographic differences in interspecific mimicry may exist because of corresponding differences in population density and coexisting model species. However, some discrepancies undoubtedly reflect superb imaginations, attempts to lengthen a list of mimics, or nondiscriminating ears. For example, Scott's (1902) credibility is reduced when he describes a duck imitating its turkey foster mother; such flexibility in vocal development among ducks has not been confirmed (D. B. Miller, personal communication).

The Appendix includes most of the references that I examined, and Table I contains a summary of the data for the Passeriformes. Some references were rejected, though, when subsequent analyses discredited them. Much of the evidence for vocal learning comes from a few authors (e.g., Vernon, Chisholm) who tried to establish lists of mimics in particular geographical regions; it is quite evident that relatively little work has been done under controlled laboratory conditions.

Most acceptable observations of vocal learning are for songbirds, though vocal imitation also occurs in the Psittaciformes, Apodiformes, and Piciformes; all four taxa are highly derived (see Nottebohm, 1972). This apparent correlation is complemented by studies of doves and chickens, whose vocalizations appear normal even without exposure to normal calls or auditory feedback during development. However, studies of many of the less highly derived orders have not been done. Studies by Caswell (1972) on Mallards (*Anas platyrhynchos*), Cosens (1981) on American Coots (*Fulica americana*), Wilkinson and Huxley (1978) on Gray-throated Rails (*Canirallus oculeus*), and Moynihan (1959) on Ring-billed and Franklin's gulls (*Larus delawarensis* and *L. pipixcan*) document very nicely the rates of change of call repertoires and their usage, and suggest that vocal imitation is not involved in vocal development. Most shorebirds produce relatively complex vocalizations for which learning seems possible (Nethersole-Thompson and Nethersole-Thompson, 1979; Oring, 1968; Miller, 1979). Sparling (1979) presents circumstantial evidence of vocal learning in Greater Prairie

TABLE I

A Summary of Data on Vocal Learning for the Passeriformes

Family	Distribution[a]	Number of species[b]	Number of species with evidence of vocal learning (percentage of family)[c]
1. Eurylaimidae	Africa, India to Borneo	14	0 (0)
2. Dendrocolaptidae	Neotropical	52	0 (0)
3. Furnariidae	Neotropical	218	0 (0)
4. Formicariidae	Neotropical	230	0 (0)
5. Conopophagidae	Neotropical	11	0 (0)
6. Rhinocryptidae	Costa Rica to Tierra del Fuego	30	0 (0)
7. Cotingidae	Neotropical	79	1?? (1.3)
8. Pipridae	Neotropical	57	0 (0)
9. Tyrannidae	New World	377	0 (0)
10. Oxyruncidae	Neotropical	1	0 (0)
11. Phytotomidae	South America	3	0 (0)
12. Pittidae	Old World tropics	24	0 (0)
13. Xenicidae	New Zealand	4	0 (0)
14. Philepittidae	Madagascar	4	0 (0)
15. Menuridae	Australia	2	2 (100.0)
16. Atrichornithidae	Australia	2	2 (100.0)
17. Alaudidae	Old World, 1 sp. to North America	79	12 (15.2)
18. Hirundinidae	Worldwide	80	1 (1.3)
19. Motacillidae	Nearly worldwide	54	3 (5.6)
20. Campephagidae	Old World tropics	70	0 (0)
21. Pycnonotidae	Africa, southern Asia	122	1 (0.8)
22. Irenidae	Oriental, East Indies	14	1 (7.1)
23. Laniidae	All continents except South America and Australia	74	7 (9.5)
24. Vangidae	Madagascar	13	0 (0)
25. Bombycillidae	Holarctic	8	0 (0)
26. Dulidae	Hispaniola	1	0 (0)
27. Cinclidae	Holarctic, South America	5	0 (0)

28. Troglodytidae	New World, 1 Holarctic	60	9 (15.0)
29. Mimidae	New World	31	6 (19.4)
30. Prunellidae	Palaearctic	12	0 (0)
31. Muscicapidae	Worldwide	1434	92 (6.4)
32. Aegithalidae	Eurasia, western North America	8	0 (0)
33. Remizidae	Eurasia, Africa, western North America	10	0 (0)
34. Paridae	Eurasia, Africa, North America	47	6 (12.8)
35. Sittidae	Widespread, except Africa and South America	26	0 (0)
36. Certhiidae	Holarctic, Africa, India	6	2 (33.3)
37. Rhabdornithidae	Philippine Islands	2	0 (0)
38. Climacteridae	Australia, New Guinea	6	0 (0)
39. Dicaeidae	Oriental Australasian	58	1 (1.7)
40. Nectariniidae	Old World tropics	117	7 (6.0)
41. Zosteropidae	Old World tropics	83	4 (4.8)
42. Meliphagidae	Australasia, southern Africa	172	1 (5.8)
43. Emberizidae	New World	560	36 (6.4)
44. Parulidae	New World	126	4 (3.2)
45. Drepanididae	Hawaiian Islands	23	0 (0)
46. Vireonidae	New World	43	4 (9.3)
47. Icteridae	New World	95	9 (9.5)
48. Fringillidae	Worldwide, except Australasia	122	26 (21.3)
49. Estrildidae	Old World tropics	127	11 (8.7)
50. Ploceidae	Eurasia, Africa	144	8 (5.6)
51. Sturnidae	Old World	111	9 (8.1)
52. Oriolidae	Warm parts of Old World	28	3 (10.7)
53. Dicruridae	Old World tropics	20	4 (20.0)
54. Callaeidae	New Zealand	3	1 (33.3)
55. Grallinidae	Australia, New Guinea	4	0 (0)
56. Artamidae	Oriental, Australasian	10	2 (20.0)
57. Cracticidae	Australasia	10	3 (30.0)
58. Ptilonorhynchidae	Australasia	17	7 (41.2)
59. Paradisaeidae	Australasia	42	0 (0)
60. Corvidae	Worldwide	106	8 (7.5)

[a] From Storer (1971).
[b] From Morony *et al.* (1975).
[c] From Appendix, pp. 311–337.

Chickens (*Tympanuchus cupido*) and Sharp-tailed Grouse (*Tympanuchus phasianellus*), members of the Galliformes along with the Domestic Fowl; this evidence needs corroboration, but does suggest that vocal learning could be more extensive than previously thought. Available data for most avian groups are simply inadequate to judge.

For several reasons, it is dangerous to generalize on the basis of the data in Appendix and Table I. Vocal learning may have accompanied rapid speciation in songbird families, and there exists a strong tendency for families in which vocal learning is documented to have the larger numbers of species (Marler, 1970a; Nottebohm, 1972; see also Chapter 7 of this volume). However, the probability of discovering vocal learning within a family is proportional to the number of species it contains. Consider the passeriform families. In the 33 where vocal learning has been found (Table I), the median number of species per family is 66; on the other hand, the median number of species is only 12 in those 25 families where vocal learning has not been documented (numbers of species are from tabulations in Storer, 1971). Do these data corroborate the intense speciation hypothesis or merely that the probability of discovering vocal learning in a family is proportional to the number of member species? Only further study will reveal the answer.

Another bias results from the largely nonoverlapping geographical distributions of bioacousticians and poorly known avian families, especially the nonoscine passerines which are heavily represented in tropical areas. Vocal learning has not been documented in 25 passerine families, but only seven have representative species (about 75 species total) in North America and Europe where behavior and natural history have been most intensively studied. In contrast, of the 33 families where vocal learning has been documented, 18 (55%) have representatives in North America and Europe.

Still another bias results from my familiarity with birds and literature of the Holarctic. Even though I have attempted to survey the literature for all faunal regions, the listing in the Appendix is certainly incomplete.

IV. WHAT IS ACTUALLY LEARNED?

Most of the species listed in the Appendix have had one to several acceptable observations of interspecific vocal learning in nature (evidence category "2"). Gathering such evidence requires only a good ear and a knowledge of local avian sounds, but verification of *intra*specific vocal learning is more difficult. Spectrographic evidence from several localities (e.g., Martin, 1979) or careful listening and detailed notes on population-unique vocal sounds from several locations (Tinbergen, 1939) are suggestive of it, but careful documentation in the laboratory is certainly the most unequivocal evidence (Marler, 1970b). In the Appendix, interspecific vocal imitation appears far more common than intraspecific vocal learning, but this merely reflects the relative ease of collecting such evidence. There are, in fact, few accomplished mimics (see Chapter 3, this volume), while evidence for intraspecific vocal learning exists for all carefully studied songbird species (Kroodsma, 1977).

Vocal learning appears ubiquitous among songbirds, but how does it come to be restricted to intraspecific learning in most cases? Vocal learning may have its advantages, but this flexibility in the developmental process is a potential hazard to birds, since they are exposed to many sounds other than conspecific song during their limited sensitive period (see Section V). Individuals of a typically nonmimicking species rarely learn the "wrong" songs, which suggests that young males (and perhaps especially females?) possess a fail-safe mechanism guiding vocal learning along species-typical pathways.

Those fail-safe mechanisms appear to vary considerably among species. In some, selective vocal learning relies heavily on an inherited set of auditory specifications for species-typical songs. Marler (1970b, 1976) has discussed some of the characteristics of this "innate auditory template." Evidence for such a template exists for several songbird species: Chaffinches (*Fringilla coelebs*) do not learn various alien songs (Thorpe, 1958); White-crowned Sparrows (*Zonotrichia leucophrys*) favor conspecific songs over Harris' Sparrow (*Zonotrichia querula*) and Song Sparrow (*Zonotrichia melodia*) songs (Marler and Tamura, 1964); and Swamp Sparrows (*Zonotrichia georgiana*) preferentially learn species-typical syllables from an array of Song and Swamp sparrow syllables which are presented in songs of varying complexity and temporal pattern (Marler and Peters, 1977). The characteristics of the "template" may prove very complex, however, for further experiments with the Swamp Sparrow have demonstrated that Song Sparrow song elements are learned more readily if they are closely associated in time with species-typical syllables (Marler and Peters, 1981).

In addition to some inherited specifications for the hearing or production of relevant sounds, or both, a social bond between juvenile and parent may promote learning of the proper vocalizations. Juvenile male Zebra Finches (*Poephila guttata*), for example, learn songs from either a conspecific or heterospecific grassfinch parent, but do not learn details of songs from visually isolated adults singing nearby or songs which are merely broadcast from a loudspeaker (Immelmann, 1969; Price, 1979). Limits to the types of sounds which juveniles imitate do exist, however (Price, 1979): if juveniles do not hear appropriate model songs, they produce a recognizable Zebra Finch song by organizing an increased number of "developmentally conservative" call notes (see Güttinger and Nicolai, 1973) into a temporal pattern resembling that of the normally learned song.

Mimicry of heterospecific sounds has been convincingly demonstrated for many other species in the laboratory. The Song Sparrow readily learns components of Swamp Sparrow song (Marler and Peters, 1977). Several icterids, including the Northern Oriole (*Icterus galbula*), Red-winged Blackbird (*Agelaius phoeniceus*), and Eastern and Western meadowlarks (*Sturnella magna* and *S. neglecta*) can imitate the songs of other species (see Appendix I). With the two meadowlarks, Lanyon (1957) found that deprivation of conspecific song models could lead to mimicry or improvisation. Eastern Meadowlarks have about 100 songs apiece while Western Meadowlarks have only about 10. Deprivation, then, is relative to the total repertoire size which must be acquired and this factor could be important in interpreting many laboratory experiments. Occasional mimicry observed in nature may result from similar natural "deprivation," where young birds are acoustically isolated from singing conspecifics during the period when song is normally acquired. This can happen among young hatched late in the singing season, at the edges of a breeding range, or among early dispersers or migrants.

Finally, in a few species there appear to be few, if any, restrictions in what is learned. Renowned mimics include free-living mockingbirds, lyrebirds, and others listed in the Appendix. The mimicry is usually not totally deceptive, though, for even the persistent mimics usually translate their imitations into a particular style which betrays the singer (see Chapter 3, this volume). Also in this class are several curious species which in capitivity are capable of imitating a great diversity of sounds; in nature, though, mimicry is rare if not absent. Examples include the Hill Myna (*Gracula religiosa*), Bullfinch (*Pyrrhula pyrrhula*), and several parrots. In nature, male and female mynas restrict their learning to several relatively simple calls from their respective sexes (Bertram, 1970). The Bullfinch is the quintessential "bird fancier's delight" (Godman, 1955), but male Bullfinches in the wild learn the song [a "very subdued mixture of warbling and creaking notes" (Peterson *et al.,* 1966)] exclusively from the male parent (Nicolai, 1959). In wild parrots, vocal imitation is evidently restricted to members of a given flock (Nottebohm, 1970).

Species recognition by song alone can become a problem under some circumstances, but many clues other than these learned vocalizations may be used by conspecifics. Thus, the strange song of the Bullfinch very likely plays no role in identifying mates (Nicolai, cited in Thorpe, 1961). Where Western and Eastern meadowlarks are sympatric and males learn heterospecific songs, females probably use more reliable calls to identify potential mates (Lanyon, 1957). Similarly, in the sibling *Empidonax* species, mate choice may be based on either calls or song, for the song is a very reliable identifier of species in which no vocal learning occurs.

It is quite possible that learned vocal signals became more variable only after they were freed from the role of species identification during mate selection (Marler, 1960). The intriguing singing behavior of several wood warblers (Parulidae; e.g., see Lein, 1978) may further exemplify this phenomenon: the primary songs which are more likely to be used during mate selection are more stereotyped and vary less geographically than do the types of songs used during male/male encounters (Gill and Murray, 1972; Kroodsma, 1981).

But such patterns fail to explain observed differences in song selectivity among some closely related species, such as the *Zonotrichia* sparrows. Perhaps Song Sparrows maintain an ability to imitate Swamp Sparrow songs so that a competitive advantage is gained in interspecific territoriality. [Rice (1978) postulates that a similar mechanism may be operating among vireos.] Alternatively, Song Sparrows may have a decreased selectivity in what is learned because they typically sing a greater diversity of song elements than do Swamp Sparrows. Similarly, male White-crowned Sparrows have only one song-type and are selective in song elements they will learn (Marler, 1970b). If this relationship is widespread, then larger song repertoires will be correlated with a lower specificity with respect to what is acceptable for learning.

Another intriguing example of selective learning may occur in the Parulidae. A hand-reared Chestnut-sided Warbler (*Dendroica pensylvanica*) did not learn 10 Yellow Warbler (*Dendroica petechia*) songs to which it was exposed (D. E. Kroodsma and R. Pickert, unpublished data), even though these sound very similar to Chestnut-sided Warbler songs (Morse, 1966). Instead, it imitated the strikingly different Common Yellowthroat (*Geothlypis trichas*) song. The Chestnut-sided Warbler may possess a finely tuned internal mechanism for discriminating the songs of *Dendroica*. Heightened levels of discrimination might be expected in a communication system where different, apparently learned songs are used in different contexts (Lein, 1978).

Vocal development in another pair of congeners, the Long-billed Marsh (*Cistothorus palustris*) and Short-billed Marsh wrens (*C. platensis*), is strikingly different and illustrates that development may be closely tied to life history. The breeding biology of these *Cistothorus* species is very similar, for both are polygynous and breed in high densities in wetland communities of low avifaunal diversity, and males may have up to 100 different songs each. Long-billed Marsh Wrens learn details of songs from surrounding males, and learn songs of both wren species in the laboratory. In contrast, neighboring male Short-billed Marsh Wrens share few song types, and in the laboratory males improvise fairly normal wild-type songs but do not imitate model songs from training tapes. This difference may be related to environmental predictability; male Long-billed Marsh Wrens return and breed in the same marshes each year, whereas Short-billed Marsh Wrens are seminomadic, an adaptation to the relative instability of the wet meadows where they breed (Kroodsma and Verner, 1978). Thus, the generalized song of the Short-billed Marsh Wren may facilitate movements of individuals from one population to another, for songs vary little throughout the geographic range of the species; on the other hand, Long-billed Marsh Wrens develop population-typical songs, match themes during countersinging, and probably respond less to the different songs of distant populations (Verner, 1975 and personal communication).

In all of the above studies, the focus has been on how birds learn particular vocalizations—as Beer (see Chapter 9, this volume) emphasizes, we have been "preoccupied with syntactics, to the neglect of semantic and pragmatic considerations . . ." Thus, male Long-billed Marsh Wrens learn their song-types, but we know little about how neighboring males develop their competence in "song-dueling" (Verner, 1975; Kroodsma, 1979). Senegal Indigo-birds (*Vidua chalybeata*) learn their songs and have local dialects, and a male may even move from one dialect area to another and change his entire song repertoire (Payne and Payne, 1977; Payne, 1981a). Different songs appear to be used in different contexts, yet how does the male know when to use each song type? Are there ritualized structural cues within each song-type that are endogenously controlled, or is it possible that context and meaning are established by convention in local populations? The same questions must be asked with regard to the Chestnut-sided Warbler and other warbler species where different learned songs are used in different contexts. In short, the study of semantics and pragmatics in the development of bird vocalizations has many questions to address.

V. TIMING OF VOCAL LEARNING

Vocal learning may occur in two distinct phases. The first involves the exposure to and "memorization" of the sound, and the second involves the reproduction of that sound so that it matches the memory (Nottebohm, 1969). These two processes can be well separated in time as in several *Zonotrichia* sparrows. Songs are memorized during the first 2 to 3 months of life, when juvenile males practice singing (i.e., subsinging), but the final adult song is not produced until the next spring when the birds are roughly 7 to 9 months old (e.g., Marler, 1970b; see also Chapter 2, this volume). In Chaffinches, though, these two processes may overlap, for some vocal learning occurs during the first spring when the males are refining their adult song (Nottebohm, 1969). Some species learn new sounds throughout life, but the two stages may still remain distinct in time; thus, one 20-year-old African Gray Parrot (*Psittacus erithacus*) acquired new sounds but did not produce them until caging conditions changed (F. Nottebohm, unpublished data).

The timing of vocal learning is known for very few species. In several species, vocal learning continues throughout life. African Gray Parrots as old as 20 years will learn new sounds (Nottebohm, 1970); Canaries (*Serinus canaria*), Greenfinches (*Carduelis chloris*), and other *Carduelis* species will continue to learn vocalizations from other adults (Nottebohm and Nottebohm, 1978; Güttinger,

1977; Mundinger, 1970); yearling and older Senegal Indigo-birds or Saddlebacks (*Creadion carunculatus*) can disperse and learn from other adults in new dialect areas (Payne, 1981a; Jenkins, 1977); and the Mockingbird (*Mimus polyglottos*) very likely imitates new sounds throughout life (Laskey, 1944).

Evidence for other species indicates that vocal learning is more restricted to a period early in life. However, the data for different species vary tremendously. Male meadowlarks acquired songs heard during (roughly) days 20 to 80 as well as days 80 to 140, but hand-reared birds did not acquire songs during the first 3 weeks of life or after the beginning of the first spring (Lanyon, 1957). In Indigo Buntings (*Passerina cyanea*) individually isolated, hand-reared birds develop highly abnormal songs, but 18-month-old males can learn new song components from loudspeakers or adjacent males (Rice and Thompson, 1968). The Red-winged Blackbird learns some songs during 3 weeks of exposure to conspecific songs between days 27 and 71 (Marler *et al.*, 1972); evidence for some learning during later years comes both from these laboratory Red-winged Blackbirds and from wild males in which the song repertoire size is highly correlated with age (Yasukawa *et al.*, 1981). In the Chaffinch and Cardinal (*Cardinalis cardinalis*), song learning normally begins during the first fall and terminates the following spring when males establish a territory (Thorpe, 1958; Dittus and Lemon, 1969).

In five well-studied species, vocal learning is nearly restricted to the first several months of life. Two male White-crowned Sparrows imitated training songs after exposure during days 8–28 and 35–56, respectively; males exposed to songs before 8 and after 56 days developed songs unlike those of the models (Marler, 1970b). Similarly, Song Sparrows learn most songs before 70 days, though some vocal learning may occur as late as 161 to 196 days of age (Mulligan, 1966). I tutored each of four male Swamp Sparrows during days 20–30, 40–50, 60–70, and in the following spring; vocal learning was restricted to the first fall and was concentrated in days 20–30 (five, two, and one songs were imitated during the three fall periods, respectively; D. E. Kroodsma, unpublished data). Song learning in the Zebra Finch terminates by about day 85, when young males are singing adult songs (Immelmann, 1969). Finally, in the Bewick's Wren, juvenile males in nature are singing song components typical of their eventual adult songs by day 60, and very little learning occurs after the year of hatching (Kroodsma, 1974).

The studies just referred to differ not only in species studied but also in all conceivable details of methodology, and cross-species comparisons are therefore next to impossible. Furthermore, if replications of some of these data were attempted, it might be discovered, as it has with imprinting, "that the outcome of an . . . experiment depends on all sorts of conditions which the experimenter may vary sometimes knowingly and sometimes unwittingly" (Bateson, 1979, p. 473).

In the Long-billed Marsh Wren, I have documented four experimental conditions which influence the timing of song learning (Kroodsma, 1978 and unpublished data; Kroodsma and Pickert, 1980). Initial experiments suggested that song learning occurred during the first 2 to 3 months of life, but I have subsequently determined that the timing and length of this sensitive period depend on (1) photoperiod, (2) the rate of exposure to different song types, (3) the quality of the stimulus (whether birds are exposed to other singing males or to songs over loudspeakers; see also Payne, 1981b), and (4) the housing and movement of birds during the experiments. Thus, four of eight juveniles which were maintained on a photoperiod simulating a hatching late in the season (August) in a Hudson River marsh (41° N latitude) learned songs from loudspeakers the following spring. However, none of 10 males that were maintained on daylengths simulating a hatching early in the season (June) learned songs under the same conditions. These four males which learned songs the next spring had also imitated or heard the fewest song types during their hatching year. Also, withholding songs from juveniles can delay termination of the learning phase during the hatching year. Novel surroundings or movement of cages during an experiment can prevent song learning. Finally, regardless of photoperiod and the number of different songs heard during the hatching year, yearlings exposed to live singing adults will learn some songs during their first spring. Thus, as Nottebohm (1969) first demonstrated with a Chaffinch, the sensitive period is not a strictly age-dependent phenomenon.

Factors influencing the learning phase provide insight into mechanisms of vocal learning. Testosterone levels during the hatching year are undoubtedly important for song development in many north temperate species of songbirds (see Arnold, Chapter 3, Volume 1). A Chaffinch that Nottebohm (1969) castrated as a juvenile did not learn songs until it was given exogenous testosterone at 2 years of age. Testosterone levels among free-living birds vary with the photoperiod, and photoperiodic differences encountered by early-versus-late hatching wrens are sufficient to influence the timing of the sensitive period for song learning. However, since singing accompanies high testosterone levels, it remains unclear whether the termination of the sensitive period occurs because of the higher hormonal levels or because of the final crystallization of adult song (Nottebohm, 1969).

For the field and evolutionary biologist, a single caveat looms above the bewildering array of data on vocal learning in the laboratory: what is actually happening in nature may not be reflected accurately by results of laboratory experiments. The complex physical and social processes occurring in nature cannot be duplicated in the laboratory. Birds may be capable of imitating songs at certain ages in the laboratory, but a lack of song learning at other ages is "negative evidence" which requires great care in interpretation. In the Long-billed Marsh Wren, for example, allowing social interactions with caged singing adults extends the song learning period considerably (Kroodsma, 1978); social pressures for songs to conform to some population-typical model may be even greater among neighboring and countersinging territorial males. Indeed, contrary to what would have been expected from laboratory data (Marler, 1970b), L. Baptista and L. Petrinovich (unpublished data) do report that one of their free-living White-crowned Sparrows altered his song pattern at 1 year of age, and that laboratory experiments unequivocally demonstrate the influence of social factors on the timing of song learning.

The timing of the song (vocal) learning period among birds varies from species to species, and questions regarding the evolution of sensitive periods can perhaps best be broached by first asking, ''Why learn?'' Nottebohm (1972) concluded that vocal learning most likely evolved under forces that (1) promoted elaboration of a simpler species-typical song or (2) favored the establishment of local dialects during microevolutionary processes. If birds are to breed during their first year, vocal learning must, of course, be accomplished during that year in order for any benefits to be derived during the first breeding season. On the other hand, vocal learning which continues throughout life could lead to age-correlated repertoire sizes; females, in an effort to select older and/or more experienced partners, could then use repertoire size as a cue in mate selection. Such a system is at least plausible in the Canary, Red-winged Blackbird, and Mockingbird (Nottebohm and Nottebohm, 1978; Kroodsma, 1976; Yasukawa *et al.*, 1981; Howard, 1974).

If the maintenance of local dialects is crucial, the timing of vocal learning must coevolve with establishment of site tenacity. A sensitive period designed to aid in restricting dispersal would occur early in life, preferably before the young have dispersed from the home locality. The early learning period among some *Zonotrichia* sparrows could contribute to the maintenance of small population sizes (see Chapter 7, this volume). Dispersal in other species may precede song learning, though; vocal learning at the site of breeding ensures that interacting conspecifics will share the same vocal signals (Kroodsma, 1974; Payne, 1975; Jenkins, 1977) and may lead to a reduced correlation between song dialect areas and populations. Dispersal distances in several songbird species are actually dependent on the hatching date (Dhondt and Huble, 1968), just as is the sensitive period for song learning in the Long-billed Marsh Wren; a sensitive period for song learning which is in part photoperiodically determined permits the dispersing juvenile some flexibility in when and where it settles and learns a local song dialect.

In general, vocal learning probably occurs at the site of breeding (irrespective of site of hatching) as a means of insuring common vocal signals among interactants. If social contexts change gradually or occasionally, songs (Payne, 1975; Jenkins, 1977) or calls (Mundinger, 1970) may be modified accordingly. Frequent changes in social context and/or a large vocal repertiore could preclude vocal changes, though, and might even select for a totally different vocal ontogeny, such as vocal improvisation (Kroodsma and Verner, 1978).

VI. CONCLUDING REMARKS

The phenomenon of vocal learning seems pervasive among birds, but it is especially pronounced among the songbirds; 290 to 294 (see below) of the 303 entries in Appendix are for oscines. If the Atrichornithidae and Menuridae are actually primitive oscines, as believed by Feduccia (1980), that leaves no solid evidence of vocal learning among the approximately 1100 sub-oscine passerines. There are two very large families in this group, the Dendrocolaptidae, with 270 species, and the Tyrannidae, with 374 species; before substantive arguments can be made concerning the role of vocal learning in rapid speciation (Marler, 1970a), further studies of vocal development in these primarily neotropical passerines are badly needed.

Unequivocal evidence for vocal learning does occur in three orders (Psittaciformes, Apodiformes, Passeriformes), while evidence for the Emerald Toucanet (*Aulacorhynchus prasinus*) of the Piciformes should be verified. Sparling (1979) would add Galliformes to the list, for his evidence of vocal learning in this order is very suggestive. Further work is clearly needed among all the non-oscines; vocalizations are usually simpler in these groups, and documenting the existence of either microgeographic variation or vocal learning may be far more difficult under such circumstances.

Determining the interplay between learned and inherited specifications continues to be a major focus of oscine song developmental studies. Marler's (e.g., Marler and Peters, 1977) refined experiments involve a search for features of vocalizations which guide selective learning in closely related sparrow species. Other comparative work is also needed in order to document how vocal development may be correlated with different life histories (e.g., Kroodsma and Verner, 1978). Thus, how does vocal development differ among migratory and nonmigratory populations of the same species, or among migratory populations of the same species living at different latitudes (and therefore under different photoperiods)? Do striking differences in repertoire sizes among some wren populations (e.g., Winter Wren, *Troglodytes troglodytes*, Kroodsma, 1980; Long-billed Marsh Wren, Verner, 1975; Kroodsma, 1979) reflect genetic differences in the capacity to learn, or is it possible that immediate environmental influences either enable or inhibit vocal learning? How might the density of conspecifics affect vocal development, or do the number of sympatric congeners influence how much can be trusted to vocal learning or how selective that vocal learning must be? Hopefully the apparent differences in song development among the oscines reflect only partially the diversity of techniques used by different investigators and reflect primarily different evolutionary strategies in vocal development. Only carefully designed studies will clarify these trends and selective pressures.

Other exciting advances will be made when we examine not only what is learned but also how birds acquire competence in using those vocalizations (see Chapter 9, this volume). Especially intriguing are those species in which different learned songs are used in different contexts. If a Chestnut-sided Warbler male learns three forms of accented-ending songs and three forms (with many variations) of unaccented-ending songs, how does he know when to use each one? And how do other males and females know how to interpret this communication system? Must this information be learned from other adults, or are there clues in the structure of each song type which dictate when and where each is to be used? While the study of semantics has predominated investigations of vocal development in songbirds, research examining the syntactic and pragmatic aspects of vocal ontogenies should prove very rewarding.

Finally, searches for sensitive periods in avian vocal learning will continue, but greater emphasis must be made on correlating the timing of vocal learning with the biology of the particular population under study. The relative timing of vocal learning and dispersal determines in large part the extent of microgeographic variation in song *and* the relationship of observed vocal dialects to population units. Documenting exactly when and where a juvenile learns his songs can be accomplished only with extensive banding and tape recording in the field. Thus, studies of the factors influencing sensitive periods in the laboratory must be complemented with field studies before the significance of the interrelationships among sensitive periods, dispersal, microgeographic vocal variation, and population structure will be understood.

ACKNOWLEDGMENT

This chapter was written while I was supported by the National Science Foundation (BNS78-02753 and BNS80-40282).

REFERENCES

Armstrong, E. A. (1963). "A Study of Bird Song." Oxford Univ. Press, London and New York.

Bateson, P. (1979). How do sensitive periods arise and what are they for? *Anim. Behav.* **27,** 470–486.

Becker, P. H. (1978). Der Einfluss des Lernens auf einfache und komplexe Gesangsstrophen der Sumpfmeise (*Parus palustris*). *J. Ornithol.* **119,** 338–411.

Bent, A. C. (1953). Life histories of North American Wood Warblers. *Bull.—U.S. Natl. Mus.* No. 203.

Bertram, B. (1970). The vocal behaviour of the Indian Hill Mynah, *Gracula religiosa*. *Anim. Behav. Monogr.* **3**(2), 79–192.

Borror, D. J. (1968). Unusual songs in Passerine birds. *Ohio J. Sci.* **68,** 129–138.

Caswell, F. D. (1972). The development of acoustical communication in the Mallard (*Anas platyrhynchos*). Master's Thesis, Univ. of North Dakota, Grand Forks.

Cosens, S. E. (1981). Development of vocalizations in the American Coot, *Fulica americana*. *Can. J. Zool.* **59,** 1921–1928.

Dhondt, A. A., and Huble, J. (1968). Fledging-date and sex in relation to dispersal in young Great Tits. *Bird-Study* **15,** 127–134.

Dittus, W. P. J., and Lemon, R. E. (1969). Effects of song tutoring and acoustic isolation on the song repertoires of Cardinals. *Anim. Behav.* **17,** 523–533.

Ewert, D. N. (1979). Development of song of a Rufous-sided Towhee raised in acoustic isolation. *Condor* **81,** 313–316.

Feduccia, A. (1980). "The Age of Birds." Harvard Univ. Press, Cambridge, Massachusetts.

Gill, F. B., and Murray, B. G., Jr. (1972). Song variation in sympatric Blue-winged and Golden-winged warblers. *Auk* **89,** 625–643.

Godman, S. (1955). "The bird fancyer's delight." *Ibis* **97,** 240–246.

Green, S. (1975). Dialects in Japanese monkeys: Vocal learning and cultural transmission of locale-specific vocal behavior? *Z. Tierpsychol.* **38,** 304–314.

Grinnell, J., Dixon, J., and Linsdale, J. M. (1930). Vertebrate natural history of a section of northern California through the Lassen Peak region. *Univ. Calif., Berkeley, Publ. Zool.* **35,** 1–594.

Güttinger, H. R. (1977). Variable and constant structures in Greenfinch songs (*Chloris chloris*) in different locations. *Behaviour* **60,** 304–318.

Güttinger, H. R., and Nicolai, J. (1973). Struktur und Funktion der Rufe bei Prachtfinken (Estrildidae). *Z. Tierpsychol.* **33,** 319–334.

Hall-Craggs, J. (1979). Sound spectrographic analysis: Suggestions for facilitating auditory imagery. *Condor* **81,** 185–192.

Heckenlively, D. B. (1970). Song in a population of Black-throated Sparrows. *Condor* **72,** 24–36.

Howard, R. D. (1974). The influence of sexual selection and interspecific communication on Mockingbird song (*Mimus polyglottos*). *Evolution* **28,** 428–438.

Immelmann, K. (1969). Song development in the Zebra Finch and other estrildid finches. *In* "Bird Vocalizations" (R. A. Hinde, ed.), pp. 61–74. Cambridge Univ. Press, London and New York.

Jenkins, P. F. (1977). Cultural transmission of song patterns and dialect development in a free-living bird population. *Anim. Behav.* **25,** 50–78.

Konishi, M. (1963). The role of auditory feedback in the vocal behavior of the Domestic Fowl. *Z. Tierpsychol.* **20,** 349–367.

Konishi, M. (1965). The role of auditory feedback in the control of vocalization in the White-crowned Sparrow. *Z. Tierpsychol.* **22,** 770–783.

Konishi, M., and Nottebohm, F. (1969). Experimental studies in the ontogeny of avian vocalizations. *In* "Bird Vocalizations" (R. A. Hinde, ed.), pp. 29–48. Cambridge Univ. Press, London and New York.

Kroodsma, D. E. (1973). Coexistence of Bewick's Wrens and House Wrens in Oregon. *Auk* **90,** 341–352.

Kroodsma, D. E. (1974). Song learning, dialects, and dispersal in the Bewick's Wren. *Z. Tierpsychol.* **35,** 352–380.

Kroodsma, D. E. (1976). Reproductive development in a female songbird: Differential stimulation by quality of male song. *Science* **192,** 574–575.

Kroodsma, D. E. (1977). A re-evaluation of song development in the Song Sparrow. *Anim. Behav.* **25,** 390–399.

Kroodsma, D. E. (1978). Aspects of learning in the ontogeny of bird song: Where, from whom, when, how many, which, and how accurately? *In* "Development of Behavior" (G. Burghardt and M. Bekoff, eds.), pp. 215–230. Garland, New York.

Kroodsma, D. E. (1979). Vocal dueling among male Marsh Wrens: Evidence for ritualized expressions of dominance/subordinance. *Auk* **96,** 506–515.

Kroodsma, D. E. (1980). Winter Wren singing behavior: A pinnacle of song complexity. *Condor* **82,** 357–365.

Kroodsma, D. E. (1981). Geographical variation and functions of song types in warblers (Parulidae). *Auk* **98,** 743–751.

Kroodsma, D. E., and Pickert, R. (1980). Environmentally dependent sensitive periods for avian vocal learning. *Nature* **288,** 477–479.

Kroodsma, D. E., and Verner, J. (1978). Complex singing behaviors among *Cistothorus* wrens. *Auk* **95,** 703–716.

Lade, B. I., and Thorpe, W. H. (1964). Dove songs as innately coded patterns of specific behaviour. *Nature (London)* **202,** 366–368.

Lanyon, W. E. (1957). The comparative biology of the meadowlarks (*Sturnella*) in Wisconsin. *Publ. Nuttall. Ornithol. Club* No. 1.

Lanyon, W. E. (1978). Revision of the *Myiarchus* flycatchers of South America. *Bull. Am. Mus. Nat. Hist.* **161,** 429–627.

Lanyon, W. E. (1979). Development of song in the Wood Thrush (*Hylocichla mustelina*), with notes on a technique for hand-rearing passerine birds from the egg. *Am. Mus. Novit.* No. 2666.

Laskey, A. R. (1944). A Mockingbird acquires his song repertory. *Auk* **61,** 211–219.

Lein, M. R. (1978). Song variation in a population of Chestnut-sided Warblers (*Dendroica pensylvanica*): Its nature and suggested significance. *Can. J. Zool.* **56,** 1266–1283.

Marler, P. (1960). Bird songs and mate selection. *In* "Animal Sounds and Communication" (W. E. Lanyon and W. N. Tavolga, eds.), Publ. No. 7, pp. 321–347. Am. Inst. Biol. Sci., Washington, D.C.

Marler, P. (1970a). Bird song and speech development: Could there be parallels? *Am. Sci.* **58,** 669–673.

Marler, P. (1970b). A comparative approach to vocal learning: Song development in White-crowned Sparrows. *J. Comp. Physiol. Psychol. Monogr.* **71,** 1–25.

Marler, P. (1976). Sensory templates in species-specific behavior. *In* "Simpler Networks and Behavior" (J. Fentress, ed.), pp. 314–329. Sinauer, Sunderland, Massachusetts.

Marler, P., and Mundinger, P. C. (1972). Vocal learning in birds. *In* "Ontogeny of Vertebrate Behavior" (H. Moltz, ed.), pp. 389–450. Academic Press, New York.

Marler, P., and Peters, S. (1977). Selective vocal learning in a sparrow. *Science* **198,** 519–521.

Marler, P., and Peters, S. (1981). Birdsong and speech: Evidence for special processing. *In* "Perspectives in the Study of Speech" (P. Eimas and J. Miller, eds.), pp. 75–112. Erlbaum, Hillsdale, New Jersey.

Marler, P., and Tamura, M. (1964). Song "dialects" in three populations of White-crowned Sparrows. *Science* **146,** 1483–1486.

Marler, P., Mundinger, P., Waser, M. S., and Lutjen, A. (1972). Effects of acoustical stimulation and deprivation on song development in Red-winged Blackbirds (*Agelaius phoeniceus*). *Anim. Behav.* **20,** 586–606.

Martin, D. J. (1979). Songs of the Fox Sparrow. II. Intra- and interpopulation variation. *Condor* **81,** 173–184.

Miller, E. H. (1979). Functions of display flights by males of the Least Sandpiper, *Calidris minutilla* (Vieill.), on Sable Island, Nova Scotia. *Can. J. Zool.* **57,** 876–893.

Morony, J. J., Jr., Bock, W. J., and Farrand, J., Jr. (1975). "Reference List of the Birds of the World." Am. Mus. Nat. Hist., New York.

Morse, D. H. (1966). The context of songs in the Yellow Warbler. *Wilson Bull.* **78,** 444–455.
Moynihan, M. (1959). Notes on the behavior of some North American gulls. IV. The ontogeny of hostile behavior and display patterns. *Behaviour* **14,** 214–239.
Mulligan, J. A. (1966). Singing behavior and its development in the Song Sparrow, *Melospiza melodia. Univ. Calif., Berkeley, Publ. Zool.* **81,** 1–76.
Mundinger, P. C. (1970). Vocal imitation and individual recognition of finch calls. *Science* **168,** 480–482.
Nethersole-Thompson, D., and Nethersole-Thompson, M. (1979). "Greenshanks." Tad Poyser, Berkhamsted, England.
Nicolai, J. (1959). Familientradition in der Gesangsentwicklung des Gimpels (*Pyrrhula pyrrhula*). *J. Ornithol.* **100,** 39–46.
Nottebohm, F. (1969). The "critical period" for song learning in birds. *Ibis* **111,** 386–387.
Nottebohm, F. (1970). Ontogeny of bird song. *Science* **167,** 950–956.
Nottebohm, F. (1972). Origins of vocal learning. *Am. Nat.* **106,** 116–140.
Nottebohm, F., and Nottebohm, M. (1971). Vocalizations and breeding behavior of surgically deafened Ring Doves (*Streptopelia risoria*). *Anim. Behav.* **19,** 313–327.
Nottebohm, F., and Nottebohm, M. (1978). Relationship between song repertoire and age in the Canary, *Serinus canarius. Z. Tierpsychol.* **46,** 298–305.
Oring, L. W. (1968). Vocalizations of the Green and Solitary sandpipers. *Wilson Bull.* **80,** 395–420.
Payne, K., and Payne, R. Large scale progressive changes in songs of humpback whales in Bermuda. *Z. Tierpsychol.* (in press).
Payne, R. B. (1981a). Population structure and social behavior: Models for testing the ecological significance of song dialects in birds. *In* "Natural Selection and Social Behavior: Recent Research and New Theory" (R. D. Alexander and D. W. Tinkle, eds.), pp. 108–120. Chiron, New York.
Payne, R. B. (1981b). Song learning and social interaction in Indigo Buntings. *Anim. Behav.* **29,** 688–697.
Payne, R. B., and Payne, K. (1977). Social organization and mating success in local song populations of Village Indigo-birds, *Vidua chalybeata. Z. Tierpsychol.* **45,** 113–173.
Peterson, R. T., Mountfort, G., and Hollom, P. A. D. (1966). "A Field Guide to the Birds of Britain and Europe." Houghton, Boston, Massachusetts.
Petrinovich, L. (1972). Psychobiological mechanisms in language development. *Adv. Psychobiol.* **1,** 259–285.
Price, P. H. (1979). Developmental determinants of structure in Zebra Finch song. *J. Comp. Physiol. Psychol.* **93,** 260–277.
Rice, J. C. (1978). Behavioural interactions of interspecifically territorial vireos. I: Song discrimination and natural interactions. *Anim. Behav.* **26,** 527–549.
Rice, J. O. (1978). Behavioural interactions of interspecifically territorial vireos. I: Song discrimination and natural interactions. *Anim. Behav.* **26,** 527–549.
Scott, W. E. D. (1902). Data on song in birds: The acquisition of new songs. *Science* **15,** 178–181.
Shiovitz, K. A. (1975). The process of species-specific song recognition by the Indigo Bunting, *Passerina cyanea,* and its relationship to the organization of avian acoustical behavior. *Behaviour* **55,** 128–179.
Slater, P. J. B. (1983). Bird song learning: Theme and variations. *In* "Perspectives in Ornithology" (G. A. Clark and A. R. Brush, eds.). Cambridge Univ. Press (in press).
Sparling, D. W. (1979). Evidence for vocal learning in Prairie Grouse. *Wilson Bull.* **91,** 618–621.
Stein, R. C. (1963). Isolating mechanisms between populations of Traill's Flycatchers. *Proc. Am. Philos. Soc.* **107,** 1–50.
Storer, R. W. (1971). Classification of birds. *In* "Avian Biology" (D. S. Farner and J. R. King, eds.), Vol. I, pp. 1–18. Academic Press, New York.

Thielcke, G. (1973). Uniformierung des Gesangs der Tannenmeise (*Parus ater*) durch Lernen. *J. Ornithol.* **114,** 443–454.

Thorpe, W. H. (1958). The learning of song patterns by birds, with especial reference to the song of the Chaffinch *Fringilla coelebs*. *Ibis* **100,** 535–570.

Thorpe, W. H. (1961). "Bird Song." Cambridge Univ. Press, London and New York.

Tinbergen, N. (1939). The behaviour of the Snow Bunting in spring. *Trans. Linn. Soc. N.Y.* **5,** 1–95.

Verner, J. (1975). Complex song repertoire of male Long-billed Marsh Wrens in eastern Washington. *Living Bird* pp. 263–300.

Wilkinson, R., and Huxley, C. R. (1978). Vocalizations of chicks and juveniles and the development of adult calls in the Aldabra White-throated Rail, *Dryolimnas cuvieri aldabranus* (Aves: Rallidae). *J. Zool.* **186,** 487–505.

Yasukawa, K., Blank, J. L., and Patterson, C. B. (1981). Song repertoires and sexual selection in the Red-winged Blackbird. *Behav. Ecol. Sociobiol.* **7,** 233–238.

2

Subsong and Plastic Song: Their Role in the Vocal Learning Process

PETER MARLER
SUSAN PETERS

I. INTRODUCTION

Story has it that Pavlov inscribed over the entrance of his laboratory the words "Description and Description." The message is familiar to ethologists and yet descriptive analysis of the developmental stages of birdsong has been largely neglected. If all oscine songs are learned, as appears to be the case (see Chapter 1, this volume), careful description and analysis of the early ontogenetic transformations to which learned songs are subject (e.g., Fig. 1) are virtually guaranteed to yield insights into the nature of the vocal learning process. There is a

ACOUSTIC COMMUNICATION IN BIRDS
VOLUME 2

ISBN 0-12-426802-1

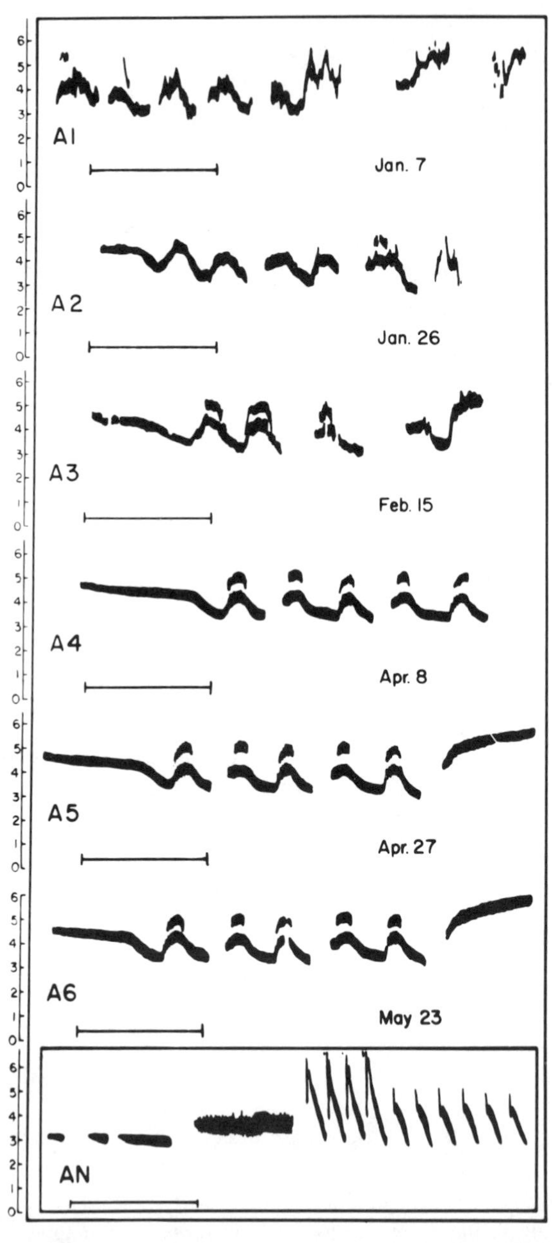

Fig. 1. A progression through subsong, plastic song, and crystallized song in a male White-crowned Sparrow reared in the laboratory and isolated from conspecific song during the sensitive period for song learning, from 10 to 50 days of age. The song dialect in his home area is also shown (AN). (From Marler, 1970. Copyright 1970 by the American Psychological Association. Adapted by permission of the publisher.)

particular need for longitudinal studies of the vocal production of particular individuals through all phases of song ontogeny. This chapter presents a case for the intensification of research on subsong and plastic song as a potential key to otherwise intractable aspects of the vocal learning process.

II. SUBSONG OF THE SONG SPARROW

An early commentary on the nature and significance of subsong is to be found in Nice's classic studies of the Song Sparrow (*Zonotrichia melodia*). Her painstaking literature review confirmed the generality of several of her Song Sparrow findings (Nice, 1943). The onset of subsong, stage 1 of song development or "warbling," as she defined it, has been recorded in a variety of songbirds between 2 and about 10 weeks of age. With the exception of Bourke's Parrot (*Neophema bourkii;* Hampe, 1939), all of her illustrations are of oscines. Subsong does not in fact seem to have been recorded in such intensively studied bird orders as Galliformes, Charadriiformes, and Columbiformes (see Nottebohm, 1972a), despite their possession of structurally elaborate patterns of crowing, calling, and cooing that are, in many respects, functional equivalents of oscine song. The crucial point appears to be that these vocal patterns are innate. Nottebohm (1972a) has assembled persuasive evidence that song learning has evolved independently in Passeriformes, Psittaciformes, and Apodiformes. To date subsong seems only to have been recorded in these three orders (e.g., Armstrong, 1963; J. W. Hardy, unpublished data; Kroodsma, 1974; Lanyon, 1957, 1960; Lemon and Scott, 1966; Nice, 1943; Nottebohm, 1975; Poulsen, 1959; Stiles and Wolf, 1979 and unpublished data; Thorpe, 1961). We may tentatively conclude that, as a preliminary stage in song ontogeny, subsong is restricted to those bird groups that learn their songs, namely, songbirds, parrots, and hummingbirds. The corollary also seems to be true; all birds that learn to sing go through a subsong stage in song ontogeny.

In her analysis of Song Sparrow song development, Nice (1943) distinguished five stages: (1) Continuous warbling; (2) Some short songs, much warbling, intervals between songs much shorter than song length; (3) Predominantly short songs (but not yet crystallized into adult form), length of intervals about equal to that of songs, some warbling; (4) Songs practically adult in form, intervals usually twice as long as songs, a period of trying out songs (adding some, deleting others); and (5) Songs adult and sterotyped except for endings restricted to the final repertoire. Variations in song duration served as useful markers for the five stages. Data on the age at which her Song Sparrows attained the different stages indicate a rough succession, although Nice was at some pains to indicate that there is not necessarily a continuous forward progression from one to the next. Moreover, there was considerable individual variation in the age at which stages were first achieved, and in the rate of progress from warbling through to full song. This was achieved by most wild males in about 5 months, but by one in an estimated 11 days. Four hand-raised Song Sparrows followed a similar sequence. They tended to achieve the different stages at a younger age than wild birds, with the first warbling at 2 to 3 weeks, and the first adult songs at just under 200 days of age. The difference may reflect no more than the difficulty in hearing and recording subsong in the field.

III. SUBSONG AND PLASTIC SONG IN THE CHAFFINCH

Another focus of research on the sequential structure of developing song is the Chaffinch (*Fringilla coelebs*). Thorpe adopted from Nicholson the term ''subsong'' for the Chaffinch equivalent of Song Sparrow ''warbling'' (Nicholson, 1927; Nicholson and Koch, 1936; Thorpe, 1955, 1958; Thorpe and Pilcher, 1958). The developmental sequence in the two species is rather similar. Chaffinch subsong is a quiet, rambling series of notes, continuous or broken up into fragments lasting 2 or 3 sec (Marler, 1956). The notes are variable, but fall into two groups; one includes a variety of chirping notes, rendered ''tchirp,'' 'tcheep,'' 'chip,'' 'seep,'' etc., the other consists of short mechanical sounding rattles. The earliest subsong consists only of the chirping notes. Later, the rattle is inserted into the chirp sequences. The characteristic form becomes 10 or 20 chirping notes, followed immediately by a rattle, then a pause before another series starts. This segmented form of subsong then changes to irregular ''plastic song'' with the chirp sequences and the terminal rattle replaced by the two components of song, the trill and the end phrase (Marler, 1956; Nottebohm, 1968; Thorpe, 1958).

In the Chaffinch the transition from plastic song to full song is a smooth one. Intermediate forms may involve a trill followed by a rattle, or a chirp sequence with an end phrase. Changes in timing and frequency modulation of the chirp sequences result in the trills of full song. The rattle becomes transformed into the end phrase of full song. Although different in details, the developmental sequences of Chaffinch and Song Sparrow song are broadly similar. Equivalent transitions from stage to stage are marked by similar changes in temporal organization and note structure. The three major categories of Chaffinch subsong, plastic song, and full song correspond well with Nice's Song Sparrow stages 1 and 2 (subsong), 3 and 4 (plastic song), and 5 (full song). Again the sequence is not obligatory, and in particular subsong may occur outside the developmental sequence, particularly in adult precopulatory courtship (Marler, 1956).

The Chaffinch does not mimic other species in its full song. However, there are sufficient records in the literature of Chaffinches incorporating imitations of alien species in subsong and plastic song to suggest that this is not an uncommon occurrence in the wild (see review in Thorpe, 1955). One captive Chaffinch, housed with Canaries (*Serinus canaria*), copied fragments of Canary song only to drop them completely with the transition to full song (Thorpe, 1958). The inclusion of sounds of other species in intermediate stages of song development, and their rejection from full song, also occurs in some other songbirds (reviewed in Güttinger, 1978). This has yet to be recorded in the Song Sparrow, however, a point of interest in light of the more recent discovery that Song Sparrows exert a degree of selectivity in rejecting heterospecific songs at an earlier stage of the learning process (Marler and Peters, 1977, 1981a,b).

Further comparisons of subsong and full song in thrushes, finches, and a few other species yielded contrasts similar to those in the Chaffinch (Thorpe and Pilcher, 1958). Figure 2 illustrates typical subsong for four male Swamp Sparrows (*Zonotrichia georgiana*). Recurring differences included *volume*, with subsong being quieter; *patterning*, with a different structure in subsong; *duration*, longer in subsong; *frequency*, with subsong tending to have a lower fundamental

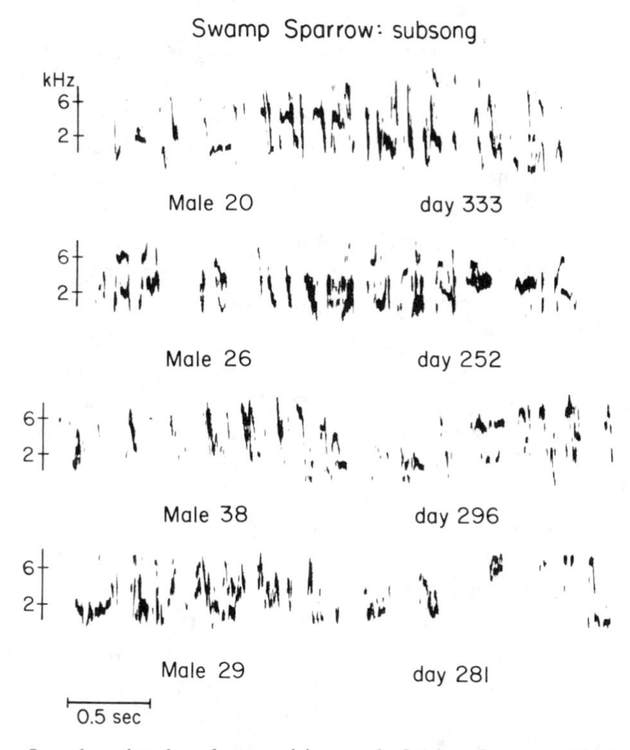

Fig. 2. Samples of early subsong of four male Swamp Sparrows illustrating its wide frequency range and amorphous structure as compared with full song (cf. Fig. 5). Consistent individual differences are much less evident in early subsong than in later stages of song development. (From Marler and Peters, 1982c.)

frequency; *frequency range,* tending to be greater in subsong; and *seasonal timing,* with subsong tending to occur earlier in the season.

With regard to the functional significance of subsong, Thorpe argued for a role as practice, assuming that the experience gained would aid in motor development. He also speculated about its perceptual significance:

> If we suppose that the animal is endowed by nature with a number of vocal motor mechanisms which enable it to utter a variety of sounds, then whenever one of these is set in action the animal hears its own voice uttering a corresponding sound. In consequence, the sense impression becomes associated with that motor mechanism. Now suppose the animal hears the same sounds uttered by another. The sound will have the same effect, namely it will have already been associated with the vocal motor mechanism (Thorpe, 1961, pp. 79–80).

The theme that auditory feedback from subsong might have significance in the development of song perception was taken up with renewed vigor by Nottebohm (1968). After deafening young male Chaffinches at various stages in the transition from subsong to full song, Nottebohm found a loss of structure after early deafening, but retention of structure once plastic song was achieved. He inferred that motor or auditory experience, or both, occurring in song ontogeny, contribute to the retention of song characteristics after severance of the auditory feedback loop. New impetus was given to the functional interpretation of subsong with Nottebohm's subsequent discovery of the lateralization of neural control of the songbird syrinx, with one side, typically the left, assuming dominance (Nottebohm, 1971, 1972b, 1977; Nottebohm and Nottebohm, 1976).

The earlier "dual-sound-source" interpretation of syringeal function (Greenewalt, 1968; Hersh, 1965) had already posed an implied ontogenetic question. How does the exquisitely precise coordination of operations on two sides of the syrinx develop? By analyzing normal call structure, studying the effects of deafening, and exploring the impact of hypoglossectomy on song development, Nottebohm showed that coordination in nestlings is quite imprecise. There is considerable improvement by the time the fledging "chirp" is produced. With the advent of subsong this trend continues and the growing dominance of the left side of the syrinx becomes discernible.

Denervation of either side of the syrinx before song crystallization leads the other side to produce loud and sustained single-trace chirps. However, early left hypoglossectomy results in an inability to group the chirp and rattle components of subsong appropriately, as though "the overly fast alternation of the two dissimilar notes cannot be mastered by one side of the syrinx alone. To this extent the rattle seems designed to ensure the harmonious participation of both hypoglossi, thus affording the bird a chance to practice the temporal integration of its two sound sources" (Nottebohm, 1972b, p. 47). This result is consistent with the notion that experience gained during performance of the subsong/plastic song/full song transitions may be directly relevant to the achievement of fully normal syringeal operation. Nottebohm also goes on to speculate, along with

Hinde (1958) and Thorpe (1954, 1961), that there may be perceptual consequences as well. ''Successive developmental stages may also help in the identification of conspecific song models'' (Nottebohm, 1972b, p. 47). By framing hypotheses in terms that permit testing, Nottebohm thus brought analysis of the functional significance of subsong and plastic song into the domain of experimental science.

IV. SONG ONTOGENY IN THE SWAMP SPARROW

In a more recent and exhaustive longitudinal analysis of song development, Swamp Sparrows were raised by hand in the laboratory and exposed to song between about 20 and 60 days of age (Marler and Peters, 1982a). The birds were isolated in soundproof boxes, on a normal upstate New York photoperiodic cycle, and recorded occasionally up to 3 months of age. Then, beginning at 90 days, each was recorded once a week for 1 hr at dawn, until they were about 400 days old. Altogether, about 15,000 samples of singing behavior were recorded. They sang sporadically in late summer and autumn, and refrained from song in winter, as Nice (1943) also noted in the Song Sparrow. Song began in earnest in the early spring. These trends are apparent in Fig. 3 which shows the number of birds recorded singing in successive weeks of the study. Autumn singing peaked at around 100 days of age, but only a quarter of the birds were recorded in song at that time, although every male probably participated at one time or another. Such

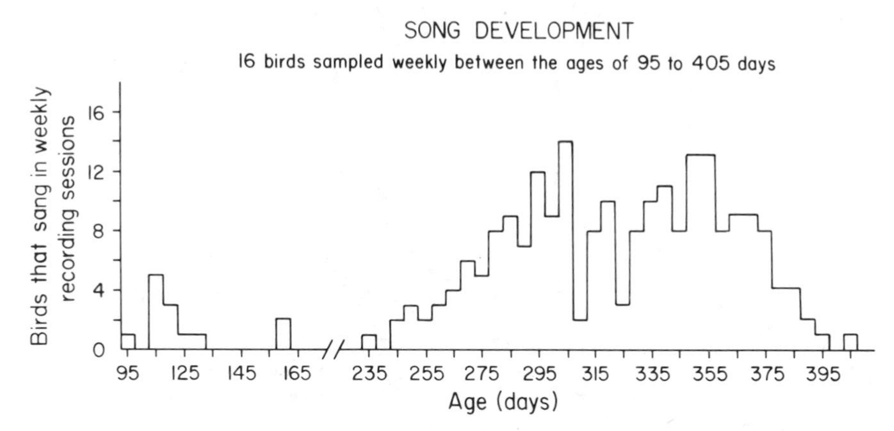

Fig. 3. Numbers of male Swamp Sparrows recorded singing in weekly early morning 1-hr recording sessions from 95 to 400 days of age. Sixteen hand-reared males kept in sound isolation chambers sang sporadically into autumn and ceased during the winter. In spring all came into frequent song, so that more than half were recorded in every session. (From Marler and Peters, 1982a.)

sporadic song started as early as 25 days of age, and peaked at approximately 100 days (Fig. 3). Then at about 250 days one male after another began to sing regularly. Most males were recorded in song every week thereafter, until at 400 days or so singing began to decline.

All singing was analyzed by real time sound spectrography (Hopkins *et al.*, 1974), and seven developmental stages were defined (Table I). These were based mainly on the structure and pattern of delivery of song syllables, which are the major components of the relatively simple, trilled song of the Swamp Sparrow (see Marler and Peters, 1982a). Stage 1 is crystallized song, stages 2–4 are phases of plastic song, stages 5 and 6 comprise subplastic song, and stage 7 is subsong. These correspond roughly to Nice's classification for Song Sparrows which relied more on song duration.

Nice noted that songs tend to shorten as male Song Sparrows progressed toward full song. This is also true of the Swamp Sparrow (Fig. 4). There was a strong and abrupt decrease in song duration and its variability around 330 days of age as males achieved stage 1 crystallized song for the first time. Data for individual males gave a very similar picture. In each case the sharp reduction in song duration and its variability coincided with other criteria for crystallization, as indicated below.

Figure 5 illustrates some of the developmental stages recorded for a single male Swamp Sparrow, with an example of stage 7 subsong at the top, and one of this male's two crystallized songs at the bottom. Some of the features of subsong noted by Thorpe and Pilcher (1958) can be detected, including the very different overall patterning of notes, the wider frequency range, and the tendency for a fundamental frequency lower than that of full song. Syllables of the original models are indicated in the box at the bottom right of the figure.

TABLE I

Stages of Song Development

Stage		Syllables		Songs	
		Morphology	Repetitions	Morphology	Duration
Crystallized song	1	Stereotyped	Clear trills	Stable order	Short
Plastic song	2	Stereotyped	Clear trills	Variable order	Short
	3	Minor variations	Clear trills	Stable order	Longer
	4	Variable	Clear trills	Variable order	Long and variable
Subplastic song	5	Rudiments	Some	Variable order	Long and variable
	6	Rudiments	None	Variable order	Variable
Subsong	7	None	None	None	Variable

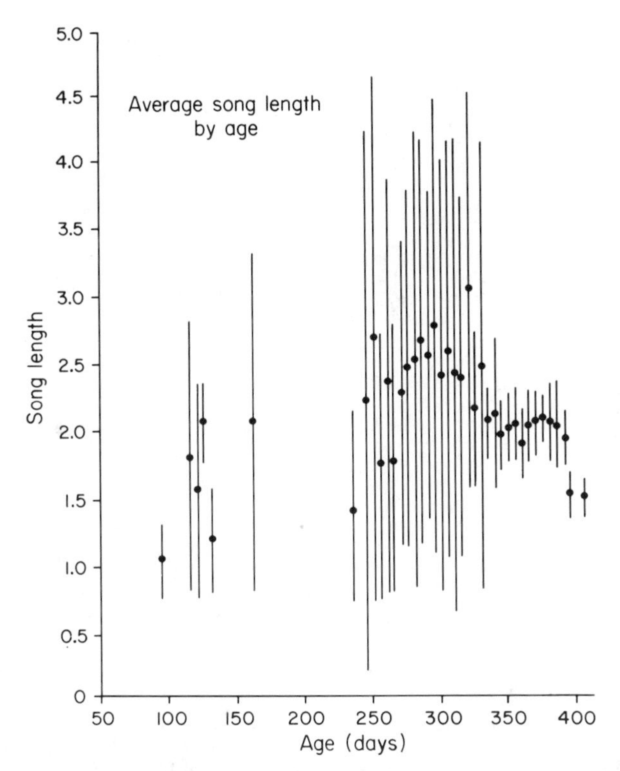

Fig. 4. Mean durations of developing songs of 16 male Swamp Sparrows in relation to age. As indicated by the standard deviation bars, variability decreases abruptly at the time of song crystallization, around 330 days of age. (From Marler and Peters, 1982a.)

Syllabic structure is at first exceedingly variable, and several syllable types may be included in a song. Syllables gradually become more stereotyped and at the time of crystallization the number of syllable types per song is reduced, typically to just one. Two-parted songs with different syllable types in each part do occur rarely in nature (D. E. Kroodsma and R. Pickert, unpublished data) and in the laboratory. The males we studied were trained with many two-parted songs, and produced some songs of this type, although one-parted songs predominated.

Nice (1943) emphasized the regularity of song development in her hand-reared Song Sparrows. To explore this question in Swamp Sparrows we simplified the classification of developmental steps by lumping stages 2–6 as plastic song. Figure 6 shows the spring onset of the three main phases of song development for all 16 male Swamp Sparrows. The onset of subsong in early spring ranged from 235 to 300 days ($\bar{X} = 270$). Prior to this all males had also given subsong in late

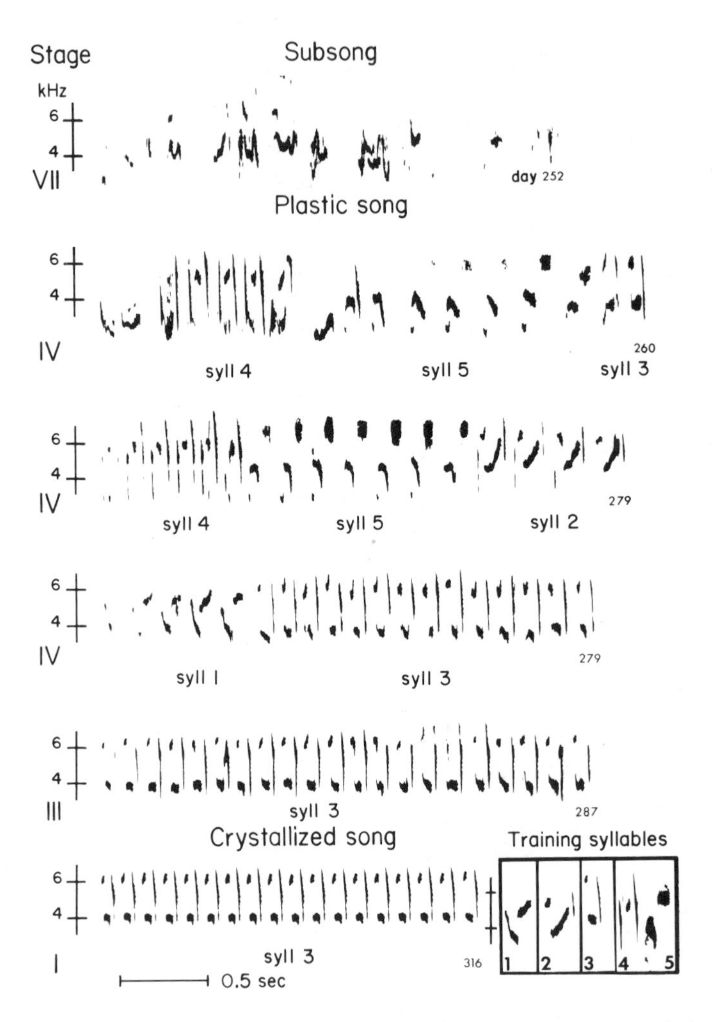

Fig. 5. Song development in a male Swamp Sparrow (No. 26). Four of the seven developmental stages are illustrated, beginning with a sample of subsong (stage 7) from day 252 and culminating with full song (stage 1) on day 316. Some of the training syllables experienced between days 22 and 62 are shown on the bottom right. Corresponding numbers identify imitations of these syllables above. The samples of stage 4 plastic song illustrate early syllabic imitations produced following a 7-month interval since the last exposure to the models. Note the amorphous, asyllabic structure of subsong, the gradual achievement of stable syllables as plastic song proceeds, and the culling of syllable types as song crystallizes.

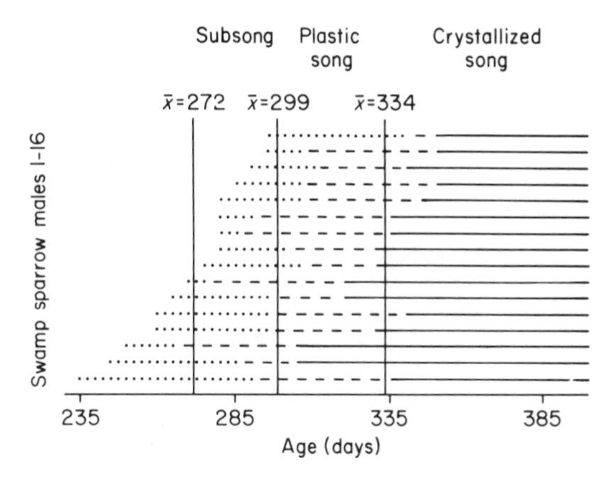

Fig. 6. The spring onset of subsong (dotted), plastic song (dashed), and crystallized song (solid) in hand-reared male Swamp Sparrows. Vertical bars mark the mean age at the onset of each stage. (From Marler and Peters, 1982a.)

summer and early fall, but had produced no plastic song whatsoever. Plastic song first emerged during a 74-day period between 266 and 340 days of age. Fully crystallized song followed predictably at around 10–12 months of age, slightly later than under field conditions in upstate New York.

Thus song development in captive Swamp Sparrows follows a regular sequence. Up to about 8 months of age only stage 7 subsong is produced, and this is interrupted by a silent period in midwinter. After the advent of plastic song at about 10 months, song development then proceeds to completion in 1 to 2 months. As in the Song Sparrow, individual Swamp Sparrows regress and recycle repeatedly, and several stages may be recorded on any one day. Despite this variability, males progress through the developmental sequence with strong regularity (Fig. 7). Such a progression agrees with Nice's (1943) description of song development in the Song Sparrow, and coincides with more qualitative reports for the Chaffinch, and a few other songbirds. Details of the structural transformations that occur must vary widely across species, considering the range of complexity which exists in crystallized song.

A proviso must be added. The Swamp Sparrows were trained in infancy with an abundance of acceptable song models. Swamp Sparrows untrained in infancy remain unstudied. White-crowned Sparrows (*Zonotrichia leucophrys*) that go untrained, or are trained with unacceptable models, exhibit a very extended developmental sequence, with crystallization occurring later than in trained birds (Marler, 1970).

Programs for song development may prove to vary considerably according to the natural variation of individual histories. Kroodsma has shown that in Long-billed Marsh Wrens (*Cistothorus palustris*) song development proceeds very differently in birds born late in the season, or exposed prematurely to shortening

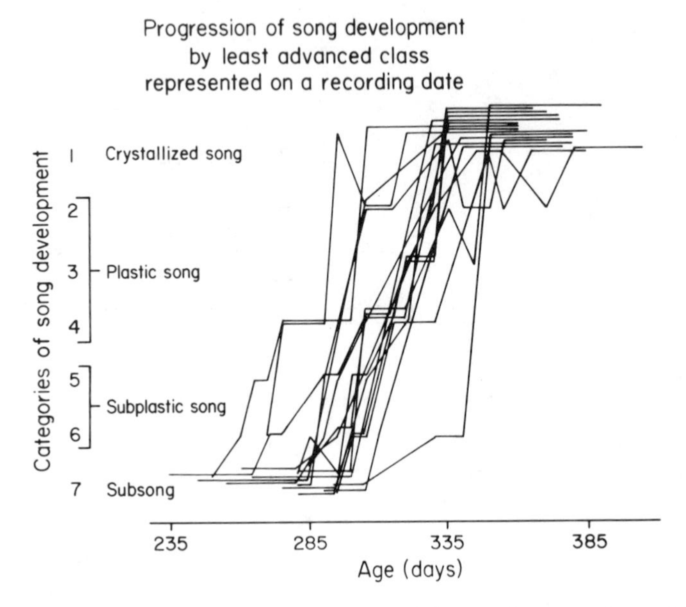

Fig. 7. A graph of the ages at which hand-reared Swamp Sparrows first achieved the seven stages of song development. Although backtracking occurred, the tendency to progress through the series is clearly evident. (From Marler and Peters, 1982a.)

days (Kroodsma and Pickert, 1980). Such variations may offer clues as to the functional significance of the developmental steps in song ontogeny.

What then is the functional significance of these various developmental stages? Are they just manifestations of a process of maturation, more or less dictated by changing androgen titers, or does the completion of one stage have direct functional implications for the next? Given the correlation between the incidence of subsong and of song learning, it is natural to assume a functional connection with the production of imitations. Moreover, it is hard to imagine song learning proceeding without a stage of vocal experimentation. How, otherwise, could the male songbird achieve the accuracy of imitation that is commonplace in studies of avian vocal learning?

V. WHEN DO THE CHARACTERISTICS OF CRYSTALLIZED SONG FIRST APPEAR?

Analysis of the exceedingly variable developmental stages of song is not an easy undertaking. One way to begin is to work backward from crystallized song to see when its particular characteristics are first detected. In the Swamp Sparrow, the distinctive syllabic morphology of the crystallized songs of the 16 intensively studied males (33 songs, 45 syllable types) was first detected at an average age of 302 days, 32 days before crystallization ($\bar{X}$ = 334 days). Despite their varied and unstable structure, even in very early renditions, the crystallized syllables of every male could be detected, but they were not the only song constituents, nor even necessarily the dominant ones. In fact, the birds produced many syllable types that were never manifest in crystallized song at all.

An analysis of all syllable types produced in stages 4 through 1 yielded an average of 12 per male, of which only three, on average, made it through crystallization. Thus song syllables are subject to attrition as development proceeds (Marler and Peters, 1982b). These trends are summarized in Fig. 8. Only a week before crystallization, males are using twice as many syllable types as afterward. The number is highest about 40 days prior to crystallization, one male achieving as many as 19 at this time. The maximum is achieved very early in plastic song development, and declines thereafter.

Male Swamp Sparrows thus produce a large amount of song material that is never used in full song. We lack the data to determine how widespread this process is, but it is not unique to the Swamp Sparrow. The strongest supporting evidence comes from Nice's Song Sparrows.

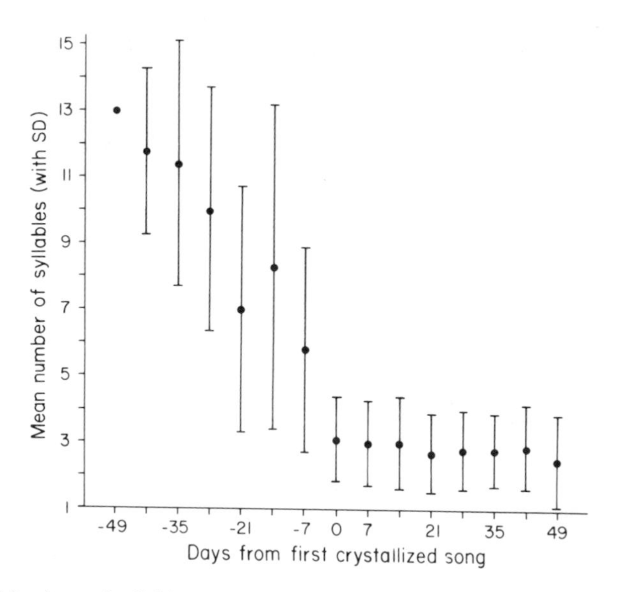

Fig. 8. Numbers of syllable types produced by hand-reared male Swamp Sparrows in the transition from plastic song to full song. One standard deviation on each side of the mean is indicated in relation to the day of song crystallization. Sample sizes ranged from 1–16. (From Marler and Peters, 1982b.)

Despite unavailability of the sound spectrograph, Nice gathered some remarkable data on structural changes that take place in the transition from subsong (''warbling''—stage 1) to full song (stage 5). In particular she detected culling of song material in the transitions through her stages 3 and 4 to full song, as indicated by the following quotation.

> Again and again I have heard young males in Stage 3 repeating exactly the songs of their adult rivals in territory establishment activities. Many of these 'imitations' are dropped, while others are adopted into the repertoire, usually in somewhat changed form. Many young birds, however, appear entirely uninfluenced by the singing of their neighbors as far as the form of their songs is concerned. Each young Song Sparrow has a large fund of potential songs, as is clear from listening to the rambling warblings of juveniles. Some individuals, before their singing becomes set in the adult mold, are capable of singing the songs of various other song sparrows hurled at them by territorial rivals. Perhaps what seems to be imitation is rather a calling forth of songs already in the potential repertoire, most of them being later lost through disuse (Nice, 1943, p. 139).

Similarly in laboratory-raised Red-winged Blackbirds (*Agelaius phoeniceus*), ''It is quite common for subsong and plastic song patterns to be lost from the repertoire before they crystallize into full song'' (Marler *et al.*, 1972, p. 588). In the White-crowned Sparrow, several hand-reared males ''passed through a phase of development in which two somewhat variable song types were used. As full song was attained and the patterns crystallized, one of these dropped out, leaving a single theme as the individual repertoire'' (Marler, 1970, p. 19). As already noted, there are frequent references in the literature to imitations of other species incorporated into early developmental stages that are rejected in the transition to full song (Lanyon, 1957; Poulsen, 1959; Thorpe, 1955, 1961). Thus the development of more song material in plastic song than is necessary for full song appears to be widespread among songbirds.

VI. LEARNING TO SING FROM MEMORY

It is common for birds to learn songs early in life and to come into full song themselves much later. In this sense they often appear to learn to sing from memory (Lanyon, 1957, 1960). However, many observers have noted imitations in various phases of song ontogeny. Early imitations would serve to bridge the gap between song learning and the production of full song, making the feat of memory less impressive than it otherwise appears. Our Swamp Sparrow studies permit us to measure this time interval between experience and rehearsal for the first time (Marler and Peters, 1982c).

Swamp Sparrows begin singing as early as the first month of life. Some males, with their last exposure to tape recorded songs between 55 and 65 days of age ($\bar{X}$ = 62 days), began singing even before training ended. On analysis, all of the

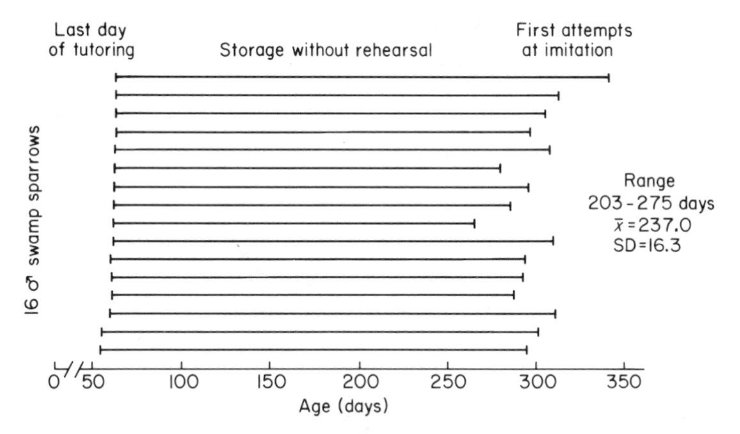

Fig. 9. Durations of periods from the last day of song exposure to the first production of an identifiable imitation, in plastic song in 16 individually isolated male Swamp Sparrows. (From Marler and Peters, 1982c. Reprinted with permission of the publisher.)

early singing proved to be subsong. No imitations could be detected until about 8 months later ($\bar{X}$ = 237 days). Values for this time interval between the first rehearsal of a learned song and the time that song was last heard are shown in Fig. 9. It is thus clear that continued access to a model is not necessary for imitation to occur. The production of imitations can be guided entirely from memory, after an interval without exposure of up to 8 months or more.

When syllable imitations appear in plastic song, they are not perfectly formed (Fig. 5). Imitations emerge gradually, beginning as approximate and variable renditions and gradually stabilizing on close copies of syllables that may have been neither heard since infancy nor rehearsed. The sequence of transformations that each undergoes is what might be expected if the experience of plastic song aids the bird in learning to control vocal operations by ear. Presumably the bird detects the errors made in early renditions and gradually corrects them by repeated practice. This continues until precise stereotypy and replication are achieved, somewhere late in plastic song development.

Although it seems reasonable to interpret plastic song as providing practice in the production of vocal imitations, this is hardly proven, and there are some puzzling features. One is the very early emergence of recognizable syllables in plastic song. If the male Swamp Sparrows first embark on syllable imitations purely by trial and error, it is surprising not to find a longer period of experimentation before identifiable syllables appear. Interestingly, the same observation has been made in the White-crowned Sparrow, in which effects of earlier training are evident very early in plastic song (Marler, 1970, p. 20).

Skill in guiding vocal production by ear may actually be refined during subsong, and only brought to bear on producing precise imitations later, in plastic song. The immediacy with which syllabic units appear in plastic song and the emergence of syllables independently of their original acoustic context suggest that learned material is committed to memory in a particular form, with syllables, or trills of the same repeated syllable, forming coherent and somewhat separate units.

VII. THE ROLE OF IMPROVISATION

While we have emphasized imitation as a major determinant of song development, many songbirds fail to produce precise imitations of models. Processes other than imitation are clearly involved. This proved to be the case in the Swamp Sparrow (P. Marler and S. Peters, unpublished data). When we compared syllables produced in the laboratory (N = 199) with those used for training in infancy only about one-third (N = 59) qualified as copies. Only 19 of these survived the process of crystallization to become part of the final complement of 45 syllable types. Thus a majority of crystallized syllables failed to qualify as imitations. What can we say about the other kinds of processes involved here?

When a musician takes a melody and subjects it to a series of experimental transformations, we speak of them as variations on that theme. There is widespread evidence of a similar process in birdsong, especially during plastic song development (Hall-Craggs, 1962; Messmer and Messmer, 1956; Sotovalta, 1956; Thorpe, 1961). Since the term "variation" is subject to other interpretations, we propose to adapt the term "improvisation" for this process. Although not strictly according to dictionary definition, the term is often used in music and by some ornithologists (e.g., Lemon, 1975) as synonymous with the composition of variations on a theme. As such, it is a component in what has been dubbed "cultural mutation" (Jenkins, 1977; Slater, 1979). We propose to restrict the term "invention" to generation of new themes that appear to emerge suddenly rather than as progressive variations on an imitated theme. Although the distinction is not always easily made, and borderline cases are frequent (see Thorpe, 1961), there are unequivocal cases of improvisation in birdsong as defined here.

There are examples of birds memorizing a particular pattern of sounds, replicating that pattern at some future date, and then in the course of plastic song subjecting it to a series of transformations, as though assuaging an appetite for novelty. An example is provided by song development in laboratory-reared Red-winged Blackbirds trained with songs of their own species and those of the Northern Oriole (*Icterus galbula;* Marler *et al.*, 1972).

As shown in Fig. 10, one male crystallized a song that was oriole-like in quality, although its patterning bore little detailed resemblance to the model to which the bird had been exposed. Investigation of earlier versions of that theme

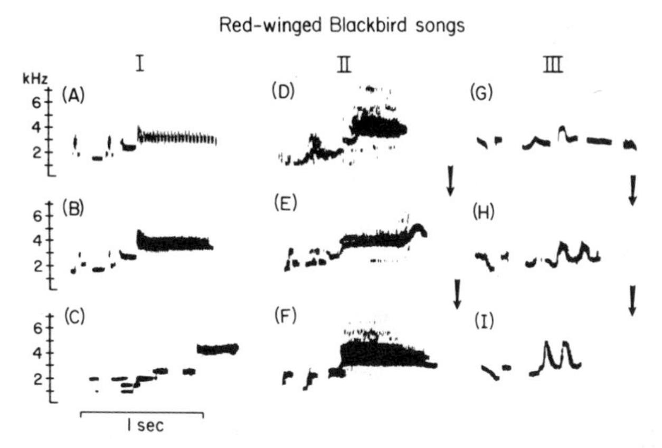

Fig. 10. Song learning, invention, and improvisation in the Red-winged Blackbird. (I) This male was trained with natural song A during infancy. At maturity he produced a copy of this song, B, and a second song, C, unlike any of the training songs and apparently invented. It is typical for this species that B closely resembles the introduction and general structure of the model, A, but deviates from the model in the fine structure of the second trill portion of the song. (II) The metamorphosis from subsong, D, through plastic song, E, to fully crystallized song, F, is illustrated for a song of one male. Processes of improvisation appear to be involved in the genesis of this song. (III) Red-winged Blackbirds will learn taped songs of the related Northern Oriole (e.g., song G) as readily as those of their own species. Imitations in full song are often imperfect copies, however (compare G and I). A precursor of I in plastic song, H, was closer to the model but was then modified by progressive improvisation. (From Marler *et al.*, 1972. Adapted and reprinted with permission of the publisher.)

in plastic song revealed that the match to the model had been more precise. The theme could be followed through a series of transformations arriving at the rather different pattern that was crystallized. The term "improvisation" seems fully appropriate here.

In captive Swamp Sparrows a certain proportion of the "noncopied" syllables developed by trained birds were explicable as cases of "copy error." About one-quarter of the noncopied syllables produced by the cohort of 16 males studied were in fact classified as poor copies rather than as noncopies. At least two processes are involved in the genesis of the poor copies. Occasionally, improvisation seemed to play a part. As illustrated in Fig. 11, early renditions of some syllables conformed more closely to a model in early rather than in late plastic song. Intermediate stages suggest that a series of progressive variations on a theme had taken place. However, this occurred rarely and, under the conditions of this study, improvisation appeared to make only a minor contribution to Swamp Sparrow song development.

Some Swamp Sparrow syllables appeared to be only approximate imitations of

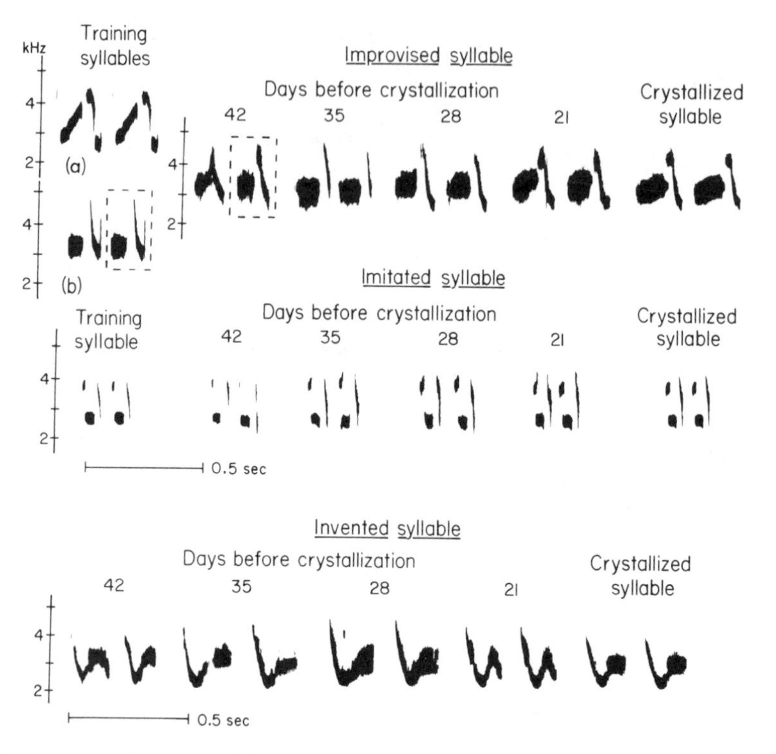

Fig. 11. Development of three Swamp Sparrow song syllables: one arising by improvisation (top), one by imitation (middle), the third by invention (bottom). Pairs of syllables are shown for each date, beginning at 42 days before crystallization. Of the two training syllables shown at the top left, the lower one, B, is believed to be the model for the improvised syllable. An approximate match is seen on day 42 (the boxed syllable). Then the second note is gradually transformed, finally coming to bear a slight resemblance to the second note of the upper training syllable, A. No other training syllables bear even a remote resemblance to the crystallized form of this syllable, which is thus classified as an improvised version of B. For comparison the development of an imitation syllable is also shown. Finally, an illustration of Swamp Sparrow syllable invention is shown. The male at the bottom was trained in youth with the 16 syllable types shown in pairs in Fig. 13. The invented syllable does not match any of the training syllables, even at early stages. However, it proceeds through the same developmental stages as improvisations and imitations, with the final crystallized form shown on right. (From P. Marler and S. Peters, unpublished data.)

a model from the outset. These could result from a failure to memorize the syllabic structure of a model accurately, or from a failure to produce a good match. In either case errors in the imitative process are involved (see Lemon, 1971, 1975; Payne, 1973). These also appear to be relatively rare in the Swamp Sparrow (P. Marler and S. Peters, unpublished data). The great majority of noncopied syllables were classifiable as "inventions," seemingly with no precursors among any of the models with which the males had been trained.

Improvisation can create not only new acoustic elements, i.e., "elemental improvisation"—but also new temporal rearrangements of existing acoustic elements—"combinatorial improvisation." Although good data are lacking, something of the latter kind must take place in the generation of the many song-types produced by such species as the Mistle Thrush (*Turdus viscivorus;* Isaac and Marler, 1963). In various thrushes (Marler, 1959), the Cardinal (*Cardinalis cardinalis;* Lemon and Scott, 1966), and other species (Lemon, 1975), one finds distinct stable themes in full song that consist of different combinations of the same basic song syllables, and which probably originated in ontogeny by combinatorial improvisation (Fig. 12).

Recombinations of learned syllables occur in the plastic song of both Swamp and Song sparrows. These sometimes demonstrate the intrusion of yet another process. The separation and rearrangement of learned syllables can result in a pattern that is innately realized, as when a Swamp Sparrow learns a syllable from a complex, multipartite song but reproduces it as a simple trill (Marler and Peters, 1977). Here the changed patterning is not attributable to an appetite for novelty, but rather to the use of learned syllables in a pattern that is in some degree genetically preordained.

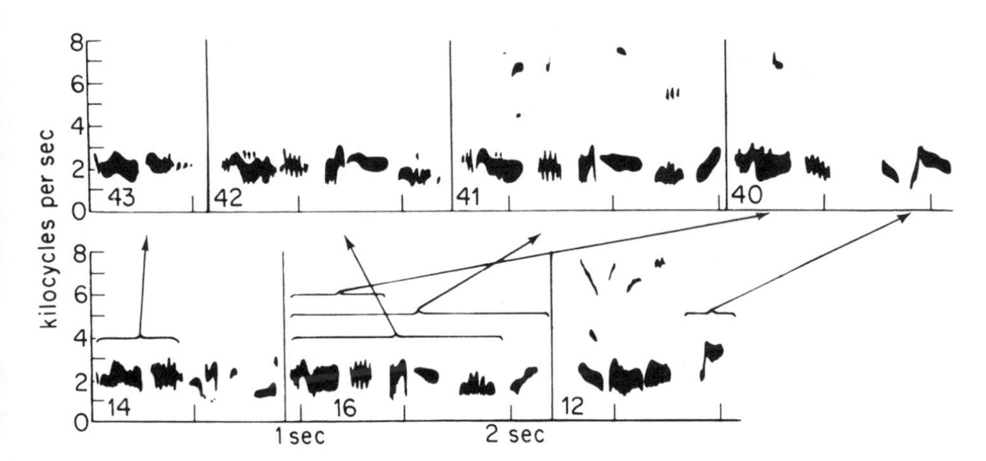

Fig. 12. An example of syntactical rearrangement of syllables in birdsong. The last four songs in a long sequence given by a Mistle Thrush are compared with earlier songs in the sequence to show how new themes must have been produced by "combinatorial invention." Numbers represent the order in the song sequence. (From Marler, 1959. Reprinted with the publisher's permission from P. R. Bell, ed., "Darwin's Biological Work," Cambridge University Press.)

VIII. THE ROLE OF INVENTION

How then should one view the large proportion of syllable types generated by Swamp Sparrows that do not seem to be created either by imitation, improvisation, or as copy errors? Inspection of their morphology reveals that they are variable and complex, more so than the simple one-note frequency inflections that constitute typical innate songs of a male Swamp Sparrow reared with no conspecific training. As illustrated in Fig. 11, such "invented" syllables seem to have a genesis similar to that of imitated ones. They begin as variable, imperfectly repeated renditions and then gradually become stereotyped.

At first sight, the sound spectrograms of invented syllables appear similar to those of copied syllables. However, a quantitative analysis of the number of distinct notes from which each syllable is constructed reveals an interesting difference. For wild birds we obtained an average syllable/note count of 2.8 (*N*

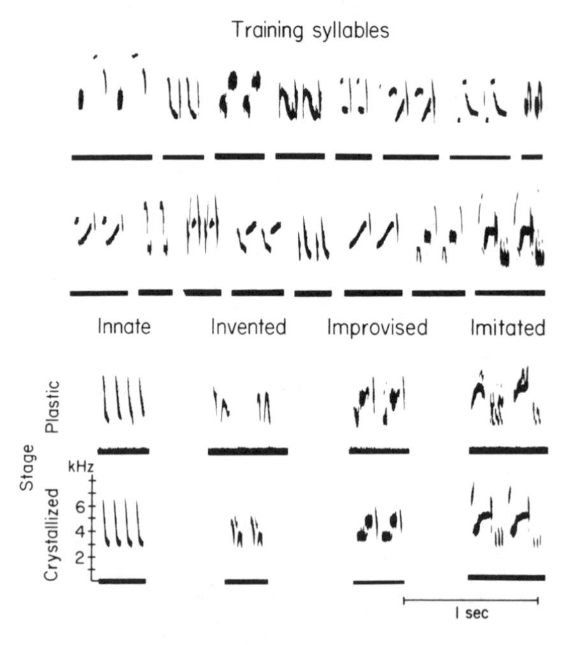

Fig. 13. Pairs of 16 song syllables used for training are shown above. Also shown are three pairs of syllables, in both plastic and crystallized versions, developed by males exposed to the training syllables. These three syllables were classified as inventions, improvisations, and imitations. Training syllables used as models are shown immediately above the improvised and imitated syllables. For comparison, a pair of innate syllables developed by an untrained Swamp Sparrow appears on the left. (From Marler and Peters, 1981b. Reprinted from *Science* 213, 780–782. Copyright 1981 by the American Association for the Advancement of Science.)

= 25 birds, 60 syllables). Similarly, in 16 males trained in captivity the note count for their copied syllables was 2.8 (N = 10 birds, 19 syllables). This was no different from that in the wild, as expected since that is where the models come from. However, the note/syllable count for a sample of the crystallized syllables judged to be invented was 1.42 (N = 10 birds, 19 syllables), significantly lower than that for copied syllables, yet still greater than that for syllables developed in the innate song of untrained males (N = 3 males, 6 syllables, $\bar{X}$ = 1.1 notes per syllable). Figure 13 gives examples of all four syllable classifications.

These data are still scanty, but they suggest vocal invention is stimulated by the acoustic stimulation that trained birds receive, even though invented syllables fail to achieve the complexity of imitated syllables. This speculative interpretation would help in understanding observations on some other species, such as the Yellow-eyed Junco (*Junco phaenotus;* Marler, 1967). As a result of training with conspecific song, males of this species develop more complex songs than in isolation, both in syllabic structure and in gross organization, even though they do not seem to specifically imitate the particular songs heard. There are difficulties of interpretation here. A junco might learn the general organization of a training song and yet build its own version of that organization with invented syllables. The potential interplay between imitation and invention is fascinating and complex, and clearly deserves closer attention in future research.

Studies of subsong and plastic song have thus revealed a number of different and sometimes conflicting processes contributing to the epigenesis of full song. We can now begin to glimpse how species differences in the manifestation and relative emphasis on these different kinds of ontogenetic processes, as well as interspecies variation in the parts of song in which they are evident, may result in very different developmental endpoints.

IX. CONCLUSIONS ON THE FUNCTIONAL SIGNIFICANCE OF SUBSONG AND PLASTIC SONG

We have suggested several functions for the ontogenetic stages through which a bird proceeds in the transition from the earliest efforts of subsong through the emergence of the crystallized themes of full song. There are compelling reasons to think that subsong aids in the acquisition of skill in performing new motor coordinations, especially insofar as these require precise coordination of the two sides of the syrinx (Nottebohm, 1972b). Furthermore, subsong and plastic song are probably important in perfection of the new skill of guiding the voice by ear rather than by proprioception or by endogenous motor programming.

It is hard to imagine the production of vocal imitations without some degree of prior experimentation, at least in early stages of development of the skill. Having achieved a certain level of proficiency, with full command of all of the vocal operations necessary to generate a range of species-specific sounds, then a new pattern could presumably be produced with virtually no experimentation at all, as long as the option for learning new sounds was still open. Hall-Craggs (1962) records a case in a mature Blackbird (*Turdus merula*), and there are other examples. However, at earlier stages immediate reproduction of a sound heard seems inconceivable. Finally, we have evidence that improvisation and invention make significant contributions to song ontogeny. Once more it seems only reasonable to expect that vocal experimentation will be a necessary step, whether the ontogenetic changes proceed in small increments or in large saltations.

Although various lines of evidence suggest that subsong and plastic song play a functional role in song ontogeny there remain some confusing paradoxes. One is the sudden emergence of imitations of memorized material early in the plastic song, with little evidence of prior experimentation at the syllabic level. It may be that experience with subsong serves to refine the capacity to guide vocal production by ear so that ultimately little prior experimentation is needed.

Another source of confusion is the recurrence of subsong and plastic song after the first year. During the annual spring recrudescence of singing behavior typical of adult, male, temperate songbirds, the emergence of crystallized song is preceded by some of the earlier ontogenetic stages. The Swamp Sparrow is a case in point.

Adult Swamp Sparrow song remains virtually unchanged after the first year. Of 33 song types produced by a group of males, 31 persisted unchanged in the second season of singing. One song was dropped, and in one song some high-frequency components were lost, as though from some syringeal abnormality. Indications from field data are very similar (D. E. Kroodsma and R. Pickert, unpublished data). The relative frequency of use of different song-types may change, but the song repertoire appears to be fixed for life.

We followed the entire process of song development in four male Swamp Sparrows in the second year. There was virtually no subsong. Plastic song did occur, though for a shorter period than in the first year. Without exception, all of the syllables produced in plastic song matched syllables used in first-year plastic song, although there was a certain amount of attrition. The four birds used 51 syllable types in first-year plastic song and only 25 in the second year. There were some shifts of emphasis, tending to favor learned syllables over invented ones, but no additions occurred. The four birds crystallized the same 12 syllable types as in year 1, all of them well rehearsed in plastic song. Thirteen plastic song–syllable types were rejected upon crystallization, as they had been in the first year.

Recurrence of subsong and plastic song after the first year has been recorded in other songbirds (see review in Lanyon, 1960, p. 332). It occurs in the Chaffinch, where plastic song commonly precedes the emergence of crystallized themes in the second and subsequent years (Marler, 1956; Nottebohm, 1968; Poulsen,

1951). Subsong may also occur, though it is infrequent, and it is not a necessary precursor to the emergence of crystallized song (Marler, 1956, p. 86). Plastic song appears to occur in Song Sparrows early in the second and subsequent seasons, although subsong does not (Mulligan, 1966; Nice, 1937). Male White-crowned Sparrows coming into song in their second or third years of life produce subsong, though they pass through the stages to full song more quickly than first-year males (Marler, 1970).

What is the functional significance of the recrudescence of early development stages of song in later years, after the patterns of adult singing are set? At this stage of life, subsong and plastic song may simply be epiphenomena, occurring as an accompaniment of low androgen levels. The rapidity with which the transition to crystallized song often occurs after the first year, the erratic occurrence of subsong, and the rapidity with which crystallized song themes can emerge in several species all agree with this interpretation. However, other viewpoints are possible.

In discussing selective attrition of syllable types in song ontogeny we have suggested that a perceptual function may be involved. We hypothesize that reproduction of syllable types serves to fix them more effectively in memory so that they may be recognized subsequently if used by other birds. Insofar as rejected syllable types were learned from companions at a particular phase of the life cycle, motor rehearsal may nevertheless help a male to remember those song patterns and to employ that knowledge in his future relationships with others. If this were the case, rehearsal in subsequent years might serve to reinforce such memories, helping them to persist and to retain functional significance throughout life.

Alternatively, subsong and plastic song may play an additional role in species in which there is a significant change in the song repertoire from year to year. The list of species for which we know this to be true is growing, including the Indigo Bunting (*Passerina cyanea:* Rice and Thompson, 1968), the Red-winged Blackbird (Marler *et al.*, 1972; Yasukawa *et al.*, 1980), the Canary (Marler *et al.*, 1973; Marler and Waser, 1977; Nottebohm and Nottebohm, 1976), and the Saddleback (*Creadion carunculatus:* Jenkins, 1977). Plastic song in particular may provide a reservoir of new material out of which additional songs can be created. For this and other reasons, it is important to determine whether repertoire additions derive from new learned material, as Jenkins' (1977) data on Saddlebacks suggest, or whether some or all of the additions stem from the excess developed in plastic song in the first year.

Given the diversity of programs for song development in birds it would surely be a mistake to assume that the various stages of song ontogeny always serve the same function. It seems inevitable that subsong and plastic song must serve a variety of roles. It remains to be seen whether these variations involve differences in kind, or whether species differences arise from shifts of emphasis and timing imposed on the same fundamental ontogenetic mechanisms. If the evidence on other aspects of song learning is any guide, the latter is likely to prove to be the more appropriate viewpoint (Marler, 1981). Perhaps here too we are coming closer to the scientist's ideal of a universal model for all birdsong learning, such that an endless variety of complex and intricate behaviors can be generated by adjustment of the nature, timing, and emphasis on a relatively limited number of ontogenetic mechanisms.

ACKNOWLEDGMENTS

This research was supported by USPHS Grant MH 14651 to P. Marler and USPHS SO7 RR-076512 to the Rockefeller University. We are indebted to V. Sherman for aid in recording and analyzing songs, to M. Searcy for art work, and to Don Kroodsma and Ted Miller for their good counsel and patience.

REFERENCES

Armstrong, E. A. (1963). "A Study of Bird Song." Academic Press, New York.

Greenewalt, C. H. (1968). "Bird Song: Acoustics and Physiology." Smithson. Inst., Washington, D.C.

Güttinger, H. R. (1978). Verwandtschaftsbezeihungen und Gesangsaufbau bei Stieglitz (*Carduelis carduelis*) und Grünlingsverwandten (*Chloris* spec.). *J. Ornithol.* **119,** 172–190.

Hall-Craggs, J. (1962). The development of song in the Blackbird, *Turdus merula. Ibis* **104,** 277–299.

Hampe, H. (1939). Zur Biologie des Bourkesittichs, *Neophema bourkii. J. Ornithol.* **87,** 554–567.

Hersh, G. (1965). Bird voices and resonant tuning in air/helium mixtures. Ph.D. Thesis, Univ. of California, Berkeley.

Hinde, R. A. (1958). Alternative motor patterns in Chaffinch song. *Anim. Behav.* **6,** 211–218.

Hopkins, C., Rossetto, M., and Lutjen, A. (1974). A continuous sound spectrum analyzer for animal sounds. *Z. Tierpsychol.* **34,** 313–320.

Isaac, D., and Marler, P. (1963). Ordering of sequences of singing behaviour of Mistle Thrushes in relationship to timing. *Anim. Behav.* **11,** 178–188.

Jenkins, P. F. (1977). Cultural transmission of song patterns and dialect development in a free-living bird population. *Anim. Behav.* **25,** 50–78.

Kroodsma, D. E. (1974). Song learning, dialects, and dispersal in the Bewick's Wren. *Z. Tierpsychol.* **35,** 352–380.

Kroodsma, D., and Pickert, R. (1980). Environmentally dependent sensitive periods for avian vocal learning. *Nature (London)* **288,** 477–479.

Lanyon, W. E. (1957). The comparative biology of the meadowlarks (*Sturnella*) in Wisconsin. *Publ. Nuttall Ornithol. Club* No. 1.

Lanyon, W. E. (1960). The ontogeny of vocalization in birds. *In* "Animal Sounds and Communication" (W. E. Lanyon and W. N. Tavolga, eds.), Publ. No. 7, pp. 321–347. Am. Inst. Biol. Sci., Washington, D.C.

Lemon, R. E. (1971). Differentiation of song dialects in Cardinals. *Ibis* **113,** 373–377.

Lemon, R. E. (1975). How birds develop dialects. *Condor* **77,** 385–406.

Lemon, R. E., and Scott, D. M. (1966). On the development of song in young Cardinals. *Can. J. Zool.* **44,** 191–199.

Marler, P. (1956). Behaviour of the Chaffinch. *Behaviour, Suppl.* No. 6.

Marler, P. (1959). Developments in the study of animal communication. *In* "Darwin's Biological Work" (P. R. Bell, ed.), pp. 150–206. Cambridge Univ. Press, London and New York.

Marler, P. (1967). Comparative study of song development in sparrows. *Proc. Int. Ornithol. Congr.* **14,** 231–244.

Marler, P. (1970). A comparative approach to vocal learning: Song development in White-crowned Sparrows. *J. Comp. Physiol. Psychol.* **71,** Suppl., 1–25.

Marler, P. (1981). Birdsong: The acquisition of a learned motor skill. *Trends Neurosci.* **4,** 88–94.

Marler, P., and Peters, S. (1977). Selective vocal learning in a sparrow. *Science* **198,** 519–521.

Marler, P., and Peters, S. (1981a). Birdsong and speech: Evidence for special processing. *In* "Perspectives on the Study of Speech" (P. Eimas and J. Miller, eds.), pp. 75–112. Erlbaum, Hillsdale, New Jersey.

Marler, P., and Peters, S. (1981b). Sparrows learn adult song and more from memory. *Science* **213,** 780–782.

Marler, P., and Peters, S. (1982a). Structural changes in song ontogeny in the Swamp Sparrow, *Melospiza georgiana. Auk* **99,** 446–458.

Marler, P., and Peters, S. (1982b). Developmental overproduction and selective attrition: New processes in the epigenesis of birdsong. *Dev. Psychobiol.* **15,** 369–378.

Marler, P., and Peters, S. (1982c). Long-term storage of learned birdsongs prior to production. *Anim. Behav.* **30,** 479–482.

Marler, P., and Waser, M. S. (1977). Role of auditory feedback in Canary song development. *J. Comp. Physiol. Psychol.* **91,** 8–16.

Marler, P., Mundinger, P., Waser, M. S., and Lutjen, A. (1972). Effects of acoustical stimulation and deprivation on song development in the Red-winged Blackbird (*Agelaius phoeniceus*). *Anim. Behav.* **20,** 586–606.

Marler, P., Konishi, M., Lutjen, A., and Waser, M. S. (1973). Effects of continuous noise on avian hearing and vocal development. *Proc. Natl. Acad. Sci. U.S.A.* **70,** 1393–1396.

Messmer, E., and Messmer, I. (1956). Die Entwicklung der Lautausserungen und einiger Verhaltensweisen der Amsel. *Z. Tierpsychol.* **13,** 341–441.

Mulligan, J. A. (1966). Singing behavior and its development in the Song Sparrow, *Melospiza melodia. Univ. Calif., Berkeley, Publ. Zool.* **81,** 1–76.

Nice, M. M. (1937). Studies in the life history of the Song Sparrow. I. A population study of the Song Sparrow and other passerines. *Trans. Linn. Soc. N.Y.* **4.**

Nice, M. (1943). Studies in the life history of the Song Sparrow. II. The behavior of the Song Sparrow and other passerines. *Trans. Linn. Soc. N.Y.* **6.**

Nicholson, E. M. (1927). "How Birds Live." Williams & Norgate, London.

Nicholson, E. M., and Koch, L. (1936). "Songs of Wild Birds." H. F. & G. Witherby, London.

Nottebohm, F. (1968). Auditory experience and song development in the Chaffinch, *Fringilla coelebs. Ibis* **110,** 549–568.

Nottebohm, F. (1971). Neural lateralization of vocal control in a passerine bird. I. Song. *J. Exp. Zool.* **177,** 229–262.

Nottebohm, F. (1972a). The origins of vocal learning. *Am. Nat.* **106,** 116–140.

Nottebohm, F. (1972b). Neural lateralization of vocal control in a passerine bird. II. Subsong, calls and a theory of vocal learning. *J. Exp. Zool.* **179,** 35–49.

Nottebohm, F. (1975). Vocal behavior in birds. *In* "Avian Biology" (D. Farner, ed.), Vol. 5, pp. 287–332. Academic Press, New York.

Nottebohm, F. (1977). Asymmetries in neural control of vocalization in the Canary. *In* "Lateralization in the Nervous System" (S. Harnad, R. W. Doty, L. Goldstein, J. Jaynes, and G. Krauthamer, eds.), pp. 23–44. Academic Press, New York.

Nottebohm, F., and Nottebohm, M. (1976). Left hypoglossal dominance in the control of Canary and White-crowned Sparrow song. *J. Comp. Physiol.* **108 A,** 171–192.

Payne, R. (1973). Behavior, mimetic songs and song dialects, and relationships of the Parasitic Indigobirds (*Vidua*) of Africa. *Ornithol. Monogr.* **11,** 1–33.

Poulsen, H. (1951). Inheritance and learning in the song of the Chaffinch (*Fringilla coelebs* L.). *Behaviour* **3,** 216–227.

Poulsen, H. (1959). Song-learning in the domestic Canary. *Z. Tierpsychol.* **16,** 173–178.

Rice, J. O., and Thompson, W. L. (1968). Song development in the Indigo Bunting. *Anim. Behav.* **16,** 462–469.

Slater, P. J. B. (1979). Cultural evolution in Chaffinch song. *Behavior* **71,** 146–166.

Sotovalta, O. (1956). Analysis of the song patterns of two Sprosser Nightingales, *Luscinia luscinia* L. *Ann. Bot. Soc. Zool. Bot. Fenn. Vanamo* **17,** 1–31.

Stiles, F. G., and Wolf, L. L. (1979). Ecology and evolution of lek mating behavior in the Long-tailed Hermit Hummingbird. *Ornithol. Monogr.* **27,** 1–78.

Thorpe, W. H. (1954). The process of song learning in the Chaffinch as studied by means of the sound spectrograph. *Nature* (*London*) **173,** 465–469.

Thorpe, W. H. (1955). Comments on "The Bird Fancyer's Delight," together with notes on imitation in the subsong of the Chaffinch. *Ibis* **97,** 247–251.

Thorpe, W. H. (1958). The learning of song patterns by birds, with especial reference to the song of the Chaffinch, *Fringilla coelebs*. *Ibis* **100,** 535–570.

Thorpe, W. H. (1961). "Birdsong: The Biology of Vocal Communication and Expression in Birds." Cambridge Univ. Press, London and New York.

Thorpe, W. H., and Pilcher, P. M. (1958). The nature and characteristics of subsong. *Br. Birds* **51,** 509–514.

Yasukawa, K., Blank, J. L., and Patterson, C. B. (1980). Song repertoires and sexual selection in the Red-winged Blackbird. *Behav. Ecol. Sociobiol.* **7,** 233–238.

3

Avian Vocal Mimicry: Its Function and Evolution

JEFFREY R. BAYLIS

I. INTRODUCTION

Although the study of the functional aspects of avian vocal mimicry is a fairly recent development, the mechanism by which mimetic elements are acquired was a central issue in one of the earliest extended works on the Darwinian

ACOUSTIC COMMUNICATION IN BIRDS
VOLUME 2

ISBN 0-12-426802-1

evolution of birdsong (Witchell, 1896). Witchell argued that mimicry was clear evidence some birds learned their song, and could acquire new elements of song from the animate and inanimate world around them. This gave him a single mechanism to account for the transmission of a song from one generation to the next, and to provide the variation on which natural selection could act. For the next 40 years, the study of vocal mimicry emphasized that it was learned, while there was relatively little discussion of its function.

Recent technological developments have produced increasingly sophisticated methods for the description and observation of bird vocalizations. It has become clear that avian vocal mimicry is widespread, often precise, and occurs in a variety of social situations (Sections III and IV). These observations have produced an interest in the functional aspects of vocal mimicry (Section II). However, much of the functional literature has lost sight of the implications of the mechanism by which mimicry is acquired (i.e., vocal learning). The peculiarities of vocal mimicry are that it is fundamentally a behavior pattern, and is acquired and passed on in a cultural manner.

In this chapter I examine aspects of the biological significance of vocal mimicry in birds. First I discuss some of the conceptual issues involved and establish a terminology. I then survey the literature for well-documented examples of vocal mimicry. Possible functions of vocal mimicry are presented, and critical tests of some hypotheses are suggested. Finally, the evolution of vocal mimicry and its implications for the evolution of vocal signaling in model and mimic species are discussed.

II. SOME CONCEPTUAL ISSUES

> In the voices of animals, hereditary similarity, involuntary or passive imitation, and voluntary or active mimicry, blend insensibly from one to the other. The influence of resemblance is to be observed throughout the animate world. Mere physical resemblances to things or to organisms are of vital importance to vast numbers of creatures which they continually protect (Witchell, 1896, p. 159).

A. Definitions

The College Edition of Webster's New World Dictionary (Guralnik and Friend, 1964) defines *mimicry* as "close resemblance of one organism to another or to some object in its environment." I define vocal mimicry as the resemblance of one or more vocalizations of an individual bird of one species either to the vocalizations typical of individuals of another species or to some environmental sound. "Mimicry" will be used in this chapter as a contraction for "avian vocal mimicry."

"*Song* is a vocal display in which one or more sounds are consistently repeated in a specific pattern. It is produced mainly by males, usually during the breeding season. All other bird vocalizations are collectively termed call notes or, simply, calls." The above definition of song is from Pettingill (1970, p. 319). Although it is not an ideal definition of song, Pettingill's is both descriptive and consistent with the implicit definitions used by most of the authors cited in this chapter; hence I consider it to be a suitable working definition. Where an authority is cited for the use of mimicry in a bird's song I accept the author's judgment as to what constitutes mimicry and song, unless the situation described is clearly at variance with the above definitions.

B. Descriptive versus Functional Definitions

The definitions given in Section II,A carry no implication of mechanism or function. Since birdsong is fundamentally a behavior pattern, it must be described in ways that do not imply function or mechanism, if these aspects are to be investigated objectively (Hinde, 1970; Marler and Hamilton, 1966). This point has been disregarded in reviews which have advocated functional definitions of mimicry (Dobkin, 1979; Vernon, 1973). Moreover, functional definitions are often incomplete, as they do not allow for chance or dysfunction as explanatory null hypotheses.

For example, Dobkin (1979) assumed that "vocal copying" must have a function, and he erected a classification scheme on this premise. Dobkin asserted that he found confusion and contradiction in the use of the term "mimicry" in the literature on avian vocal behavior. I surveyed much the same material as he and found that prior to 1966 the literature on mimicry was quite clear and consistent in using mimicry as a *descriptive* term, not as a *functional* term (Armstrong, 1963; Berger, 1961; Chisholm, 1950; Marshall, 1950; Pettingill, 1970; Robinson, 1974; Saunders, 1929; Thorpe, 1961; Thorpe and North, 1966; Vernon, 1973; Welty, 1975; Witchell, 1896). Hence Dobkin found that mimicry was tangled in contradictory hypotheses; but a single phenomenon may be explained by many functional hypotheses, some competing and some not (van der Steen, 1972). This should not be disturbing if the phenomenon can be described in nonfunctional terms. It is only when we define a mammalian forelimb as a limb used for walking, and we are confronted with the flipper of a whale, the wing of a bat, and the human arm that the phenomenon of a forelimb becomes confusing and contradictory.

The difficulty that a functional definition creates was obvious in Dobkin's attempt to classify the indigo-birds (*Vidua* spp.) as "vocal imitators." Indigo-birds are brood parasites on other sympatric weaver-finch species. The male indigo-birds sing a courtship and territorial song that contains mimetic elements copied from the song of the male host parent (Payne, 1973, 1976).

The following definitions are quoted directly from Dobkin (1979, pp. 349–350): "Vocal imitation is the use of species specific calls or songs of conspecifics. Vocal imitation is used intraspecifically." "Vocal mimicry is the use of calls or songs of other species that are predominantly aggressive, predatory, or otherwise noxious. Vocal mimicry functions interspecifically." Dobkin was reluctant to call indigo-birds "vocal mimics," because the presumed function of the song containing copied elements is intraspecific. He classified the indigobirds as "vocal imitators" because of the intraspecific function of the song, but the use of elements not belonging to conspecifics clearly violates his own definition. In fact, indigo-birds do not fit in any of the four functional categories suggested by Dobkin (see Section IV,C).

Witchell (1896) used both the terms "imitation" and "mimicry" in describing resemblance of elements in a bird's song to other sounds. He was generally consistent in using mimicry to describe a bird of one species employing the song or calls typical of another species. This use of mimicry as a descriptive term has remained remarkably consistent in the avian literature (Armstrong, 1963; Pettingill, 1970; Welty, 1975). "Vocal imitation" has generally been used to describe cases where an individual acquires the call or song of a conspecific individual. Both terms have been used in a descriptive sense, and many functional hypotheses have been suggested for the phenomena to which they apply. "Vocal resemblance" would certainly be a better term for the general phenomenon than "vocal mimicry," but mimicry is a term already clearly and consistently in use in the literature, and has been used in a descriptive rather than a functional manner. It is only because of recent attempts to define mimicry in terms of "deception of a receiver" (Wickler, 1968) that the issue has become confused, as inevitably it must if we depend on a functional definition of a behavior pattern.

C. Scope

I have given emphasis to cases of mimicry where the mimetic elements occur in the principal or most conspicuous vocalizations of the mimicking species. Calls are discussed, but are not considered in detail for reasons given below. The occurrence of mimicry in "subsong," "whisper song," or "recording song" is not considered because of ambiguity in terminology and the difficulty of evaluating the criteria used for establishing mimicry in such rambling and plastic vocalizations.

III. A SURVEY OF MIMICS

> To make a complete study of the Mocker's powers of imitation, one would need to know: First, songs and calls of birds now extinct in this section of the world, which may have been

imitated by the ancestors of the present Mocker; second, songs and calls of all contemporary species in this part of the United States; third, the songs and the calls of all Mockingbirds now singing (Mayfield, 1934, p. 17).

A. Criteria

Mayfield overstated the case a bit, but his quote hints at the enormous practical problems involved in the study of mimicry. Obviously there are operational difficulties with the use of the term "resemblance" in the definition I have given. Gramza (1972) observed that the vast majority of studies have relied on the human ear to spot auditory similarity. He suggested increased reliance on instrumentation to reduce or make explicit the criteria used by the human observer for making the judgment that two sounds resemble each other (see Bertram, 1970, for an approach to this problem). While this admonition for future studies is laudable and aids in the evaluation of some past studies, it is of little practical help in evaluating most published reports of mimicry. If current instrumentation were up to the task of identifying mimetic elements, someone would have marketed a reliable speech-controlled typewriter and I would not now be writing at a pace limited by my clumsiness. We must always face the problem that "resemblance" is as much a property of how a sound is perceived as it is of physical properties of the sound signal itself. A trained observer will have the final word, and will always be the "gold standard" against which instrumental judgments will be tested. Even this ignores the problem that a bird may not perceive resemblance where a human observer does. The chief advantage of instrumentation is that the criteria used for judgment can be clearly and operationally stated.

The criterion I have used for accepting a species as a mimic is the judgment of an author. In each case the model species must be clearly identified. There are implicit criteria used by authorities for reporting cases of mimicry. The obvious points are that the species involved must be sympatric for at least part of the year, and that the vocalization in question is atypical of the mimicking species and typical of the model species. Cases where vocalizations of two species both converge on a new sound in zones of sympatry have not been included, although some of the cases of mimicry reported here may involve character convergence (Brown, 1977; Cody, 1969; Dobkin, 1979; Moynihan, 1968). Cases where conspecifics acquire sounds from each other are considered examples of vocal imitation rather than mimicry and I have not included them (Armstrong, 1963; Bertram, 1970; Dobkin, 1979; Thorpe and North, 1966; Vernon, 1973).

B. The Survey

Past attempts to draw broad patterns from the occurrence of mimicry have been confined to Australian birds. Chisholm (1932) examined the natural history of 19 species reputed to be mimics. They included species that foraged and sang mostly on the ground (9 species), species that used both trees and the ground (5 species), and species that used only trees (5 species). The mimic species Chisholm studied inhabit either tropical or temperate areas of Australia, and a few inhabit both. None of the mimics is migratory, and most are "old" forms in genera endemic to Australia.

Marshall (1950) examined the life history patterns of many of the same species as Chisholm, and questioned whether Australia has an unusually large number of mimics. He found that 53 of the approximately 360 passerines found in Australia are mimics; this compares with 30 mimics in 160 passerine species in Great Britain. Marshall concluded that it is not the proportion of mimics but rather the high quality and intensity of mimicry by Australian birds that is remarkable. Most Australian mimics inhabit scrub, heath, or thick forest where lack of visibility would favor sound as a distance signal. Marshall felt that mimicry was favored as a consequence of the need to produce a continual long-range advertisement by voice. Thus he was suggesting that mimicry is a way of increasing repertoire size and hence song diversity.

Armstrong (1963) noted that many of the world's accomplished mimics do not inhabit visually constrained environments. He suggested that mimicry could be used to allow individual distinctiveness in song. Hartshorne (1973) also reviewed the mimicry literature, and concluded that mimicry is widely distributed among families and regions. He noted that it tends to be more common among large birds and rarely occurs in migratory species. Hartshorne observed that many versatile singers are not mimics, and that it is quite possible for a species to have a large repertoire and individual distinctiveness without resorting to mimicry. Vernon (1973) surveyed the South African mimics and found that 63 of 370 passerine species were mimics. He compared the proportion of avian mimics in Britain, Australia, and South Africa, and concluded that mimics were about equally frequent in all three locations (19, 15, and 17%, respectively).

I have attempted my own survey of the natural history of species noted for their mimicry (Table I). As a result of the criteria I used in accepting an example, the survey consists of only eight cases. The species (or genus in the case of *Vidua*) must have been the object of one or more investigations aimed specifically at some aspect of the species' song. The song or calls must have been demonstrated to contain mimetic elements, and at least some general data on the life history of the species or population under study must be available.

If mimicry is a unitary phenomenon with a single function common to all or most mimics, then the results of the survey should present us with a pattern or syndrome of natural history characteristics of mimic species (Table I). Several aspects of their natural history are shared by all eight examples. All eight use mimicry in their song, a finding that is hardly surprising given the nature of the survey. In all cases only the male is reported as singing during the breeding season, and all are intraspecifically territorial during the breeding season, although in the lyrebirds (*Menura* spp.) and the indigo-birds (*Vidua* spp.) the male's territory is used only for display and copulation. Except for the use of mimicry in song, all these similarities are exactly what we would expect to see in any random assemblage of passerine species (Brown, 1975).

The first column in Table I following the species names contains information on what Hartshorne (1956) called the "continuity" of the song. This is the proportion of time within a bout of singing in which the bird is actually producing sound. Six of the eight species in Table I were ranked by Hartshorne (1973) as "superior" singers, and received his score for continuity. This score ranges from a low of 1 to a high of 9. Hartshorne (1973) also used a term he called "song scope," which he also ranked from 1 to 9. Birds that have a large repertoire and frequently change song patterns receive a high score for song scope. Sexual dimorphism was scored for a species if male and female plumage differ during the breeding season.

The number of model species is the average number of species whose song elements appear in songs of a single individual; an individual may copy more than one song element from a given species, but that will not be reflected in this number. The numbers in parentheses in this column represent Hartshorne's (1973) score for vocal imitativeness ("skill in mimicry") for ranked species. The last column indicates whether or not the species is described as being interspecifically aggressive.

The breeding habitats of species listed in Table I are diverse. The two lyrebird species inhabit and breed on the forest floor (Chisholm, 1932). The African robin-chats (*Cossypha* spp.) inhabit forest crown (Harcus, 1977). The indigo-bird males defend prominent song perches in open grassy scrub (Payne, 1973). The Marsh Warbler (*Acrocephalus palustris*) breeds in reeds and brush along the edges of bodies of water (Voous, 1960). The Mockingbird (*Mimus polyglottos*) inhabits parkland, while the Gray Catbird (*Dumetella carolinensis*) prefers thick brush and hedgerows (Bent, 1948). Among these eight examples of mimicry, there is certainly no obvious correlation of mimicry with habitat.

There is a clear tendency for the mimics to be versatile singers. Six of the eight species were rated as superior singers by Hartshorne, and all six were ranked high in song continuity. Similar high rankings occur for song scope, with indigo-birds having the smallest repertoire size. There is no clear relationship between sexual dimorphism and mimicry: three cases show marked dimorphism, five show none. Not surprisingly, mating systems follow the same relationship as sexual dimorphism.

There is only one seedeater among these mimics; all the rest are at least partially insectivorous. Most appear to be unusually generalized in their food habits, although thorough studies have not been done on diet and foraging habits in any of these species. Only two of the mimics are migratory. Although the

TABLE I

Selected Aspects of the Natural History of Known Mimics[a]

	Song continuity[b]	Song scope[c]	Sexual dimorphism	Mating system[d]	Diet[e]	Migratory	Number of model species[f]	Interspecific aggression	Authorities[g]
Australia									
Menuridae (Lyrebirds)									
Menura novaehollandiae	+(9)	(9)	+	P	I	–	15(9)	–	1,2,3,4
M. alberti	+(9)	(9)	+	P	I	–	(9)	–	1,2,4
Africa									
Turdinae (thrushes)									
Cossypha dichroa	+(9)	(9)	–	M	I;F	–*	12(7)	–	5,6,7
C. natalensis	+(8)	(8)	–	M	I;F	–*	6?(7)	–	5,6,8
Ploceidae (Viduinae)									
Vidua spp.	?	18	+	P	S	–	1	–	5,9,10
Europe									
Sylviinae (warblers)									
Acrocephalus palustris	+(9)	90(8)	–	M	I	+	76(7)	–	11,12,13,14

North America									
Mimidae (Mimid thrushes)									
Mimus polyglottos	+(9)	150(8)	–	M	I;F	–*	20(7)	+	15,16,17
Dumetella carolinensis	+	175	–	M	I;F	+	?	+?	15,18,19,20

[a] A plus (+) denotes the presence of a trait, and a minus (–) denotes its absence. A question mark (?) means that the entry may not be accurate or the information is not available. An asterisk (*) denotes that some small proportion of the population may exhibit the trait. See text for additional discussion.

[b] A species was scored a plus for song continuity if one or more authorities stated that it sings continuously. The number in parentheses is Hartshorne's (1973) score for continuity, which ranges from a low score of 1 to a high score of 9.

[c] The number outside of parentheses in the song scope column is the average song repertoire size per bird. The number in parentheses is Hartshorne's (1973) score for song scope (1–9).

[d] Mating systems were classed as polygynous or promiscuous (P), or monogamous (M).

[e] Diet was scored as insectivorous (I), frugivorous (F), seed-eating (S), or some combination of these.

[f] The number of model species is the number of different model species mimicked within the song repertoire of one bird. The number in parentheses is Hartshorne's (1973) score for vocal imitativeness (1–9).

[g] 1, Chisholm (1932); 2, Marshall (1950); 3, Bell (1976); 4, Robinson (1974, 1975); 5, Mackworth-Praed and Grant (1963); 6, Newman (1968); 7, Harcus (1977); 8, Farkas (1969); 9, Nicolai (1964); 10, Payne (1973, 1976); 11, Voous (1960); 12, Vernon (1973); 13, Dowsett-Lemaire (1979); 14, Peterson *et al.* (1974); 15, Bent (1948); 16, Wildenthal (1963); 17, Howard (1974); 18, Thompson and Jane (1969); 19, Boughey and Thompson (1976); 20, Fletcher and Smith (1978).

extent of the Gray Catbird's mimicry is unknown, the Marsh Warbler may well be the most extensive mimic in the world and undergoes a very long migration (Dowsett-Lemaire, 1979). The number of model species is highly variable from case to case, and only one species is definitely aggressive toward other species.

The number of conclusions that can be drawn from Table I is limited. Good mimics tend to have large song repertoires and sing continuously. There does not appear to be a strong relationship between the occurrence of polygyny and mimicry. While mimics tend to be year-round residents, at least some mimics are migrants. Only one (possibly two) of the well-studied mimics defends an interspecific territory. The examples given here inhabit almost as many habitats as there are genera in the list. My conclusion from this survey is that mimicry is unlikely to have a single, simple function. It is not a unitary concept; rather it is many things whose only common aspect is vocal resemblance. We can expect the reasons for the resemblance to be diverse.

The appendix to this volume contains a much less selective sampling of the world's mimics. This is a list of 123 species that have been classed as mimetic by one or more authorities. Mimicry here is defined to include the use of mimetic elements as calls, in subsong, or in song. Within the classification, a distinction is made between three grades of mimicry. A *persistent* mimic is defined as a species where most or all individuals are described as mimicking regularly. A *consistent* mimic is a species where some individuals are reported to mimic. *Casual* mimics are species where only isolated individuals are reported to mimic. Many of the species in this list were listed by Hartshorne (1973) as superior singers.

Some familial consistency is obvious in the appendix. In the Old World, larks (Alaudidae), thrushes (Turdinae), and warblers (Sylviinae) contain mimetic members wherever they are represented. The larks are an interesting case, for almost every species recorded as a mimic is described as using mimetic song during a flight display. Shrikes (Laniidae) and drongos (Dicruridae) also include mimics wherever they are found. The paucity of mimics in Asia probably reflects my limited knowledge of the avian literature of that region. The bulk of Australian mimics are indeed accounted for by primitive groups and endemics, as Chisholm (1932) observed. The Starling (*Sturnus vulgaris*) is a mimic everywhere it is now found.

Except for the introduced Starling, North America has no mimics in any family shared with the Old World. Mimicry may be universal in the mimid thrushes (Mimidae). South America has few reported mimics, and only one belongs to a family occurring in the Old World (Turdinae). As with North America, South America's mimics are endemic to the New World. One cannot predict mimicry from knowing that a species is a superior singer; nor can one predict a superior singer from knowledge that a species is a persistent mimic.

A striking feature of the mimics listed in the Appendix is their restriction to the passerines. Indeed, only four species of sub-oscines are listed. It is tempting to conclude that only species that learn their vocal repertoire mimic, and vocal learners are largely confined to the passerines. However, some non-passerines are vocal imitators, yet are not known to mimic in nature (Thorpe and North, 1966; see also Chapter 1, this volume). Nonpasserines are frequently used as models by mimics. These facts suggest that some special aspect (or aspects) of the ecology, physiology, or social context in which vocalizations are used has given passerine birds an advantage as mimics. One obvious factor is the possession of a true syrinx. However, lyrebirds lack this feature, yet are mimics of stunning skill and range. So far vocal learning is the only factor clearly common to all mimics.

IV. POSSIBLE FUNCTIONS OF VOCAL MIMICRY

> If the young of some song-birds acquire every danger-cry and call-note solely by imitating the voices of their parents, and keep strictly to these models, and the notes are common to all birds of the same species respectively, a similarity between those notes in allied species is just as valuable an indication of a common origin as it would have been if the perpetuation of the notes had resulted from inheritance (Witchell, 1896, p. 90).

Given the diversity of species and ecological situations in which mimicry has been described it is not surprising that many functions have been suggested. In this section I examine some of these hypotheses and review the evidence for them. I have tried to make the list reasonably exhaustive, but have avoided the broad functional and conceptually exhaustive categories suggested by Dobkin (1979). I confine my discussion to the functions of mimicry for which there is at least some strong circumstantial evidence in the literature.

A. Null Hypotheses

1. The Resemblance Is Due to Chance

The first step in arguing that a case of mimicry requires a functional explanation is to demonstrate that the resemblance observed is not due to chance (Brown, 1977; Gramza, 1972). This becomes especially important in a versatile singer with a large repertoire, like the Brown Thrasher (*Toxostoma rufum*). It has been argued that the Brown Thrasher does not actually mimic but improvises so freely that coincidence alone produces mimetic resemblance to other species (Townsend, 1924). As Armstrong (1963) cautioned, "loquacious birds" are apt to mimic by chance. In light of Kroodsma and Parker's (1977) estimate of a repertoire size of about 2000 song units on one individual Brown Thrasher, Townsend's admonition seems apt.

Chance alone is a very difficult hypothesis to disprove, yet it is the logical null hypothesis against which any functional hypothesis must compete. Ideally we should have some quantitative knowledge of the range of motor and perceptual abilities of birds to produce and detect various sounds. If we also have some knowledge of the repertoire sizes of the species in question, the number of species in the same location together with their repertoire sizes and relative abundance, etc., then we may be able to put some sort of estimate on the number of calls, notes, or songs we would expect to find shared by two sympatric species on the basis of chance alone. This would be next to impossible. We must rely on the judgment of field biologists to decide what constitutes a significant resemblance, yet we should always expect some chance resemblances (Brown, 1977).

2. Shared Environmental Factors

A second hypothesis is that the resemblance is due to some environmental factor common to the two species but not directly involving any interaction between them. For example, if acoustical properties of a shared environment render certain sounds more conspicuous or more easily propagated, two species may evolve similar vocal signals without either species having influenced the evolution of the other. This is a hypothesis that may be tested by inference from the comparative method (Hailman, 1977).

There is at least one case where clearly unrelated species are allopatric yet live in similar habitats, and have converged toward similar optical and acoustical signals. Hailman (1977) cited Emlen's observation of an African pipit that resembles the Eastern Meadowlark (*Sturnella magna*) of North America in plumage and song, and inhabits a similar environment. Further, the pipit responded to playback of the Eastern Meadowlark's song. It would seem absurd to treat this as a case of vocal mimicry, yet the case might have been reported as mimicry if the two had been sympatric.

A general example is the case where the species in question share a common predator such as a hawk. An alarm call is given by many species of passerine birds when a hawk flies overhead (Marler, 1959), and the calls of many passerines, whether sympatric or allopatric, strongly resemble each other. We need not suppose that one passerine species has influenced another to explain its evolution. The call has a common function which has resulted in a similar form. An equivalent case may be made for the striking resemblance found in the mobbing calls of passerine birds.

These observations are important for the interpretation of cases of mimicry. The Eastern Meadowlark and the African pipit provide a clear example of convergence where direct interaction or shared ancestry may be excluded. It may be very difficult to reach this conclusion for sympatric species. In the case of the alarm calls, it is clear that one can explain the resemblance in functional terms on the basis of each species individually, without reference to other passerines. This is why I am reluctant to treat examples of mimicry involving alarm calls in any great depth. The most ubiquitous form of mimicry is where one species is said to mimic the alarm calls of another (Armstrong, 1963; Chisholm, 1950; Marshall, 1950; Robinson, 1974, 1975; Vernon, 1973; Witchell, 1896). It is also an ambiguous category without knowledge of the call repertoire of each species both in areas where it is sympatric with and in areas where it is allopatric with the suspected model. One is inclined to wonder how often the observer reporting mimicry is simply more familiar with the alarm call repertoire of the model than with that of the mimic.

3. *Deprivation of Conspecific Songs*

The third null hypothesis is to suppose that the resemblances in song represent "mistakes" that are actively selected against. Saunders (1923, 1929) suggested one way an individual whose song is learned might acquire song elements of another species. If a species is invading a new area, the number of conspecific teachers available from which to learn song may be very low, and a young bird might copy song elements from an individual of another species. Similarly, a bird hatched late in a season might miss the song peak of its own species, but overlap the peak of another species and learn the wrong song. Dowsett-Lemaire (1979) has suggested the latter as the mechanism by which the Marsh Warbler acquires much of its large repertoire of mimicked song elements. It may not be disadvantageous for a bird to acquire alien song elements when colonizing a new area, if use of such sounds aids in interspecific competition (Brown, 1977; Cody, 1969; Dobkin, 1979; Moynihan, 1968).

Another possibility is that some cases of mimicry result when parents of one species foster young of another. Such interspecific "helping" occurs in birds, and is at least as common in the literature as the instances of casual mimicry in the Appendix (Skutch, 1961). If the mimetic feature was learned through a situation where only the foster parents were present, then we might expect the mimicking individual never to sing a normal song. Such a case has been reported for a Field Sparrow (*Spizella pusilla;* Short, 1966). Experiments with cross-fostering could establish whether or not a fostered individual would sing in a manner consistent with the casual mimics. Such experiments have been carried out for over 200 years, and what few results there are show consistency with this "foster parent" hypothesis (Armstrong, 1963; Eberhardt and Baptista, 1977; Kroodsma, 1976; Witchell, 1896).

B. Intraspecific (Social) Functions

> In the evolution of bird-song, mimicry as well as invention have through sexual selection played a part (Townsend, 1924, p. 541).

Several authors have observed an inverse relationship between song quality and conspicuous plumage in birds; the world's best songsters tend to be drab

(Armstrong, 1963; Hartshorne, 1973; Witchell, 1896). This observation suggests that in some species, complex and conspicuous vocal displays have replaced gaudy plumage as a vehicle for sexual selection (Armstrong, 1963; Hartshorne, 1973; Kroodsma, 1977).

Authors have also noted a correlation between complexity and continuity in birdsong; those species which tend to sing for the largest percentage of the time have the largest repertoires and change song elements the most often (Hartshorne, 1956, 1958, 1973; Hunt, 1922; Witchell, 1896). Hartshorne (1956, 1973) argued that this relationship between virtuosity and continuity is based on the biological principle of habituation. In a species having a simple song, any tendency to increase stimulation by simply increasing singing time would be counteracted by the receiver's habituation to the signal, unless the complexity of the signal was also increased. The empirical correlation Hartshorne (1973) claimed to have found has been challenged (Dobson and Lemon, 1975), but Kroodsma (1978) showed that the challenge was based on a misinterpretation of Hartshorne's measure of virtuosity.

The suggested reasons why it might be of advantage for a bird to increase the stimulation value of its song are sexual selection, operating either through competition between males or through female choice (Howard, 1974; Krebs, 1977; Kroodsma, 1977; Robinson, 1974); sequestering of resources (Krebs, 1977; Kroodsma, 1977); and the stimulation of mates to bring them into reproductive condition (Kroodsma, 1976). Only the last hypothesis has been tested in any definitive manner, where it was found that complex Canary (*Serinus canaria*) songs stimulated female Canaries into repoductive condition earlier than simple songs (Kroodsma, 1976).

The significance of mimicry for the last hypothesis is that it provides a mechanism for increasing individual repertoire size, and hence song diversity (Armstrong, 1963; Dobkin, 1979; Hunt, 1922; Marshall, 1950; Nottebohm, 1972; Townsend, 1924; Witchell, 1896). However, any such tendency to elaborate through mimicry would be counteracted to some unknown extent by the pressure to preserve species identity (Dobkin, 1979; Hunt, 1922; Marshall, 1950; Witchell, 1896).

If mimicry has been a mechanism for song elaboration for the purposes described above, I would expect that the role of mimetic song elements in habituation reduction would work only if the mimic did not imitate species that were likely to be singing during its own breeding season. Model songsters would devalue the mimicked signal by using it themselves. A mimic should tend to copy rare species for the same reasons. Actually, the reverse of both of these is true in the case of the Mockingbird (Laskey, 1935; Michener and Michener, 1935; Richardson, 1906; J. R. Baylis, unpublished data) and two species of robin-chats (Farkas, 1969; Harcus, 1977).

The two lyrebirds of Australia do fit this prediction. Robinson (1974, 1975) noted that these birds are outstanding vocal mimics, but they do not breed during the primary breeding season, when most of their models sing. Here is a clear case where habituation (or confusion) due to the model's song is avoided by the mimic. These species are polygynous and sexually dimorphic, strengthening the argument that the mimetic elements have been favored due to sexual selection for large repertoire size (Kenyon, 1972). Another potential example of sexual selection favoring mimicry is provided by the Marsh Warbler, which breeds in Europe but mimics African models. The potential role of sexual selection is less clear in the latter species, as it is sexually monomorphic and monogamous, and males do not sing after mating (Dowsett-Lemaire, 1979). Howard (1974) claimed to have demonstrated a positive relationship between early mating date and repertoire size in male Mockingbirds, but the ages of the birds were unknown and the methods used to estimate repertoire size were questionable.

In many mimic species repertoire size probably increases with age (Laskey, 1944). This means that all males can potentially possess a large repertoire and hence will have an equal chance in breeding competition with respect to repertoire at a given age. Such a trait cannot be said to be a response to sexual selection, because it varies *within* an individual with ontogeny and is not a heritable property of individuals (Wade and Arnold, 1980). Thus a simple demonstration of a positive relationship between repertoire size and mating success in a songbird is not sufficient. One must show such a relationship with repertoire size between males of the same age class.

For species with polygynous or promiscuous mating systems, intrasexual variance in mating success is fairly easily measured, and has an obvious source. Monogamous mating systems can also produce sexual selection, which may either be mutual, as in the Great Crested Grebe (*Podiceps cristatus;* Huxley, 1914), or differential, as is posited in serially monogamous birds where only the male sings. A character cannot be said to have evolved in response to sexual selection unless the variation in the trait in question is heritable, and individual variation with respect to the trait results in individual variation in mating success within a sex through intrasexual competition for access to mates, and/or intersexual mate choice (Darwin, 1874; Wade and Arnold, 1980). Wade and Arnold (1980) give a method for calculating the intensity of sexual selection, based on variation in mating success. Some potential sources for such variation in mating success in serially monogamous birds are:

(1) Differential rates of remating in males and females may result in one sex exhibiting a higher variance in reproductive success than another. Thus a bird may be strictly monogamous within a breeding episode, but if males switch mates and start sooner on the next breeding episode than their old mates, differential sexual selection could occur.

(2) A biased adult sex ratio could produce differential sexual selection. A male-biased sex ratio has been suggested in some monogamous birds, where

mortality related to reproduction is higher in females than males (Selander, 1972).

(3) "Cheating" or copulating outside the pair bond may occur, and would be likely to produce a greater variance in male reproductive success.

(4) Males that mate earlier in the season than other males might show a greater reproductive success, through greater fledging success (Darwin, 1874; Howard, 1974; Fisher, 1958). This argument assumes that there is *no heritable component* to the tendency of females to mate early. Howard was aware of this assumption in his Mockingbird study, but offered no empirical evidence that this assumption was met by his female Mockingbirds (Howard, 1974). All of the above factors could lead to differential sexual selection on male song repertoire. However, in each case it is necessary to show that song repertoire is instrumental in intrasexual competition for mates or for female choice of males. Thus far only the lyrebirds and probably the bowerbirds and indigo-birds appear to be clear candidates for sexual selection favoring mimesis as a mechanism for enlarging repertoire size. Other cases, while suggestive, must await further evidence.

1. *Identification of Potential Mates*

Two cases have emerged in recent years that have generated hypotheses for the function of mimicry that seem counterintuitive. In both instances, it appears that mimetic song elements are used for the identification of potential mates.

The clearest examples of this use of mimicry are the parasitic indigo-birds (Nicolai, 1964; Payne, 1973, 1976). The viduine finches are brood parasites of the firefinches (*Lagonosticta* spp.) of Africa. The genus *Vidua* contains many similar species, each specific to a single host. The courtship songs of male indigo-birds contain mimetic elements copied from a single host species' calls and songs. Females are attracted to the male's singing site, and choose mates who are singing songs of the appropriate host species. Although the sounds produced by the mimic are used in pair-bond maintenance in the host species, they evoke no overt host response when mimics sing in large aviaries (Sullivan, 1976).

The critical test of the hypothesis that this mimetic song functions in potential mate identification would be to cross-foster individuals of the parasite species under several species of host, and see if as breeding adults the parasite species chose mates on the basis of the mimetic elements of the song, rather than the nonmimetic elements. The evidence already gathered on mate choice appears consistent with the hypothesis (Payne, 1973, 1976). It is possible that similar mimicry occurs in other bird parasite–host complexes (Mundy, 1973), but should be expected only in cases where the parasite is highly host specific. The Brown-headed Cowbird (*Molothrus ater*) is less specialized in its parasitic habits and appears to encode species identification in courtship song differently (King and West, 1977).

Another case where a mimetic song may serve a role in species isolation is the Marsh Warbler. A study of about 20 birds in Belgium disclosed that the average individual repertoire of about 100 song elements was at least 80% mimetic (Dowsett-Lemaire, 1979; Lemaire, 1977). Of the mimetic elements, 60% were African and 40% were European. The unidentified 20% of the song elements are probably mimicry of African species. She concluded that it is too early to suggest a function for the incredible amount of mimicry in the song of the Marsh Warbler, other than suggesting a possible role in species isolation with the Reed Warbler (*Acrocephalus scirpaceus*). Sexual selection is possible but has no obvious role in this monogamous species, as song declines markedly after pairing (Catchpole, 1980; Dowsett-Lemaire, 1977).

Another possible function of mimicry is based on mate choice. In a migratory species such as the Marsh Warbler, a good portion of each year is spent in Africa. Such a species is probably closely adapted to specific migration routes and wintering grounds as well as to its breeding grounds. However, neighbors in the breeding habitat may go to very different areas upon migration (Fretwell, 1972). If individuals are highly adapted to a migration route and wintering habitat, then individuals should be favored who selectively mate with others that have followed the same route and gone to the same location. Mimetic vocal learning that encompasses the migration season and the start of the winter residence would allow for mate selection on the breeding ground based on auditory history, since many of the species on the route are local in distribution (Dowsett-Lemaire, 1979). Blyth's Reed Warbler (*Acrocephalus dumetorum*) is sympatric with the Marsh Warbler over the eastern portion of its range. Blyth's Reed Warbler overwinters in the Indian subcontinent rather than Africa, and it would be interesting if species isolation in the region of sympatry is enforced by Blyth's Reed Warbler mimicking Indian species (Voous, 1960). A third sibling species, the African Reed Warbler (*A. baeticatus*), resides in South Africa and also mimics. It is easily told from the European species only by song (Vernon, 1973). It does seem that the mimetic song of the Marsh Warbler could serve a function involving species if not population identification. A conclusion must await tests to see if females choose males who match their own auditory history. The population isolation hypothesis predicts that mates would overwinter in locations with a similar avifauna. My purpose in putting forth this hypothesis is to encourage others to look for similar cases of mimicry where a migrant mimics the calls and songs of species on its wintering grounds. Mimicry of this sort could very easily go unnoticed unless it is specifically sought out.

2. *As a Threat*

Robinson (1975) noted that many of the sounds mimicked by lyrebirds and bowerbirds in Australia are calls of aggressive species or predators. These sounds are employed more often by the singer when involved in fights with an intruding male or when a potential predator intrudes on the display ground. He suggested that the mimicry may function as an intraspecific and interspecific threat. This hypothesis may be tested by comparing edited song in playback studies. Mimetic songs containing only those elements suspected of being from aversive models can be tested against similar songs lacking these elements, and responses compared.

C. Interspecific (Ecological) Functions

1. *Parasite/Host Relationships*

The parasitic indigo-bird examples involve not only an intraspecific function of mimicry, but potential interspecific functions. The female indigo-bird may use the host song to locate potential nests in which to deposit her eggs. It is clear that the evolution of song in the parasite and host species will not be independent; the parasite's mimicry of the host will have some effect on the form of host song. Assuming that there is a disadvantage to being parasitized, selection should favor those elements of the host song that are difficult to mimic, rendering inappropriate matings more common in the parasite species. As in any evolving signal system, the motor and sensory systems concerned with song of parasite and host can be expected to coevolve. The mimicry system has an interspecific function in that it allows the indigo-birds to specialize on a host by isolating subpopulations that have succeeded in parasitizing a particular species.

2. *Mobbing*

Many cases involve the mimicry of alarm notes or mobbing calls of one species by another. Morton (1976) observed this behavior in the Thick-billed Euphonia (*Euphonia laniirostris*). Females used the mobbing calls of other species when the nest was threatened, and males also used these calls in their song (Remsen, 1976). Individuals of the model species were attracted to the mimic's calls, and joined the mimic in mobbing the predator. Similar situations have been described for Australian bowerbirds and lyrebirds (Marshall, 1950; Robinson, 1974, 1975) and the White-eyed Vireo (*Vireo griseus;* Adkisson and Conner, 1978).

No deception needs to be invoked to explain this form of mimicry. The individuals attracted were from species nesting nearby. It may be to the advantage of all involved to drive off potential nest predators. Mimetic calls are simply a convenient way to widen the response to individuals of other species, and need not be interpreted as exploitation. Similar hypotheses have been used to explain the evolution of mobbing calls without resorting to genetic relatedness (Armstrong, 1963). Since relatedness is not required, the same explanations can serve to describe the evolution of an interspecific response. The fact that a response occurred does not imply that the individuals of the model species were deceived. It would be interesting to see if the mimicking species in such cases will also respond to the model's mobbing calls.

3. *Interspecific Threats*

Much of the mimicry of lyrebirds and bowerbirds is copied from aggressive or predatory species. Since such calls are used more frequently when a human intruder enters the male's display site, or by a female when her nest is disturbed, then one of the functions of the mimicry might be as vocal threat (Robinson, 1974). This would be difficult to test, but the observation that the Spotted Bowerbird's (*Chlamydera maculata*) mimetic calls are given with increasing frequency the closer one approaches to the nest suggests a possible method (Chisholm, 1932). Individuals of many species will attack potential predators that approach their nest. A tethered live or a stuffed "predator" could be introduced at various distances from a bowerbird nest, and the frequency of mimetic calls recorded at each distance along with the number of attacks on the stimulus predator. If the mimetic calls function as a threat, then their frequency of occurrence should increase with decreasing distance, along with the probability of overt attack. It is important that the frequency of mimetic calls be recorded as a proportion of all calls given at each distance. We might expect the absolute number of mimetic calls to rise if the bowerbird simply vocalizes more frequently as an intruder nears the nest.

4. *Prey*

Some cases of mimicry, such as shrikes mimicking the calls of their prey, may lure prey within reach (Armstrong, 1963; Marshall, 1950; Witchell, 1896). This interpretation seems unlikely, because such a strategy would impose selection on the prey species for detection of the ploy (Armstrong, 1963). However, just such a luring strategy is known to be used by female fireflies of the genus *Photuris*, and hence it is a possible function of mimicry (Lloyd, 1965).

5. *Interspecific Territoriality*

The possibility that vocal mimicry functions as a means of defending or maintaining an interspecific territory has been proposed in several forms. Hartshorne (1973) observed that some versatile mimics are remarkably vigorous and generalized in the defense of their territories. He suggested that mimicry could serve to discourage other species from approaching the territory of the mimic. If a given song inhibits invasion by conspecifics, a similar sound sung by a heterospecific may have a similar effect, without deceiving a bird into thinking a conspecific holds that territory, but he did not specify the mechanism of exclusion. Rechten (1978) suggested that vocal mimicry functions as an interspecific "Beau Geste" territorial exclusion strategy (Krebs, 1977).

Dobkin (1979) argued that it is unlikely that a model would mistake a mimic's song for that of a conspecific. All mimic species impart some characteristic qualities to the song that are discernible to a skilled human observer, and these qualities might also be used by a model species to detect the mimic. Since many territorial species can recognize conspecific individuals by their song, surely they would be able to tell whether a singer was a conspecific (Dobkin, 1979). Abilities for individual recognition are not *de facto* evidence that a species can or will not respond to a mimic. Several experiments done with altered song have shown that even some rather artificially manipulated song will elicit a response that is indistinguishable from response to normal song (Boughey and Thompson, 1976; Emlen, 1972; Fletcher and Smith, 1978).

Howard (1974) rejected the hypothesis that the mimetic song of the Mockingbird was involved in interspecific territoriality. His test was based on reasoning that if interspecific territoriality were important, all the Mockingbirds on his study site should have shared mimicked vocalizations as the most common elements in their song repertoire. Dobkin (1979) pointed out that this was not a reasonable deduction. There is no reason to suppose that interspecific functions of a song would be more important than intraspecific functions. Indeed, it is remarkable that in a bird with as large a repertoire as the Mockingbird, 2 of the 14 syllable patterns used most frequently by all birds were imitations of Blue Jays (*Cyanocitta cristata;* Howard, 1974).

It is unnecessary to assume that mimicry would function in interspecific territoriality only if it involved deception (Section V,A below), or even an *a priori* avoidance by the model species. Nor is it reasonable to suppose that it could function in this regard only if mimetic elements were the most common ones in the resultant song. Hartshorne's (1973) statement of how mimicry might function in interspecific territoriality is realistic. Let us assume the existence of a bird such as the Mockingbird, which aggressively defends its territory from a variety of individuals of other species as well as its own (Bent, 1948; Moore, 1978; Mueller and Mueller, 1971). Its song and display flights warn other conspecifics away from the territory, but it will attack a variety of other intruders that are potential competitors for food (Moore, 1978). By mimicking the songs of these other species, the Mockingbird will be able to capture the attention of the individuals it mimics. It is not deception that the Mockingbird is practicing, but rather the phrasing of a warning in the language of each species that might intrude. Most birds probably do not pay attention to the songs of other species; rather they use a two-valued system (Hailman, 1977). The singer belongs to "my" species, or he is "another." "Another" will usually be ignored. The Mockingbird may circumvent this tendency to ignore and be ignored by other species by mimicry. This is not deception, as the Mockingbird will attack individuals of many species. The above should be considered when designing playback experiments to test hypotheses about mimicked song. In cases such as the

Mockingbird's, it may be that only birds that have interacted with Mockingbirds will avoid a Mockingbird mimicking their song. A naive bird may ignore it.

The hypothesis described above requires an assumption: The mimic must be interspecifically aggressive during the breeding season, as most sympatric passerines would not respond to song when not breeding. Does this adequately describe the Mockingbird? Anecdotal evidence for Mockingbird territoriality is plentiful, but detailed observations are confined to the winter territories. Mueller and Mueller (1971) described the winter aggressive behavior of a Mockingbird at a feeder. They recorded attacks on seven species of birds while other species were ignored. They noted that the species attacked learned to flee with the mere arrival of the Mockingbird at the feeder. Moore (1978) also observed winter territories of Mockingbirds, and recorded the number of attacks on each species as well as the number of times each species intruded into a Mockingbird's territory. Of 176 attacks observed, the highest proportion were delivered to individuals of species likely to compete with the Mockingbird for food. He did not report how many attacks were directed at other Mockingbirds.

I studied Mockingbirds in the months of March through September of 1975, and March through August of 1976 in Dutchess County, New York. During the breeding season I recorded every observed attack made by six banded male birds within their territories on individuals of their own and other species. Only attacks where both the attacker and the species of the bird attacked could be reliably identified were included. Although I do not have information on the relative frequency of intrusion, the data presented in Table II give an estimate of the relative effort devoted to intra- and interspecific territorial defense by these breeding male birds. Fellow Mockingbirds were attacked more than any other single species, but it is clear that the sum of the next two most frequently attacked species exceeds the total number of attacks on Mockingbirds. As far as overt attacks are concerned, it may be concluded that a Mockingbird in New York spends more time chasing other species than he spends chasing his own. This would appear to verify the anecdotal literature describing this species as interspecifically territorial in breeding as well as in winter territories.

Howard (1974) implied that if interspecific competition were important to Mockingbirds, the mimetic song elements should be the most commonly used notes and shared by all individuals. Dobkin (1979) rejected this line of reasoning, but did not suggest how frequently mimetic elements should be used. I think there are at least two possible predictions that can be made here. If Hartshorne's principle of monotony and habituation is correct for birdsong, a mimic should not use a copied phrase more frequently than the model would, or the model species will habituate to it. A second and simpler hypothesis is that the mimic should use the song elements based on its frequency of encounters with the model species. This would be easily accomplished by song matching with the model species, or using its song or call when an individual of that species is seen.

TABLE II

Species Attacked by Six Territorial Male Mockingbirds during the 1975 and 1976 Breeding Seasons at Millbrook, New York

Species attacked	Attacks observed
Mockingbird *Mimus polyglottos*	42
Brown Thrasher *Toxostoma rufum*	26
American Robin *Turdus migratorius*	21
Blue Jay *Cyanocitta cristata*	18
Gray Catbird *Dumetella carolinensis*	17
Starling *Sturnus vulgaris*	11
Eastern Kingbird *Tyrannus tyrannus*	11
Song Sparrow *Zonotrichia melodia*	8
Common Flicker *Colaptes auratus*	6
Cardinal *Cardinalis cardinalis*	5
Red-winged Blackbird *Agelaius phoeniceus*	4
Cedar Waxwing *Bombycilla cedrorum*	3
American Kestrel *Falco sparverius*	2
Killdeer *Charadrius vociferus*	2
Rufous-sided Towhee *Pipilo erythrophthalmus*	2
Total number of attacks observed	178

A third prediction is that mimicry should be highest during the establishment phase of territorial defense in the early breeding season, and decline during the nonbreeding season, when other birds are not singing and may not respond to song. This has been observed in the Mockingbird, which sings beyond the season of most passerines (Miller, 1938).

There is a considerable body of anecdotal evidence to suggest that Mockingbirds do countersing with other species of birds. Interspecific countersinging was described by Gander (1929), Halkin (1977), Hatch (1967), Howard (1974), Laskey (1944), and my own unpublished observations. In one instance of countersinging, I observed a Cardinal (*Cardinalis cardinalis*) and a Mockingbird in full view of one another and singing from treetops less than 30 m apart. There is evidence that Mockingbirds associate the appearance of birds with their specific songs and call notes. Laskey (1944) described an episode where a captive Mockingbird gave the call note of a Common Flicker (*Colaptes auratus*) upon seeing a silent Flicker; Townsend (1924) described a similar episode. I would have given little weight to such observations, except that I have observed instances when American Robins (*Turdus migratorius*) and Blue Jays were the visual stimuli for Mockingbird mimicry. Such an association may not be peculiar to Mockingbirds; Vernon (1973) cited a similar instance in an African drongo. It appears that Mockingbirds do respond to the song of other species of birds, and there is evidence of visual association with the sound. Individuals of other species respond to Mockingbirds as well, and other species countersing to Mockingbird mimicry (Davis, 1940; Halkin, 1977; Hatch, 1967; J. R. Baylis, unpublished data). Whether such responses can be shown to occur with a frequency and under circumstances to effect territorial exclusion is the remaining question. If interspecific territoriality can be effected by song mimicry, it is likely to occur only among the Mimidae, Dicruridae, and Irenidae, as Hartshorne (1973) suggested. These families are notably aggressive to other species as well as mimetic in full song.

6. *Interference*

Harcus (1977) suggested that vocal interference was a possible function of mimicry in the Chorister Robin-chat (*Cossypha dichroa*). Singing individuals of this species reply to other species during the dawn chorus, and even though the Chorister Robin-chat is unaggressive, the simple act of mimicry may interfere with the social signals of other species and hence give an advantage to the robin-chat. Another mechanism by which this may occur is exemplified by a flock of Smooth-billed Anis (*Crotophaga ani*) put to flight by a Mockingbird who sang their alarm call (Davis, 1940).

V. THE EVOLUTION OF VOCAL MIMICRY

A. Mimicry Need Not Imply Deception

Earlier I objected to a functional definition of vocal mimicry, and chose to treat it as a descriptive term. It is obvious from the examples above and the various selective forces that have been implicated in the evolution of mimicry that it is not a unitary phenomenon. It is rarely necessary to resort to the notion of deception to explain the mimetic relationship. The indigo-birds are mimetic by any intuitive criterion, yet there is no clear deception or even disadvantage to any receiver of the mimicked signal. Similar instances occur in the Marsh Warbler and lyrebirds. To exclude these as cases of mimicry because they may not happen to copy the calls of aversive species, or because they do not involve the deception of a receiver, would be folly.

Avian vocal mimicry is a phenomenon very different from other modes of mimicry that have been examined by biologists. Our modern conception of mimicry is based largely on trophic interactions between birds and insects. It would be a mistake to attempt to press all examples of mimesis into this single mold. There are two key points which suggest that vocal mimicry in birds will involve relationships between model and mimic very different from those usually considered in visual mimicry systems.

1. Generation Times

In the case of Batesian visual mimicry involving insect prey and vertebrate predators, the generation times of the prey and predator are disparate. The insect has the advantage of a shorter generation time, and deceptive mimicry can keep ahead of an adaptation to improve detection on the part of the predator. In the case of vocal mimicry, the generation times of the model and mimic are very similar if not identical. It would be difficult for a deceptive mimic to achieve lasting success.

2. Plasticity of Response

In the case of insect Batesian mimicry, an insect bears a resemblance to a *class* of other individuals. In the case of avian vocal mimicry, the resemblance is learned and in some cases highly specific (Tretzel, 1966). For example, a Mockingbird probably does not learn Cardinal song in the abstract, but learns a Cardinal song from an *individual* Cardinal. In principle, one could probably identify the individual Cardinal from whom the Mockingbird learned the song. The types of relationships available to vocal mimics and their models can be highly plastic, and yet be highly specific to individual models.

These are not the only aspects of vocal mimicry that differ from the more classical cases, but these alone are sufficient to suggest that the functional basis of vocal mimicry may not be confined to cases where there is "the inability of signal-receivers to distinguish between legitimate and illegitimate signals" (Dobkin, 1979, p. 350). Vocal mimics can be involved in complex competitive, cooperative, and parasitic relationships within and between species that cannot be easily equated to the simpler predator–prey relationships involved in classical mimicry.

B. Mimicry and The Cultural Transmission of Information

> Involuntary imitation, such as that of young birds learning the notes of their parents, is generally perpetuated by inherited tendency; and this, together with the protection afforded by imitation, no doubt chiefly perpetuates some of those common features by which each species of animal is distinguished (Witchell, 1896, p. 160).

Virtually every author who has reviewed the subject of vocal mimicry has noted that only birds which learn their songs mimic to any extent, although current evidence is not exhaustive. Witchell (1896) was the first to formulate this principle. Others have suggested that mimics inherit their mimetic songs, but the notion has never been popular (Dickey, 1922; Visscher, 1928). The Starling has been introduced throughout the world, and everywhere it mimics only local species (Saunders, 1923; Townsend, 1924). Unlike Batesian mimicry, the model and mimic often interact directly in avian vocal mimicry; hence it is difficult to imagine how inherited mimicry could occur in birds if there were any disadvantage to the model. Even if the model's call were also genetically determined, the mimic must always follow in a race with generation times too close to accurately track the model.

Little has been published on the responses of conspecifics to song in mimetic species, but available information suggests that some persistent mimics engage in bouts of matched countersinging with their conspecific neighbors (Halkin, 1977; Harcus, 1977; Howard, 1974). Countersinging may be correlated with an extended learning period allowing individuals to learn their neighbors' songs (Armstrong, 1963). This might represent an antecedent condition for the evolution of vocal mimicry through song learning, for it favors an ability to learn and develop a large repertoire size. It would be a small evolutionary step for an ecological generalist to broaden the learning and countersinging responses to include other species, if countersinging is indeed a form of ritualized territorial aggression. The more general case is that countersinging favors prolonged vocal learning and large repertoires; any selective advantage that accrues to an individual that does not confine its learning to conspecifics will favor mimicry. The proximate mechanisms may be manifold.

If both the model and the mimic learn their songs, the information is purely cultural in the sense that it is transmitted only by example from one generation to the next; vocal mimicry is the interspecific transfer of cultural information. This makes the mimic a potential social parasite of the model. If this characterization of mimicry is accurate, then Mayfield's statement on the problems involved in studying mimicry could be correct; perhaps the 85% of Mockingbird song consisting of unidentified elements represents extinct songs of other species, living or dead.

The above characterization of Mockingbird song as mostly mimetic is surely unrealistic, because it is based on misapprehension of how cultural evolution would operate. Cultural evolution is frequently characterized as being Lamarckian (Simon, 1980). Cultural information has the potential of being transferred at rates much faster than generation times and through paths other than descending kinship lineages. Such modes of transfer are not possible in Lamarckian systems. A second misapprehension is that the acquisition of new traits occurs with every generation.

The simplest way to think of this system is to treat each event of primary mimicry (i.e., where a mimic copies a vocal element from a model) as a *cultural mutation*. The mimetic element may then be passed on to offspring through vocal imitation, without the offspring ever having to hear the model (Saunders, 1923). Thus a mimic may have a core repertoire of its own species' unique culturally transmitted sounds, and a repertoire of mimetic sounds faithfully copied from conspecifics. Acquisition of mimetic elements by primary mimicry may actually be a fairly rare event. This may explain why Mockingbirds are said to mimic more in the northern part of the United States, where their range is extending. Because of the scarcity of conspecifics, primary mimicry (''mutation'') may be more common (Hartshorne, 1973).

C. Mimicry and the Coding of Information

Mimicry poses the obvious problem of preserving the species specificity of signals. I doubt that this is as great a problem as has been suggested. The songs copied by mimics may indeed have a courtship or territorial function, but they are long-distance signals (Marler and Hamilton, 1966) and they are not the only signals used by the species. Time would be wasted by an inappropriate response, but time may or may not be valuable. Visual displays and species-specific action patterns would prevent heterospecific matings or damaging fights over resources where no real competition exists. This is an issue only where a long-distance signal would serve as the sole basis for an inappropriate response. This would be most likely in situations such as mobbing or interspecific territoriality (exclusion).

Employment of other signal modalities could circumvent confusion in long-distance acoustical signaling. The mimetic larks all seem to use mimicry during a conspicuous, looping display flight when ample visual cues are available from plumage and action patterns (Section III,B); it is unlikely that a conspecific would fail to recognize the source of the song. Similarly, unpaired male Mockingbirds in full song utilize conspicuous song perches and punctuate their song by periodic leaps into the air, powered by 2–3 wing beats, and a swooping return to the song perch (J. R. Baylis, unpublished data). The bird is in full song during the ''jump'' display, and the white wing and tail patches characteristic of the species are conspicuously flashed. Mockingbirds also sing at night, when their models are mostly silent. Other mimics may have similar mechanisms for avoiding confusion.

The possibility of interspecific confusion should exert some selection pressure against mimicry unless some mechanism exists to code species specificity in spite of mimetic elements in the song. Virtually every study of mimicry dealing with this question has some statement that the mimic has a characteristic manner of singing (Boughey and Thompson, 1976; Dowsett-Lemaire, 1979; Farkas, 1969; Howard, 1974; Payne, 1973; Wildenthal, 1965), and that species specificity is coded not by the elements of the songs, but rather by the way the elements are arranged. Thus, there is an order of song structure higher than the elements themselves.

A mimic's song may resemble human speech in that the tokens representing the elements of the song (''phonemes'' in human speech) have meaning only relative to each other; each token by itself has no meaning. This gives great versatility to the song, for the elements in such a system can be recombined into new arrangements. Within the family Mimidae in eastern North America, the Gray Catbird sings each token in its song once, the Brown Thrasher sings each twice, and the Mockingbird sings each four to six times. The token sung and its species of origin may not matter (Boughey and Thompson, 1976). This higher level coding of information may be basic to all vocal mimics that breed at the same season as their models.

A simpler form of coding may occur in the songs of consistent and casual mimics. Witchell (1896) and Hunt (1922) both noted that many birds that mimic appear to sing the mimetic elements at the end of a species-specific song. This may represent a simpler form of coding that can preserve species specificity while allowing an individual to acquire notes from another species.

If deception is involved in the mimicry system, the model might be expected to have a simpler song organization. The model may be typified by short, stereotyped songs where song elements have individual importance; each element should function as a social releaser (Lorenz, 1935). An example of such a song is found in Indigo Buntings (*Passerina cyanea*), which respond to song artificially manipulated in the order of elements or number of repetitions, but do not respond unless the elements themselves remain intact (Emlen, 1972). A precondition for being a mimic may be the coding of information on a level of song organization higher than simply the notes of which the song is composed or their order of appearance.

This is especially suggestive because the most ubiquitous form of mimicry reported is the copying of calls. Calls tend to be highly stereotyped and are more easily characterized as releasers than song. If deceptive mimicry were to occur, mimicry of calls is the most likely case.

D. Are Some Species Easier to Mimic than Others?

I surveyed the species mimicked by the consistent and persistent mimics of North America [Appendix: Mockingbird—Halkin (1977), Laskey (1944), Mayfield (1934), Townsend (1924), Visscher (1928), and Wildenthal (1963); Gray Catbird—Townsend (1924); Yellow-breasted Chat (*Icteria virens*)—Cook (1935) and Townsend (1924); White-eyed Vireo—Adkisson and Conner (1978); Starling—Townsend (1924)]. The American Robin and the Blue Jay were mimicked by four of the five species. All of the Mockingbirds described by the above authors mimicked the above two models as well as the Cardinal and the Common Flicker. Some species are very commonly copied; either these species are easy to copy, or their songs and calls are easily recognized by the human observers, or they are of special significance to the mimicking species.

The American Robin and Blue Jay countersing with conspecifics, and the Cardinal song-matches. I suspect that a countersinging model would facilitate learning in a mimic that countersings, and may play a role in the preferential copying of some species.

VI. CONCLUSION

> We may safely lay it down as an axiom that vocal utterance is always subject to variation (Witchell, 1896, p. 140).

Avian vocalizations are highly variable at all levels from within an individual to between species. Given such a range of variation, it is not surprising that we occasionally observe instances of vocal resemblance between individuals belonging to different species. Such resemblance may be due to chance, a shared phylogenetic history, the result of natural selection favoring a common form of vocalization, or the copying of one species' vocalizations by another. The latter two categories of resemblance are functional in the sense that they represent evolutionary responses to natural selection. Section IV examined some of the suggested sources of natural selection favoring mimesis. However, there is not yet a single study which has subjected one of these hypotheses to a critical test. This is rather surprising, since the review in Section III revealed that mimetic species occur everywhere there are songbirds, and they usually make up about 15% of the total number of resident species. Mimesis is widespread, fairly common, and raises fundamental evolutionary issues. Why has it failed to produce definitive studies?

Certainly part of the problem is the difficulty of objectively identifying mimesis. It has clearly created difficulties in species with large mimetic repertoires. This has been a problem because the researcher has been tempted to study the *entire* repertoire in the mimic species. Far greater progress might be made by selecting a specific mimic–model relationship, and examining it alone.

For example, Mockingbirds clearly mimic Cardinal song, and the copied song elements are easily detected. By confining a study of song mimesis to these two species a variety of questions can be readily answered. Do individual Mockingbirds command as large a repertoire of Cardinal songs as individual Cardinals? Do Mockingbirds and Cardinals interact during singing bouts? Is a Mockingbird more likely to sing Cardinal song immediately after hearing a Cardinal? Do Mockingbirds whose territories overlap or are adjacent to Cardinal territories sing Cardinal song more often? Such questions would be extraordinarily difficult to answer if one attempted to deal with all of Mockingbird mimesis at once; but such questions can be dealt with by restricting the study of mimesis to specific questions involving specific species pairs. The questions asked should be selected in such a way as to differentiate between the functional hypotheses examined in Section IV, always bearing in mind that mimicry can simultaneously serve many functions.

Dobkin (1979) maintained that a confusion existed in the literature with respect to the function of mimicry, and that this had produced a lack of definitive rigor in testing hypotheses. I concur with his opinion. In Section IV I attempted to clearly state some hypotheses, with the assumptions of each and the conditions under which it may operate. The first step in any investigative study of mimicry must be a clear statement of the hypothesis. Other possibly competing hypotheses must also be examined, and a critical set of predictions must be generated that will discriminate between the two. Any assumptions made by the hypotheses as limiting conditions must be shown to be valid for the system under study. Only then can the findings lead to the acceptance or rejection of the hypothesis. This level of rigor has yet to be applied to most cases of vocal mimicry, although suitable systems abound. The number of competing hypotheses and the complex relationships between two or more species will require great ingenuity on the part of experimentalists and field observers, but the evolutionary issues involved are important, basic, and well worth a substantial effort.

A further problem that has seriously hampered the study of avian mimicry is that we have no real appreciation of its prevalence and scope in most species. There is really little or no excuse for this, for the phenomenon has been recognized for hundreds of years and served a central role in the early argument that birds learn their songs. Witchell (1896) wished to apply Darwinian thinking to the evolution of birdsong, and recognized that in order to do so he had to demonstrate the existence of variation on which Darwinian selection could act, and document the sources of that variation. Lacking a twentieth century knowledge of genetics, he could not use mutation or sexual recombination as a source of variation. He sought his source of variation in mimicry. As an appendix to his chapter on mimicry, Witchell included tabular data on grouped individuals of each species known for mimicry. Here he recorded the number of songs he heard, the number that were mimetic, and the frequency with which various model species occurred in each mimic's repertoire. One may fault Witchell's nineteenth century field methods, but here was a clear attempt to make quantitative observations on how frequently mimesis occurred in each species, and the frequency with which various species were used as models. If subsequent publications on mimicry had been as formal in their approach, the conclusions we could have drawn from the survey in Section III might have been much more extensive. Avian vocal mimicry is a phenomenon that has raised many questions. Very few have been definitively answered.

ACKNOWLEDGMENTS

I am extremely grateful to Peter Marler for providing the opportunity for a hard-core fish ethologist to learn about birds and birdsong. Donald Kroodsma, Myron Baker, and Michael Gochfeld assisted Professor Marler in introducing me to the literature and field techniques of ornithology. Jack P. Hailman and Sylvia Halkin read an earlier draft of this chapter, and their comments and ideas resulted in substantial improvements as well as continuing my remedial education. All of the individuals above should receive credit for what knowledge this chapter contains; I alone am to blame for any displays of ignorance or omissions. Jessica Egerton and Tom Gorman ably assisted me in collecting the field data on Mockingbirds. These and other data were gathered as part of a research project on Mockingbird song funded by NIMH Grant MH14651 to Peter Marler. His support and the support of The Rockefeller University are gratefully acknowledged.

REFERENCES

Adkisson, C. S., and Conner, R. N. (1978). Interspecific vocal imitation in White-eyed Vireos. *Auk* **95,** 602–606.

Armstrong, E. A. (1963). "A Study of Bird Song." Oxford Univ. Press, London and New York.

Bell, K. (1976). Song of the Superb Lyrebird in southeastern New South Wales, Australia with some observations on habitat. *Emu* **76,** 59–63.

Bent, A. C. (1948). Life histories of North American nuthatches, wrens, thrashers, and their allies. *Bull.—U.S. Natl. Mus.* **195.**

Berger, A. J. (1961). "Bird Study." Wiley, New York.

Bertram, B. (1970). The vocal behaviour of the Indian Hill Mynah, *Gracula religiosa. Anim. Behav. Monogr.* **3,** 79–192.

Boughey, M. J., and Thompson, N. S. (1976). Species specificity and individual variation in the songs of the Brown Thrasher (*Toxostoma rufum*) and Catbird (*Dumetella carolinensis*). *Behaviour* **57,** 64–90.

Brown, J. L. (1975). "The Evolution of Behavior." Norton, New York.

Brown, R. N. (1977). Character convergence in bird song. *Can. J. Zool.* **55,** 1523–1529.

Catchpole, C. K. (1980). Sexual selection and the evolution of complex songs among European warblers of the genus *Acrocephalus. Behaviour* **74,** 149–166.

Chisholm, A. H. (1932). Vocal mimicry among Australian birds. *Ibis* **13,** 605–625.

Chisholm, A. H. (1950). Further notes on vocal mimicry. *Emu* **49,** 232–234.

Cody, M. L. (1969). Convergent characteristics in sympatric populations: A possible relation to interspecific competition and aggression. *Condor* **71,** 222–239.

Cook, H. P. (1935). The song of the Yellow-breasted Chat. *Wilson Bull.* **42,** 297–298.
Darwin, C. (1874). "The Descent of Man and Selection in Relation to Sex," 2nd ed. Henneberry, Chicago, Illinois.
Davis, D. E. (1940). Social nesting habits of the Smooth-billed Ani. *Auk* **57,** 179–218.
Dickey, D. R. (1922). The mimetic aspect of the mocker's song. *Condor* **24,** 153–157.
Dobkin, D. S. (1979). Functional and evolutionary relationships of vocal copying phenomena in birds. *Z. Tierpsychol.* **50,** 348–363.
Dobson, C. W., and Lemon, R. E. (1975). Reexamination of monotony threshold hypothesis in bird song. *Nature* (*London*) **257,** 126–128.
Dowsett-Lemaire, F. (1979). The imitation range of the song of the Marsh Warbler, *Acrocephalus palustris,* with special reference to imitations of African birds. *Ibis* **121,** 453–468.
Eberhardt, C., and Baptista, L. F. (1977). Intraspecific and interspecific song mimesis in California (U.S.A.) Song Sparrows. *Bird-Banding* **48,** 193–205.
Emlen, S. T. (1972). An experimental analysis of the parameters of bird song eliciting species recognition. *Behaviour* **41,** 130–171.
Farkas, T. (1969). Notes on the biology and ethology of the Natal Robin, *Cossypha natalensis. Ibis* **111,** 281–291.
Fisher, R. A. (1958). "The Genetical Theory of Natural Selection." Dover, New York. (Orig. publ., 1930.)
Fletcher, L. E., and Smith, D. G. (1978). Some parameters of song important in conspecific recognition by Gray Catbirds. *Auk* **95,** 338–347.
Fretwell, S. D. (1972). "Populations in a Seasonal Environment," Monograph in Population Biology, Vol. 5. Princeton Univ. Press, Princeton, New Jersey.
Gander, F. F. (1929). Notes on bird mimicry with special reference to the Mockingbird (*Mimus polyglottos*). *Wilson Bull.* **16,** 93–95.
Gramza, A. F. (1972). Avian vocal mimicry; the phenomenon and its analysis. *Z. Tierpsychol.* **30,** 259–265.
Guralnik, D. B., and Friend, J. H. (1964). "Webster's New World Dictionary: College Edition." World Publ., New York.
Hailman, J. P. (1977). "Optical Signals." Indiana Univ. Press, Bloomington.
Halkin, S. (1977). A study of the natural history and counter-singing behavior of eastern Mockingbirds, *Mimus polyglottos polyglottos,* in the Lincoln, Massachusetts area. B.A. Honors Thesis Biol., Harvard Univ., Cambridge, Massachusetts.
Harcus, J. L. (1977). The functions of mimicry in the vocal behavior of the Chorister Robin. *Z. Tierpsychol.* **44,** 178–193.
Hartshorne, C. (1956). The monotony threshold in singing birds. *Auk* **73,** 176–192.
Hartshorne, C. (1958). Some biological principles applicable to song behavior. *Wilson Bull.* **70,** 41–56.
Hartshorne, C. (1973). "Born to Sing. An Interpretation and World Survey of Bird Song." Indiana Univ. Press, Bloomington.
Hatch, J. J. (1967). Diversity of song of Mockingbirds (*Mimus polyglottos*) reared in different auditory environments. Ph.D. Thesis, Duke Univ., Chapel Hill, North Carolina.
Hinde, R. A. (1970). "Animal Behaviour." McGraw-Hill, New York.
Howard, R. D. (1974). The influence of sexual selection and interspecific competition on Mockingbird song (*Mimus polyglottos*). *Evolution* **28,** 428–438.
Hunt, R. (1922). Evidence of musical "tastes" in the Brown Towhee. *Condor* **24,** 193–203.
Huxley, J. S. (1914). The courtship-habits of the Great Crested Grebe (*Podiceps cristatus*); with an addition to the theory of sexual selection. *Proc. Zool. Soc. London* **35,** 491–562.
Kenyon, R. F. (1972). Polygyny among Superb Lyrebirds. *Emu* **72,** 70–76.

King, A. P., and West, M. J. (1977). Species identification in the North American Cowbird: Appropriate responses to abnormal song. *Science* **195,** 1002–1004.

Krebs, J. R. (1977). The significance of song repertoires: The Beau Geste hypothesis. *Anim. Behav.* **25,** 475–478.

Kroodsma, D. E. (1976). Reproductive development in a female songbird: Differential stimulation by quality of male song. *Science* **192,** 574–575.

Kroodsma, D. E. (1977). Correlates of song organization among North American Wrens. *Am. Nat.* **111,** 995–1008.

Kroodsma, D. E. (1978). Continuity and versatility in birdsong: Support for the monotony-threshold hypothesis. *Nature (London)* **274,** 681–683.

Kroodsma, D. E., and Parker, L. D. (1977). Vocal virtuosity in the Brown Thrasher. *Auk* **94,** 783–784.

Laskey, A. R. (1935). Mockingbird life history studies. *Auk* **52,** 370–381.

Laskey, A. R. (1944). A Mockingbird acquires his song repertoire. *Auk* **61,** 211–219.

Lemaire, F. (1977). Mixed song, interspecific competition and hybridization in the Reed and Marsh warblers (*Acrocephalus scirpaceus* and *palustris*). *Behaviour* **63,** 215–240.

Lloyd, J. E. (1965). Aggressive mimicry in *Photuris:* Firefly femmes fatales. *Science* **149,** 653–654.

Lorenz, E. K. (1935). Companions as factors in the bird's environment. *In* K. Lorenz, "Studies in Animal and Human Behavior," Vol. 1. Harvard Univ. Press, Cambridge, Massachusetts, 1970.

Mackworth-Praed, C. W., and Grant, C. H. B. (1963). "Birds of the Southern Third of Africa," Vol. 2. Longmans, Green, New York.

Marler, P. (1959). Developments in the study of animal communication. *In* "Darwin's Biological Work," (P. R. Bell, ed.), pp. 150–206. Cambridge Univ. Press, London and New York.

Marler, P., and Hamilton, W. J. (1966). "Mechanisms of Animal Behavior." Wiley, New York.

Marshall, A. J. (1950). The function of vocal mimicry in birds. *Emu* **50,** 5–16.

Mayfield, G. R. (1934). The Mockingbird's imitation of other birds. *Migrant* **5,** 17–19.

Michener, H., and Michener, J. R. (1935). Mockingbirds, their territories and individualities. *Condor* **37,** 97–140.

Miller, L. (1938). The singing of the Mockingbird. *Condor* **40,** 216–219.

Moore, F. R. (1978). Interspecific aggression: Toward whom should a Mockingbird be aggressive? *Behav. Ecol. Sociobiol.* **3,** 173–176.

Morton, E. S. (1976). Vocal mimicry in the Thick-billed Euphonia. *Wilson Bull.* **88,** 485–487.

Moynihan, M. (1968). Social mimicry: Character convergence versus character displacement. *Evolution* **22,** 315–331.

Mueller, H. C., and Mueller, N. S. (1971). Flashes of white in wings of other species elicit territorial behavior in Mockingbirds. *Wilson Bull.* **83,** 442–443.

Mundy, P. J. (1973). Vocal mimicry of their hosts by nestlings of the Great Spotted Cuckoo and Striped Crested Cuckoo. *Ibis* **115,** 602–604.

Newman, K. (1968). "Garden Birds of South Africa." Am. Elsevier, New York.

Nicolai, J. (1964). Der Brutparasitismus der Viduinae als ethologisches Problem. Prägungsphänomene als Faktoren der Rassen- und Artbildung. *Z. Tierpsychol.* **21,** 129–204.

Nottebohm, F. (1972). The origins of vocal learning. *Am. Nat.* **106,** 116–140.

Payne, R. B. (1973). Behavior, mimetic songs and song dialects, and relationships of the parasitic indigobirds (*Vidua*) of Africa. *Ornithol. Monogr.* **11,** 333.

Payne, R. B. (1976). Song mimicry and species relationships among the West African Pale-winged Indigobirds. *Auk* **93,** 25–38.

Peterson, R. T., Mountfort, G., and Hollom, P. A. D. (1974). "A Field Guide to the Birds of Britain and Europe." Houghton, Boston, Massachusetts.

Pettingill, O. S. (1970). "Ornithology." Burgess, Minneapolis, Minnesota.

Rechten, C. (1978). Interspecific mimicry in birdsong: Does the Beau Geste hypothesis apply? *Anim. Behav.* **26,** 304–312.

Remsen, J. V. (1976). Observations of vocal mimicry in the Thick-billed Euphonia. *Wilson Bull.* **88,** 487–488.

Richardson, C. H. (1906). Birds whose notes are imitated by the western Mockingbird. *Condor* **8,** 56.

Robinson, F. N. (1974). The function of vocal mimicry in some avian displays. *Emu* **74,** 9–10.

Robinson, F. N. (1975). Vocal mimicry and the evolution of bird song. *Emu* **75,** 23–27.

Saunders, A. A. (1923). Mimicry in bird songs. *Condor* **25,** 68–69.

Saunders, A. A. (1929). "Bird Song," Handbook No. 7. New York State Mus., Albany.

Selander, R. K. (1972). Sexual selection and dimorphism in birds. *In* "Sexual Selection and the Descent of Man, 1871–1971" (B. Campbell, ed.), pp. 180–230. Aldine, Chicago, Illinois.

Short, L. L., Jr. (1966). Field Sparrow sings Chipping Sparrow song. *Auk* **83,** 665.

Simon, H. A. (1980). The behavioral and social sciences. *Science* **209,** 72–78.

Skutch, A. F. (1961). Helpers among birds. *Condor* **63,** 198–226.

Sullivan, G. A. (1976). Song of the finch (*Lagonosticta senegala*), interspecific mimicry by its brood parasite (*Vidua chalybeata*), and the role of song in the host's social context. *Anim. Behav.* **24,** 880–888.

Thompson, W. L., and Jane, P. L. (1969). Analysis of Catbird song. *Jack-Pine Warbler* **47,** 115–125.

Thorpe, W. H. (1961). "Bird Song," Cambridge Monographs in Experimental Biology, No. 12. Cambridge Univ. Press, London and New York.

Thorpe, W. H., and North, M. E. W. (1966). Vocal imitation in the tropical Bou-Bou Shrike (*Laniarius aethiopicus major*) as a means of establishing and maintaining social bonds. *Ibis* **108,** 432–435.

Townsend, C. W. (1924). Mimicry of voice in birds. *Auk* **41,** 541–552.

Tretzel, E. (1966). Spottmotivprädisposition und akustiche Abstraktion bei Gartengrasmücken (*Sylvia borin borin* (Bodd.)). *Zool Anz., Suppl.* No. 30, 333–343.

van der Steen, W. J. (1972). Ecology, evolution, and explanatory patterns in biology. *J. Theor. Biol.* **36,** 593–616.

Vernon, C. J. (1973). Vocal imitation by southern African birds. *Ostrich* **44,** 23–30.

Visscher, J. R. (1928). Notes on the nesting habits and songs of the Mockingbird. *Wilson Bull.* **40,** 209–216.

Voous, K. H. (1960). "Atlas of European Birds." Nelson, London.

Wade, M. J., and Arnold, S. J. (1980). The intensity of sexual selection in relation to male sexual behavior, female choice, and sperm precedence. *Anim. Behav.* **28,** 446–461.

Welty, J. C. (1975). "The Life of Birds." Saunders, Philadelphia, Pennsylvania.

Wickler, W. (1968). "Mimicry in Plants and Animals." McGraw-Hill, New York.

Wildenthal, J. L. (1963). Structure in primary song of the Mockingbird, *Mimus polyglottos.* Master's Thesis, Univ. of Kansas, Lawrence.

Wildenthal, J. L. (1965). Structure in the primary song of the Mockingbird, *Mimus polyglottos. Auk* **82,** 161–189.

Witchell, C. A. (1896). "The Evolution of Bird Song." Adam & Charles Black, London.

4

The Ecological and Social Significance of Duetting

SUSAN M. FARABAUGH

I. INTRODUCTION

In most bird species song is restricted to males. Female song, when it does occur, does not represent a reversal of sexual roles; rather, both males and females sing. In most cases, they sing at the same time, often with precise temporal coordination, in a duet. Ornithologists have long been intrigued by duetting, particularly with the elaborate antiphonal songs of many tropical species where mates exchange notes in such a regular manner that unless the listener is standing between the two singers, he may think that only one bird is singing.

ISBN 0-12-426802-1

Yet, the precision of some duets is only one interesting aspect of the general phenomenon known as duetting.

In this chapter, I examine the ecological and social significance of duetting. However, before this can be done, we must decide what a duet is. Once duetting is defined, the selection pressures that favored the evolution of vocal duets are examined by considering the ecological and social characteristics that are shared by duetting species from various unrelated taxa, and that distinguish these species from their nonduetting kin. The functions of duets are discussed in relation to these characteristics. Advantages gained both by female participation in song and by coordination of female and male vocalizations are considered. Finally, the structure of vocal duets is discussed in relation to their function.

II. WHAT IS A DUET?

A. Toward an Objective Definition of Duetting

Many papers, including a number of major review papers (Payne, 1971; Thorpe, 1972; von Helverson, 1980), have been written about duetting, but the phenomenon remains inadequately defined. Nevertheless, there are several characteristics that most authors would consider essential to any definition of duetting.

The first characteristic involves the identity of the duetters. Duets are performed by the members of a mated pair or extended family group. Therefore, countersinging by neighboring territorial males is not considered to be duetting even if the timing of the interchange is strikingly precise. If more than two family members are involved, the term "chorus singing" is sometimes used (Payne, 1971). For simplicity, however, I prefer not to distinguish between two versus three or more birds in this review; therefore, when I discuss duetting, I am referring to participation of two or more birds.

The second characteristic refers to the type of behavior that is used in a duet. Although mutual displays in modalities other than sound can occur, duetting refers only to mutual acoustic displays. However, it must be mentioned that duets are often accompanied by coordinated visual displays in many species (Kunkel, 1974; Baptista, 1978; Wickler, 1980; Todt and Fiebelkorn, 1980; S. M. Farabaugh, unpublished data), and these displays can be more elaborate than the vocal duet in some species. While there are restrictions on the modality involved, there are virtually no restrictions on the types of sounds used. In this way, possible biases toward songbirds are eliminated. Duets can involve songs, calls, or nonvocal sounds such as bill clattering of storks (Haverschmidt, 1947) and drumming of woodpeckers (Kilham, 1959). Again, for simplicity, when I use the term "vocalization" throughout the chapter I do not mean to exclude nonvocal sounds.

The third characteristic of duetting, involving the temporal coordination of the two birds' vocalizations, is the most complex. There are many facets of temporal coordination, but in the past authors have dealt with only one aspect: precision of timing. Thorpe (1961, 1963, 1972) has suggested that true duets are precisely timed. However, some of the species that he classified as having precise antiphonal vocalizations have recently been shown to be imprecise; they often overlap their calls or call irregularly (Payne, 1971; Huxley and Wilkinson, 1979). I do not mean to suggest that these species are not true duetters, but rather, that precision of timing may vary in some degree.

My gestalt view of duets is that, for any species, bouts of certain elements in the repertoire of one bird frequently overlap with bouts of certain elements in the repertoire of its mate. For instance, male song bouts may frequently overlap with female song bouts, or male song bouts may overlap with bouts of female "churr" calls, or bouts of male bill clattering may overlap with bouts of female bill clattering. Further, there is some organization of both participants' elements within the region of overlap. This view can be expressed in terms of three variables which can be measured for any species: one measure of bout overlap, i.e., the percentage of bouts that overlap with bouts of the mate; and two measures of organization of male and female elements within overlapping bouts, i.e., the precision of timing and the sequential ordering of elements.

The term "bout" must be defined before bout overlap can be discussed. Vocalizations, like other behaviors, are clumped rather than randomly distributed in time. These clumps are called bouts. Bouts are separated in time from each other by intervals that are significantly longer than the intervals between behaviors within bouts. Determination of what is and is not a bout is a thorny question for theoretical statisticians (see Machlis, 1977), but usually a working definition can be arrived at for any particular species.

The first variable, percentage of bouts that overlap, can be measured once bouts of each element type in the repertoire of each member of the pair are determined. The percentage of male bouts that overlap with female bouts, and vice versa, can be calculated for each type of bout (bouts of song, bouts of each call type, etc.). If the percentage overlap is high, these overlapping bouts may be duets. For example, for a pair from hypothetical species X, we may find that, in a sample of 200 male song bouts and 100 female song bouts, there are 100 overlapping bouts. Therefore, 50% of all male song bouts overlap with female song bouts, while 100% of all female song bouts overlap with male song bouts. In a sample from hypothetical species Y, there may be 200 male song bouts and 200 female song bouts arranged in 200 overlapping bouts. For species Y, both male overlap of female and female overlap of male are 100%. On the basis of these overlap percentages, species Y might be considered a stronger duetter than species X. Yet, both species X and Y would score 100% overlapping, since at least one bird's bouts overlap 100% with its mate's bouts. Differences between the overlap percentages for each mate indicate a difference in the amount of participation of each mate in this particular type of vocalization. In the example from species X, the female sang less than the male did, but when she did sing she always overlapped with the male.

Within overlapping bouts, organization of male and female elements can be viewed in terms of temporal precision, the second variable. Temporal precision can be described by calculating the coefficient of variation after measuring the variance of the interval from the onset of each male element to the onset of the next female element. But if the female's vocal element begins just after the male's element ends, and male elements vary in duration, then the interval from the end of each male element to the onset of the following female element should be measured instead. In either case, the variance about the mean interval length can then be calculated. Variance is not a good statistic to use for a comparison of precision in samples with different means, because variance is generally positively correlated with the size of the mean. For this reason, precision should be measured using the coefficient of variation (CV = standard deviation/mean) which allows comparisons of percentage variation in two populations, independently of the magnitude of their means (Sokal and Rohlf, 1973). Low CV's indicate low variation about the mean, and hence high precision of timing. For any species, those types of overlapping bouts which have low CV's may be duets. For a discussion of the uses of the coefficient of variation to describe behavior sequences including variability of the pattern as a whole and of its constituent elements, see Schleidt (1974).

The third variable involves the order of male and female elements within the area of overlap. An ordered pattern can be distinguished from a random pattern using the test of serial randomness of nominal categories (Zar, 1974). A sequence of like elements, bounded on either side by either an unlike element or no element, is termed a run. Two types of deviation from randomness can be indicated by the numbers of runs in a particular sequence: clumped sequences yield low numbers of runs while alternating sequences yield high numbers of runs. Random sequences yield intermediate numbers of runs. Sequences yielding high numbers of runs may indicate duetting. Letting N_1 be the total number of elements of the first category, N_2 the total number of elements of the second category, and U the number of runs, the critical U value $[U(0.05, N_1, N_2)]$ can be read from Table 34D in Zar (1974). For example, in the sequence "FM-MFFMFMMMFFMMFFFMFMMM" there are 10 "F" elements and 12 "M" elements arranged in 12 runs. For this sequence $U(0.05,10,12) = 7, 17$; that is, if U is greater than 7 and less than 17, the sequence is considered random. In this case the sequence is random, but a sequence of 10 male elements and 12 female elements arranged in 18 or more runs would be considered highly alternating and possibly a duet. The highest possible number of runs for such a sequence is 21

(FMFFMFMFMFMFMFMFMFMFMF) while the lowest number of runs is 2 (MMMMMMMMMMFFFFFFFFFFF). Of these two nonrandom sequences only the former should be considered a possible duet.

The number of runs can be used to generate a continuous variable: the percentage alternation, which equals the number of runs (U) divided by the maximum number of runs [($U_{max} = 2 (N_1) + 1$, if $N_1 < N_2$ or $U_{max} = 2 (N_1)$, if $N_1 = N_2$] times 100 [(percentage alternation $= 100 \times U/(U_{max})$]. The number of runs is not a useful continuous variable, because it is proportional to the sizes of N_1 and N_2. For example, a sequence of 6 male songs and 5 female songs arranged in the following sequence (MFMFMFMFMFM) has 100% alternation; yet, a sequence of 16 male songs and 5 female songs (MMMFMFMMMMMFMMMFMFM-MM) also has 100% alternation. Clearly the former sequence is more highly alternating than the latter. When N_1 does not approximately equal N_2, percentage alternation should be calculated by counting the number of times that the onset of a female element occurs between two male element onsets, dividing this number by the number of male onsets minus one, and multiplying this by 100. Using this method, the former sequence above would still be 100% alternating, but the latter would be only 33% alternating.

All three of the foregoing variables are descriptive. In and of themselves they cannot define what is and is not a duet. To remedy this, one possibility would be to set arbitrary critical values that would define limits for duetting. However, since these variables have been measured for few if any species, there is not yet sufficient information for setting such limits. Another possibility would be to use discriminant analysis techniques to sort duetters and nonduetters using the three continuous variables just described. It may be possible to collect samples of vocal bouts from species which have already been classified as duetters and calculate percentage overlap, coefficients of variation of interval lengths, and percentage alternation, for both duet bouts and overlapping bouts that are not considered to be duets (such as mobbing calls, begging calls of young, etc.). Using these three calculated variables in a discriminant analysis for two classes (duets or nonduets), one could determine whether these measures yield two significantly different groups. If so, the same discriminant technique could then be applied to sort other species into the duetting or nonduetting group.

In summary, overlapping bouts of vocal or nonvocal sounds given by the members of a mated pair or extended family group can be considered duets providing that a high percentage of such bouts given by one bird overlap with a particular type of bout given by the other bird. The organization of elements within these overlapping bouts should be such that the coefficient of variation of intervals between elements is low or percentage alternation of elements is high, or both. Critical limits defining duets using these three variables (percentage overlap, coefficient of variation of intervals, and percentage alternation) cannot be set until these variables are measured for more species. Alternatively, duets could be sorted from nonduets using these variables in a discriminant analysis for two classes.

B. Classification of Duets

Duetting, as defined here, includes many different types of vocal interactions, from the elaborate pair-specific antiphonal song of Buff-breasted Wrens (*Thryothorus leucotis*) to the overlapping calling bouts of American Oystercatchers (*Haematopus ostralegus*). Four variable characteristics of duets may be useful in classifying different types of duets: (1) pattern of elements, (2) relative participation of mates in vocalizations used in duets, (3) precision of timing, and (4) type of sound used by each mate.

Pattern of elements refers to the pattern of male and female element overlap within bouts. This can vary from antiphony to complete synchrony of male and female vocal elements. Golani (1976) described the various ways that two visual displays could occur simultaneously. These are applicable to simultaneous vocal interactions as well (Fig. 1). One or more of these cases may be applicable for any single duetting species.

The second variable statistic of duetting is a measure of the relative participation of mates in the vocalizations used in duets. As mentioned in the previous section, when a duetting pair has 100% overlap of a particular type of vocal bout, it does not necessarily mean that the mates are participating equally in the production of such vocalizations. 100% overlap merely indicates that the maximum possible overlap has been achieved; a pair in which the female sings less than her mate will be considered 100% overlapping if, whenever she does sing, her song bouts overlap with her mate's bouts. The relative participation of mates in duet vocalizations, in this case song, can be measured by the number of bouts of the less frequent singer divided by the number of bouts of the more frequent singer times 100. For example, the relative participation in song for species X and Y mentioned previously (both scored 100% overlap, but X had 200 male and 100 female song bouts arranged in 100 overlapping bouts, while Y had 200 male and 200 female song bouts arranged in 200 overlapping bouts) is 50% for species X and 100% for species Y. In species X, males sing more than do females.

Although precision of timing, as expressed by the CV of the interval between mates' elements, is one of the variables used to define duetting, there still can be some variation among species in this characteristic. There may be some species that are classified as duetters because they have a high percentage of overlap and a high percentage of alternation, but not necessarily because they are also precise. Thus having used precision as a defining characteristic of duets as such, we can also use it to classify and rank different duets relative to each other.

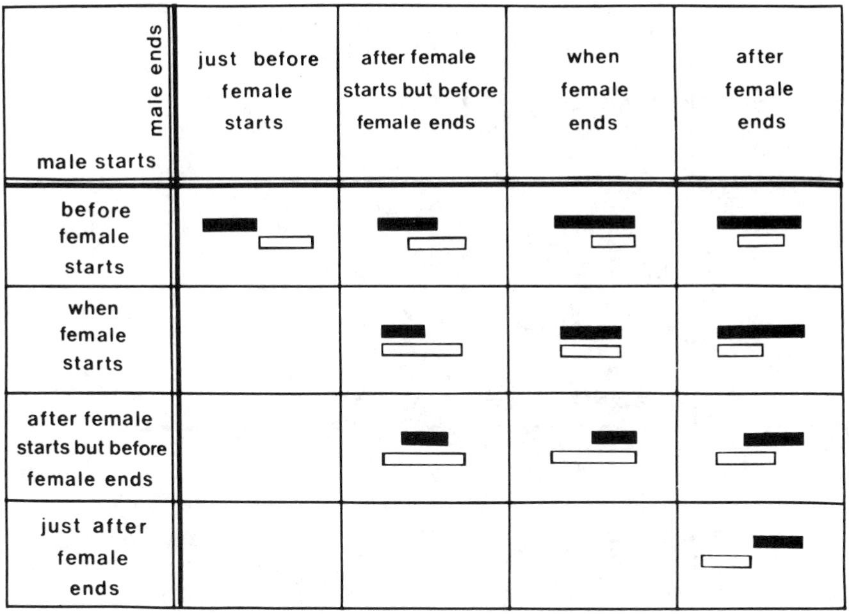

male ends / male starts	just before female starts	after female starts but before female ends	when female ends	after female ends
before female starts				
when female starts				
after female starts but before female ends				
just after female ends				

Fig. 1. Temporal relationships between "simultaneous" vocal elements. The bars represent temporal duration of elements, with each solid bar representing the male's and the open bar the female's vocal element. Time proceeds from left to right. (After Golani, 1976.)

The type of sound (song, call, or nonvocal sound) may also vary from species to species. In any one individual's repertoire, some elements may be used in duets while other elements are not, and mates may use different types of sounds in duets. For example, duetting in Carolina Wrens (*Thryothorus ludovicianus*) consists of male song and female churr calls (Fig. 2a). On the other hand, mates may use the same type of sound—both sing or both call. Even when mates use similar types of sounds there may be sexual and individual specificity. Sometimes the same motif is used by both birds, but the female's version is of slightly higher frequency, as in Rufous-and-white Wrens (*T. rufalbus;* Fig. 2b). Still other species have distinct male and female vocal elements which are joined to form pair-specific duets, as in Buff-breasted Wrens (*T. leucotis;* Fig. 2c).

Duets include a wide variety of vocal styles. In this section, four possible characteristics were discussed which could be used to describe and classify the duets of most duetting species. The significance of these interspecific differences will be discussed in Section V.

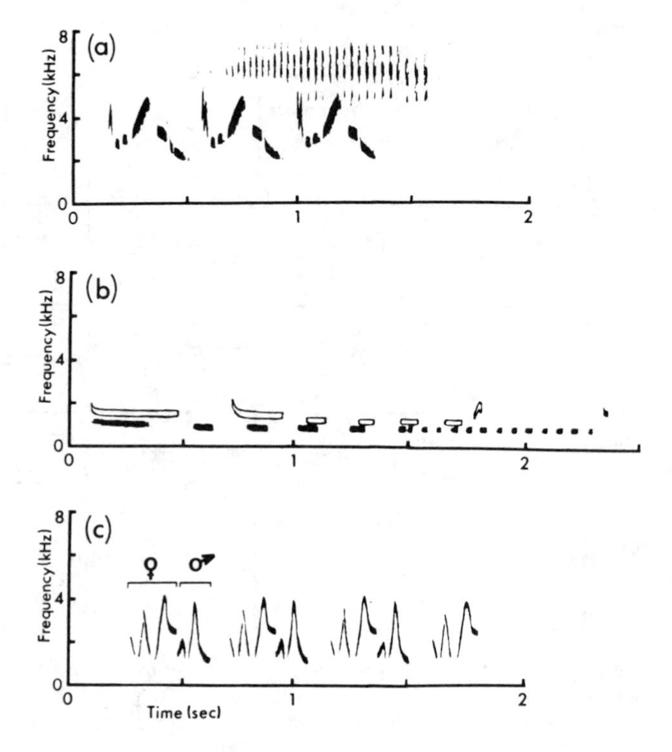

Fig. 2. Examples of the type of sounds used in the duets of three *Thryothorus* species. (a) Male song and female "churrs" of the Carolina Wren (*T. ludovicianus*). (b) Male (solid notes) and female (open notes) of the Rufous-and-white Wren (*T. rufalbus*). Songs have similar general structure except that the female's is higher in frequency. (c) Male (second repeating element) and female (first repeating element) song of the Buff-breasted Wren (*T. leucotis*). Repertoires of male and female song elements are consistently different (From S. M. Farabaugh, unpublished data.)

III. WHAT ARE DUETTING SPECIES LIKE?

Duetting species are a taxonomically diverse group; thus far, 222 species in 44 families have been reported to duet (Thorpe, 1972; Kunkel, 1974). If the broader definition presented here gains general acceptance, many more species will be added to this list. Thus far, in only 8 families are 10% or more of the species known to duet, and only 3 of these families are large, with 20 or more duetting species: in wrens (Troglodytidae) 23 of 59 species duet; in shrikes (Laniidae) 21 species of 64 duet; and in honeyeaters (Meliphagidae) 22 of 172 species duet. Duetting is not restricted to higher songbirds: although 55% of duetting species are oscines (oscines account for 48% of all bird species; see Austin, 1960), 33% of known duetting species are non-passerine (40% of all birds are non-passerine).

Duetting has evolved independently in many different taxonomic groups. Common characteristics of these unrelated duetters may indicate the selective pressures which favor the evolution of vocal or nonvocal acoustic duets. Since no single taxonomic group dominates the list of known duetters, I will consider those characteristics common to a majority of duetting species. In addition, I will compare a few duetting species to their nonduetting close relatives to see whether or not these characteristics also distinguish duetters from closely related nonduetters.

A. Common Characteristics of Duetting Species

Because the array of duetting species is so diverse, it is difficult to find characteristics common to all or even to a large portion of them. Duetters occupy many different habitats, from tropical rain forests and more seasonal tropical savannahs to highly seasonal temperate forests. Many duetters favor dense thickets and vine tangles, while others are found in the relatively open understory of mature forests or in open grassland. Some duetters are primarily insectivorous while others are omnivorous or granivorous. Nonetheless, there are several characteristics frequently mentioned as being common to most duetting species: occurrence in the tropics, sexual monomorphism, year-round territoriality, and prolonged monogamous pair bonds (Payne, 1971; Thorpe, 1972; Kunkel, 1974).

Many authors claim that duetting is primarily a tropical phenomenon (Diamond and Terborgh, 1968; Thorpe, 1972; Kunkel, 1974), yet duetting species occur in every major faunal region of the world. Over 80% of known duetting species do indeed occur in the tropics. However, this figure is confounded by the fact that there are many more bird species in the tropics; Griscom (1945) estimated that 85% of all bird species are found in the tropics. Still, 80% may be a low estimate of tropical duetting species since habits of many tropical birds have not been thoroughly studied.

One way to test whether there is a tropical bias among duetters is to compare two areas, one tropical and one temperate, that have nearly equal numbers of breeding species and for which habits of birds are known. Both Panama and North America (apart from Mexico) have approximately 700 breeding species [710 in Panama: Ridgely (1976); and 695 in North America: Robbins *et al.*, 1966]. Excluding those species which have breeding populations in both Panama and North America, there are 101 Panamanian duetters and 23 North American duetters (Table I). Clearly, there are significantly more Panamanian than North American duetting species ($\chi^2 = 42.8$, $p << 0.001$). Based on this sample, we may conclude that there are more tropical duetting species than expected, considering the geographical distribution of all species.

TABLE I

North American and Panamanian Duetting Species

Family	Species[a]	Common name	Sex morph[b]	Source
A. North America				
Anatidae	*Cygnus columbianus*	Whistling Swan	M	Johnsgard (1965)
	C. cygnus	Trumpeter Swan	M	Banko (1960; also cited in Johnsgard, 1975)
	Anser albifrons	White-fronted Goose	M	Boyd (1954, cited in Johnsgard, 1975)
	A. caerulescens	Snow Goose (blue phase)	M	Johnsgard (1965)
	A. caerulescens	Snow Goose	M	Johnsgard (1965)
	A. rossi	Ross' Goose	M	Ryder (1967, cited in Johnsgard, 1965)
	A. canagicus	Emperor Goose	M	Johnsgard (1965)
	Branta canadensis	Canada Goose	M	Collias and Jahn (1959)
	B. bernicla	Brant	M	Johnsgard (1975)
	B. b. nigricans	Brant	M	Johnsgard (1965)
Phasianidae	*Colinus virginianus*	Bobwhite	D	Stokes and Williams (1968)
	Lophortyx californica	California Quail	D	Stokes and Williams (1968)
	L. gambelii	Gambel's Quail	D	Stokes and Williams (1968)
Rallidae	*Rallus limicola*	Virginia Rail	M	Kaufmann (cited in Keith *et al.*, 1970)
Haematopodidae	*Haematopus bachmani*	Black Oystercatcher	M	E. H. Miller (unpublished data)

Strigidae	*Strix varia*	Barred Owl	M	Huxley (1919)
	Otus trichopsis	Whiskered Owl	M	Davis (1972)
Picidae	*Melanerpes erythrocephalus*	Red-headed Woodpecker	M	Kilham (1959)
	M. carolinus	Red-bellied Woodpecker	D	Kilham (1958)
	Picoides villosus	Hairy Woodpecker	D	Kilham (1960)
Corvidae	*Corvus brachyrhynchos*	Common Crow	M	Chamberlain and Cornwell (1971)
Muscicapidae	*Chamaea fasciata*	Wrentit	M	Erickson (1938)
Troglodytidae	*Thryothorus ludovicianus*	Carolina Wren	M	Diamond and Terborgh (1968)
Mimidae	*Toxostoma redivivum*	California Thrasher	M	Van Tyne and Berger (1959)
Fringillidae	*Pipilo aberti*	Abert's Towhee	M	Marshall (1960)
	P. fuscus	Brown Towhee	M	Marshall (1960)
	Cardinalis cardinalis	Cardinal	D	Lemon (1968)
B. North America and Panama				
Haematopodidae	*Haematopus ostralegus*	American Oystercatcher	M	Miller and Baker (1980)
Strigidae	*Bubo virginianus*	Great Horned Owl	M	Rodger (cited in Armstrong, 1963)
Tyrannidae	*Myiodynastes luteiventris*	Sulphur-bellied Flycatcher	M	Rising (cited in Power, 1966)
C. Panama				
Tinamidae	*Crypturellus soui*	Little Tinamou	M	E. S. Morton (unpublished data)
Cracidae	*Ortalis cinereiceps*	Gray-headed Chachalaca	M	Ridgely (1976)
Phasianidae	*Colinus cristatus*	Crested Bobwhite	D	Keith (cited in Thorpe, 1972)

(continued)

TABLE I (*Continued*)

Family	Species[a]	Common name	Sex morph[b]	Source
	Odontophorus gujanensis	Guiana Marbled Wood-quail	M	Skutch (1947)
	O. guttatus	Spotted Wood-quail	M	Ridgely (1976)
Rallidae	*Aramides cajanea*	Gray-necked Wood Rail	M	Chapman (1929)
Jacanidae	*Jacana spinosa*	Northern Jacana	M	E. S. Morton (unpublished data)
Psittacidae	*Ara ambigua*	Great Green Macaw	M	E. S. Morton (unpublished data)
	A. macao	Scarlet Macaw	M	E. S. Morton (unpublished data)
	Brotogeris jugularis	Orange-chinned Parakeet	M	Power (1966)
	Pionopsitta haematotis	Brown-headed Parrot	D	E. S. Morton (unpublished data)
	Pionus menstruus	Blue-headed Parrot	M	E. S. Morton (unpublished data)
	Amazona autumnalis	Red-lored Amazon	M	E. S. Morton (unpublished data)
	A. ochrocephala	Yellow-headed Parrot	M	E. S. Morton (unpublished data)
	A. farinosa	Mealy Amazon	M	E. S. Morton (unpublished data)
Strigidae	*Pulsatrix perspicillata*	Spectacled Owl	M	Chapman (1929)
Apodidae	*C. spinicauda*	Band-rumped Swift	M	E. S. Morton (unpublished data)
	Chaetura brachyura	Short-tailed Swift	M	E. S. Morton (unpublished data)
	Panyptila cayennensis	Lesser Swallow-tailed Swift	M	E. S. Morton (unpublished data)

Trogonidae	*Trogon massena*	Slaty-tailed Trogon	D	Davis (1972)
	T. violaceus	Violaceous Trogon	D	E. S. Morton (unpublished data)
Momotidae	*Electron platyrhynchum*	Broad-billed Motmot	M	E. S. Morton (unpublished data)
	Baryphthengus martii	Rufous Motmot	M	E. S. Morton (unpublished data)
	Momotus momota	Blue-crowned Motmot	M	Skutch (1964)
Ramphastidae	*Pteroglossus torquatus*	Collared Aracari	M	E. S. Morton (unpublished data)
	Ramphastos sulfuratus	Keel-billed Toucan	M	E. S. Morton (unpublished data)
	R. ambiguus swainsonii	Chestnut-mandibled Toucan	M	E. S. Morton (unpublished data)
Picidae	*Picumnus olivaceus*	Olivaceous Piculet	M	E. S. Morton (unpublished data)
	Melanerpes formicivorus	Acorn Woodpecker	D	E. S. Morton (unpublished data)
	Dryocopus lineatus	Lineated Woodpecker	D	E. S. Morton (unpublished data)
	Campephilus guatemalensis	Pale-billed Woodpecker	D	E. S. Morton (unpublished data)
	C. melanoleucos	Crimson-crested Woodpecker	D	E. S. Morton (unpublished data)
	C. haematogaster	Crimson-bellied Woodpecker	D	E. S. Morton (unpublished data)
Furnariidae	*Xenops minutus*	Plain Xenops	M	E. S. Morton (unpublished data)
Formicariidae	*Cercomacra nigricans*	Jet Antbird	D	S. M. Farabaugh (unpublished data)

(continued)

TABLE I (*Continued*)

Family	Species[a]	Common name	Sex morph[b]	Source
	Cercomacra tyrannina	Dusky Antbird	D	Keith (cited in Thorpe, 1972)
	Cymbilaimus lineatus	Fasciated Antshrike	D	E. S. Morton (unpublished data)
	Dysithamnus mentalis	Plain Antvireo	D	E. S. Morton (unpublished data)
	D. puncticeps	Spot-crowned Antvireo	D	E. S. Morton (unpublished data)
	Formicivora grisea	White-fringed Antwren	M	E. S. Morton (unpublished data)
	Gymnopithys leucaspis	Bicolored Antbird	D	E. S. Morton (unpublished data)
	Herpsilochmus rufimarginatus	Rufous-winged Antwren	D	E. S. Morton (unpublished data)
	Hylophylax naevioides	Spotted Antbird	D	E. S. Morton (unpublished data)
	Myrmeciza longipes	White-bellied Antbird	D	E. S. Morton (unpublished data)
	M. exsul	Chestnut-backed Antbird	D	E. S. Morton (unpublished data)
	Myrmotherula axillaris	White-flanked Antwren	D	J. Gradwohl (unpublished data)
	M. fulviventris	Checker-throated Antwren	D	J. Gradwohl (unpublished data)
	M. schisticolor	Slaty Antwren	D	E. S. Morton (unpublished data)

	Microrhopias quixensis	Dotwinged Antwren	D	E. S. Morton (unpublished data)
	Taraba major	Great Antshrike	D	S. M. Farabaugh (unpublished data)
	Thamnophilus doliatus	Barred Antshrike	D	Haverschmidt (1947)
	T. punctatus	Slaty Antshrike	D	Haverschmidt (1953)
Cotingidae	*Tityra semifasciata*	Masked Tityra	D	E. S. Morton (unpublished data)
	T. inquisitor	Black-crowned Tityra	D	E. S. Morton (unpublished data)
	Querula purpurata	Purple-throated Fruit Crow	D	E. S. Morton (unpublished data)
Tyrannidae	*Casiempis flaveola*	Yellow Tyrannulet	M	E. S. Morton (unpublished data)
	Attila spadiceus	Bright-rumped Attila	M	E. S. Morton (unpublished data)
	Conopias parva	White-ringed Flycatcher	M	E. S. Morton (unpublished data)
	Elaenia flavogaster	Yellow-bellied Elaenia	M	E. S. Morton (unpublished data)
	Megarhynchus pitangua	Boat-billed Flycatcher	M	E. S. Morton (unpublished data)
	Mionectes olivaceus	Olive-striped Flycatcher	D	E. S. Morton (unpublished data)
	Myiodynastes maculatus	Streaked Flycatcher	M	E. S. Morton (unpublished data)
	Myiopagis gaimardii	Forest Elaenia	M	E. S. Morton (unpublished data)
	Myiozetetes cayanensis	Rusty-Margined Flycatcher	M	E. S. Morton (unpublished data)

(*continued*)

TABLE I (*Continued*)

Family	Species[a]	Common name	Sex morph[b]	Source
	M. granadensis	Gray-capped Flycatcher	M	E. S. Morton (unpublished data)
	Pitangus lictor	Lesser Kiskadee	M	E. S. Morton (unpublished data)
	P. sulphuratus	Great Kiskadee	M	E. S. Morton (unpublished data)
	Sirystes sibilator	Sirystes	M	E. S. Morton (unpublished data)
	Todirostrum cinereum	Common Tody-flycatcher	M	E. S. Morton (unpublished data)
	Tyrannulus elatus	Yellow-crowned Tyrannulet	M	E. S. Morton (unpublished data)
Troglodytidae	*Campylorhynchus zonatus*	Banded Cactus Wren	M	Armstrong (1955)
	C. turdinus	White-headed Wren	D	E. S. Morton (unpublished data)
	Cyphorhinus aradus	Song Wren	D	Chapman (1929)
	Henicorhina leucophrys	Gray-breasted Wood-wren	M	Skutch (1940)
	H. leucosticta	White-breasted Wood-wren	M	E. S. Morton (unpublished data)
	Thryothorus fasciatoventris	Black-bellied Wren	M	Skutch (1940)
	T. leucotis	Buff-breasted Wren	M	Armstrong (1955)
	T. modestus	Plain Wren	M	Skutch (1940)
	T. nigricapillus	Bay Wren	M	S. M. Farabaugh (unpublished data)
	T. nigricapillus	Riverside Wren	M	E. S. Morton (unpublished data)

	T. rufalbus	Rufous-and-white Wren	M	S. M. Farabaugh (unpublished data)
	T. rutilus	Rufous-breasted Wren	M	Skutch (1940)
	T. thoracicus	Stripe-throated Wren	M	Ridgely (1976)
	Troglodytes aedon	House Wren	M	Chapman (1929)
Mimidae	*Donacobius atricapillus*	Black-capped Mocking Thrush	M	Haverschmidt (1947)
Muscicapidae (Sylviidae)	*Microbates cinereiventris*	Tawny-faced Gnatwren	M	E. S. Morton (unpublished data)
	Ramphocaenus melanurus	Strait-billed Gnatwren	M	Ridgely (1976)
Vireonidae	*Hylophilus aurantiifrons*	Golden-fronted Greenlet	M	E. S. Morton (unpublished data)
	H. decurtatus	Lesser Greenlet	M	E. S. Morton (unpublished data)
	H. ochraceiceps	Tawny-crowned Greenlet	M	E. S. Morton (unpublished data)
Icteridae	*Icterus chrysater*	Yellow-backed Oriole	M	E. S. Morton (unpublished data)
	I. mesomelas	Yellow-tailed Oriole	M	Davis (1972)
Emberizidae (Thraupinae)	*Rhodinocichla rosea*	Rose-breasted Thrush	D	Hardy (cited in Thorpe, 1972)
	Thraupis episcopus	Blue-gray Tanager	M	E. S. Morton (unpublished data)
	T. palmarum	Palm Tanager	M	E. S. Morton (unpublished data)
(Emberizinae)	*Saltator maximus*	Buff-throated Saltator	M	Thorpe (1972)
	Tiaris olivacea	Yellow-faced Grassqúit	D	E. S. Morton (unpublished data)

[a] Species names from Robbins *et al.* (1966) and Ridgely (1976), though modified where necessary to conform to the "Note on Taxonomy," pp. xix–xx.

[b] M = sexually monomorphic, D = sexually dimorphic; based on descriptions in Robbins *et al.* (1966) and Ridgely (1976).

A second characteristic that is thought to be common to most duetting species is sexual monomorphism in plumage. Again, the relative abundance of sexually monomorphic species in comparison to dimorphic species among all birds should be considered. We can test whether the numbers of monomorphic and dimorphic duetters which breed in Panama differ from the expected numbers based on the ratio for all Panamanian species. Of all breeding species in Panama, 447 (63%) are monomorphic, while 263 (37%) are dimorphic [based on descriptions in Ridgely (1976)]. As shown in Table I, 66 Panamanian duetters are monomorphic and 35 are dimorphic. These numbers do not differ significantly from the expected 63% (63.63) to 37% (37.37) ratio ($\chi^2 = 0.24$, $0.90 > p > 0.50$). Thus, it appears that monomorphism is not correlated with duetting. However, there may be some general differences between monomorphic and dimorphic duetting species. Duets of the latter group may be less elaborate, because of weaker selection favoring sexual distinctiveness in sounds if duets occur when dimorphic mates are in visual contact.

The third common characteristic of most duetting species is that they are nonmigratory, and many maintain territories throughout the year. This is true for temperate as well as tropical species. Most of the North American duetting species are year-round residents, and at least two temperate species, the Carolina Wren and the Wrentit (*Chamaea fasciata*), maintain year-round territories (Diamond and Terborgh, 1968; Erickson, 1938). Others may retain some attachments to their territories even if they do not defend them during the nonbreeding season. Common Crows (*Corvus brachyrhynchos*), which join huge winter roosts at night, return each day to the general home area to feed (Aldous, 1944).

In the tropics, where seasonal changes in food availability are not as extreme as those of the temperate zones, permanent residency on year-round territories is common. Even tropical species that join mixed-species flocks defend year-round territories (Powell, 1980; Munn and Terborgh, 1979; Gradwohl and Greenberg, 1980). Individuals join the flock as it passes through their territories. Territories of different species are highly overlapping, and flock membership depends at any particular moment on location of the flock. It is difficult to say whether year-round territoriality is proportionally more common among duetting than nonduetting species, because those species which are thought to maintain such territories are not well studied. I know of no thoroughly studied species which has been shown to have year-round territoriality that does not duet according to my definition. For this reason, I believe that year-round territoriality is probably more common among duetting than nonduetting species.

Prolonged monogamous bonds, where mates remain together for more than one breeding season, and in some species for life, are usually associated with year-round territoriality; all of the duetting species which have permanent territoriality are thought to have these bonds. Many tropical species are thought to have these bonds as well, whether territorial or not (Mayr, 1963; Kunkel, 1974), but there are few exact data because there are few long-term studies of color-banded populations. Therefore, it is also difficult to determine whether prolonged monogamous pair bonds are more common among duetters than nonduetters. The Anseriformes may at first appear to be a good example of a group of birds with lifelong pair bonds but no duetting. However, I consider the triumph ceremonies of the large anatids (geese and swans) to be duets because they involve mutual vocal displays. Many ducks have prolonged pair bonds, but very few duet. Yet, these bonds should not be considered monogamous because of the high incidence of rape in duck societies (McKinney, 1975). Thus one can make the statement that most Anseriformes that have prolonged monogamous pair bonds also duet. I believe, despite the lack of long-term data, that prolonged pair bonds probably are more common among duetters than nonduetters and that many birds thought to have these bonds may also duet.

Kunkel (1974) suggests that year-round territories function to keep the pair together. It is just as possible that prolonged monogamous bonds are only an artifact of permanent territories, because a monogamous pair would result if each bird excluded others of like sex. Even the extreme mate fidelity of some pelagic seabirds is thought to be the result of a permanent bond to the nest site (Morse, 1980). However, prolonged monogamous pair bonds are probably also correlated with duetting because many duetting species, which do not necessarily defend territories year-round, have such bonds. These species may still be involved in other types of year-round defense; for example, migratory anatids defend their young throughout the winter migration.

Of the three characteristics more common among duetting species, prolonged monogamous pair bonds and year-round territoriality seem to be the most important. The higher proportion of tropical duetting species may be related to the fact that prolonged monogamous bonds, and especially year-round territoriality, are more likely in the tropics. In Panama, 60% of passerines have permanent pair bonds and year-round territories (Morton and Farabaugh, 1980). Kunkel (1974) suggests that the few temperate species with year-round territoriality and prolonged monogamous bonds are recent intruders from the tropics which have conserved old patterns in a new environment.

In summary, three of the four characteristics mentioned are proportionally more common among duetting species: prolonged monogamous pair bonds, year-round territoriality, and occurrence in the tropics. The fourth characteristic, sexual monomorphism, is no more common among duetting than among all bird species.

B. Contrasting Related Duetting and Nonduetting Species

If these characteristics common to most duetting species are also those which distinguish particular duetting species from their nonduetting kin, then there is greater support for the contention that selection favoring duetting may be related to these characteristics. Rather than considering all duetters, I will give three examples, two contrasting duetting and nonduetting populations of the same species, and one contrasting a lone duetting species with its nonduetting congeners.

Some populations of the House Wren (*Troglodytes aedon*) duet while others do not. Breeding populations are found from Canada to Tierra del Fuego, although some authors consider the southern populations (from southeastern Mexico to Tierra del Fuego) to be a separate species, *T. musculus*. In the north temperate zone, House Wrens are polygynous, territories are maintained only during the breeding season, and females do not sing. Yet in Central America, they are monogamous and hold territories year-round, and the female's song overlaps with male song (Fig. 3). These females seem to be territorial, because they respond strongly to playbacks of female song (S. M. Farabaugh, unpublished data).

Another species which has duetting and nonduetting populations is the Red-winged Blackbird (*Agelaius phoeniceus*). It is one of the best studied polygynous species. Male song and visual displays are also well described and well studied. Male Red-winged Blackbirds are jet black with distinct red epaulets, while females resemble large sparrows. Females, though fairly aggressive near their nests, do not help to defend the territory, and territories are maintained only during the breeding season. In Cuba, however, E. S. Morton (unpublished data) found that females are jet black like the male, although they do not have the colorful epaulets. He found several pairs on territories in the Zapata Swamp, with only one female per male; and these birds were defending territories during their nonbreeding season. Morton heard both males and females sing antiphonal duets. In this tropical swamp, conditions are such that a classic polygynous species with solitary male song may have become a monogamous, permanently territorial, duetting species.

The final example concerns an unusual mallard, the African Black Duck (*Anas sparsa*). This species' display repertoire, which includes a mutual greeting display similar to the triumph ceremonies of geese and swans, is distinct from behavior of other *Anas* species (Johnsgard, 1965). McKinney *et al.* (1978) have shown that the differences between *A. sparsa*'s behavior and that of other *Anas* species are correlated with differences in its ecology. It is a river specialist rather than a pond dweller. Year-round territoriality is a profitable strategy in such a habitat, because everything the pair needs for survival and reproduction (food, nest sites, sleeping places, etc.) can be found on a defensible stretch of river, while pond dwellers have separate sites for feeding and nesting (McKinney *et al.*, 1978). Both males and females cooperate in defense, both are involved in fighting over mates and territories, and both have well-developed wing spurs. Pairs are monogamously bonded and forced copulations are not a common mat-

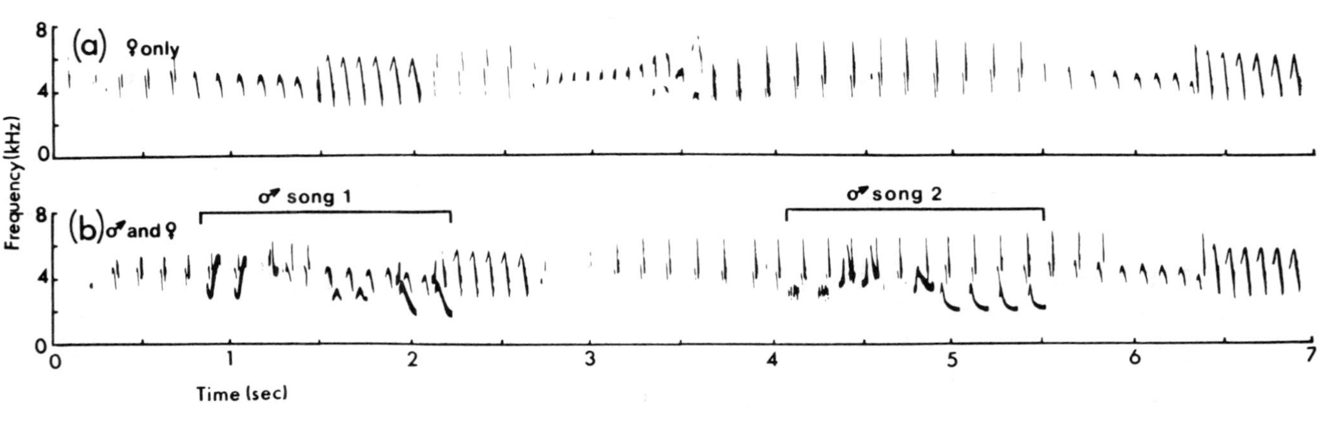

Fig. 3. Duets of House Wrens (*Troglodytes aedon*) in Panama. Female song consists of long chipping bouts that have a consistently repeated pattern of different types of chips. (a) Section of a larger duet bout where male does not overlap with female chipping song. (b) Another section of the same bout where male does overlap. Note that the male sang two songs, each of a different type.

ing strategy as in other *Anas* species. I consider *A. sparsa*'s mutual greeting display, which includes joint vocalizations (McKinney *et al.*, 1978), to be a vocal duet. This unusual duetting mallard with year-round territoriality, joint defense, prolonged monogamous bonds, and no forced copulations differs from almost all its congeners. The only other *Anas* species that are similar are also river dwellers, and may also duet.

From these three examples, it appears that year-round territoriality and prolonged monogamous pair bonds are indeed characteristics which distinguish duetters from closely related nonduetters. It is also true that duetting populations of House Wrens and Red-winged Blackbirds are tropical, although other tropical *Anas* species do not duet while *Anas sparsa* does. As for sexual monomorphism, both duetting and nonduetting House Wrens are monomorphic; duetting Red-winged Blackbirds and African Black Ducks are dimorphic, though less dimorphic than their nonduetting relatives. This reduction in dimorphism seems to be related to male and female participation in similar activities.

IV. FUNCTIONAL SIGNIFICANCE OF DUETTING

Having established a correlation between duetting, year-round territoriality, and prolonged monogamous bonds, we may now ask how the latter two characteristics favor the evolution of duets. To answer this, two different aspects of duetting must be considered: communication between pairs and communication between mates.

A. Year-Round Territoriality: Communication between Pairs

Seibt and Wickler (1977) have stated that territory maintenance is the prime function of most duets. Considering energetic costs and the risk of detection by predators, they suggest that sound intensity of the signal should correspond with the distance between sender and intended receiver. From observations of distances between duetting partners and measurements of sound pressure levels of duets for a number of species, they conclude that the intended receiver is a neighboring conspecific and not the mate. However, as Zahavi (1979) pointed out, higher intensity could communicate a higher level of arousal and indicate a greater threat to the receiver. Higher intensities could also be more threatening because the receiver may think that the sender is closer than he or she really is. In any case, even though data on sound pressure levels may not prove that the intended receiver is a neighbor rather than a mate, many other authors have reported that duets are used in territorial encounters and that they represent female and male participation in territorial defense (Hooker and Hooker, 1969;

Thorpe, 1972; Kunkel, 1974; Todt, 1975; Harcus, 1977; Todt and Fiebelkorn, 1980). The evolution of territorial duetting can be better understood by answering two questions: first, under what conditions is vocal defense by females advantageous? and second, why should the female's vocal defense be coordinated with the male's?

1. *Under What Conditions Is Vocal Defense by Females Advantageous?*

Most models of avian mating systems are based on optimal choice of mates by females. These models can be extended to explain the evolution of female participation in singing. If there are significant differences in male territory quality, females may be more successful by mating with a mated male on a superior territory than with a bachelor male on an inferior territory (Verner, 1964; Orians, 1969). If there is a significant contribution by males to female fitness (e.g., in care of young, defense of a depreciable resource, etc.), the addition of another female can lower the first female's fitness (Altmann *et al.*, 1977). It is to the male's advantage to acquire more females only if their combined reproductive output is more than that of a monogamously mated female. Yet, if the cost of territorial exclusion of additional females is less than the cost in lowered fitness, females should defend their territories from other females whether or not such exclusion benefits their mates. As Wittenberger (1976) pointed out, when female defense of mated status is profitable (*sensu* Brown, 1964), monogamy will result even though ecological conditions favor the evolution of polygyny. In birds, female defense of territory from other intruding females could include the development of female song. Such female defense of territory is particularly profitable in tropical species with year-round territories.

The maximum number of permanent residents on a year-round territory cannot exceed the number that can be supported during annual periods of low food availability. In the tropics there is a decline in insect abundance during the driest part of the year, and this decline is greatest in severe dry seasons (Janzen and Schoener, 1968; Janzen, 1973; Wolda, 1978a, and references cited therein). In areas with such severe dry seasons, there might be additional advantages for females which exclude other females from their permanent territories. However, even when there is sufficient food for more than one female during the dry season, there may not be enough to support multiple females and their young during the breeding season, because habitats with moderate dry seasons show smaller increases in insect abundance during wet (breeding) seasons. The equatorial forest of Sarawak in Borneo has only a severalfold increase in insect abundance from dry to wet seasons, in contrast to a 100-fold increase in Costa Rica and Panama, and a several thousand-fold increase in temperate woodlands (Fogden, 1972; Janzen, 1973; Wolda, 1978b).

Orians (1969) noted that a first female's fitness could be lowered by the presence of more females, but he concluded that their exclusion by the first female would be costly because of chilling or loss of eggs. He thought that it was unlikely that the adverse effects of a second female would be sufficient to select for this behavior. This is a valid argument for temperate species that are territorial only during the breeding season, but it does not apply to tropical species which are territorial year-round. During the nonbreeding season, cost of defense for females should not be greater than that for males, and during the breeding season loss of eggs due to chilling may be no more costly than loss due to increased competition for food. Tropical wren females do sing less when they have eggs in the nest, but their responses to playbacks of intruder females do not completely disappear (S. M. Farabaugh, unpublished data).

2. *Why Should the Female's Vocal Defense Be Coordinated with the Male's?*

If the exclusion of intruding females were the only factor favoring female defense, one would not necessarily predict coordination of defense between mates. In fact, exclusion of extra females may not be to the male's advantage (see above). Yet, there do not seem to be any clear examples of species wherein both mates participate in territorial defense without some degree of cooperation and coordination. Such coordination of vocalizations could be related to the pair-bonding functions of the duet, but some territorial factors may also favor it.

If it is profitable for females to invest in defense against unmated females, they may also benefit from a coordinated defense with the mate against other classes of intruders. Pair defense is probably the best defense against expansion by neighboring territorial pairs. Further, in cases of interspecific territorial confrontations, a coordinated pair may be able to defeat larger solitary intruders. For example, during occasional interspecific clashes between Buff-breasted Wrens (males weigh about 21 g and females about 18 g, n=35 mated pairs), which have precise antiphonal duets, and Rufous-and-white Wrens (males weigh about 28 g and females about 25 g, n=25 mated pairs), which have infrequent, imprecise overlapping duets, the smaller Buff-breasted Wrens are able to repel a solitary Rufous-and-white Wren (S. M. Farabaugh, unpublished data).

When mates are involved in territorial interactions with neighbors, birds could accidentally attack their mates in the confusion of battle. But, during bouts of counterduets, females usually fight females and males fight males in many species, including African Black Ducks (McKinney *et al.*, 1978), Checker-throated and White-flanked antwrens (*Myrmotherula fulviventris* and *M. axillaris;* J. Gradwohl and R. Greenberg, unpublished data), Carolina Wrens (E. S. Morton, unpublished data), and many tropical *Thryothorus* wrens (S. M. Farabaugh, unpublished data). Temporal coordination of vocalizations, and pair- and individual-specific repertoires, may aid in reducing such misdirected aggression by providing information about sexual and individual identity. Visual displays associated with the duet and high-frequency, low-intensity vocal elements which are audible only over short distances may also aid in lowering aggressiveness by appeasing the mate. Female southern House Wrens assume a head-down, wing-fluttering posture when the male is within 5 m of the female during territorial duets (S. M. Farabaugh, unpublished data). Wickler (1976) believes that the low-intensity elements in the duets of the African Slate-colored Boubou Shrike (*Laniarius funebris*) represent communication with the mate.

During the breeding season, it may be advantageous for a male to monitor his mate's contacts with other males (Parker, 1974). In some cases other mated males may be more of a threat than unmated nonterritorial males. For example, when male White Ibis (*Eudocimus albus*) are feeding away from the nest, their mates will solicit copulations from other mated males and never from unmated males (Rudegeair, 1975; also cited in Dewsbury, 1978). Birkhead (1979) found that when a male Black-billed Magpie (*Pica pica*) is confronted with a caged female, he will attack if his mate is present, but court the caged female if his mate is not present. Territorial encounters are a time when females are likely to encounter other mated males. Strong response by the male to his mate's song during the breeding season may insure against cuckoldry by informing both the mate and the neighboring male of his presence. Duetting in three north temperate quail species (Bobwhite, *Colinus virginianus;* California Quail, *Lophortyx californica;* and Gambel's Quail, *L. gambelii*) is thought to function in part to let unmated males know that a female has a mate (Stokes and Williams, 1968). For example, when a female California Quail is separated from her mate, she gives a call to which both her mate and other unmated males may respond vocally. The mated male usually responds antiphonally with a call which is used in spacing of males, but only if other unmated males also answer the female's call; otherwise the mated male answers with a call similar to the female's.

In summary, duets can function as territory-claiming vocal communication between neighboring pairs. In general, female participation in vocal defense of territory will be favored if the benefits gained by excluding intruders is greater than the cost of such exclusion. Female participation is particularly profitable in species which maintain territories throughout the year. It is advantageous for vocal defense of mates to be precisely coordinated and include pair- and individual-specific elements, because this allows an organized offense against other conspecific pairs, while reducing the possibility of misdirected aggression toward the mate. A coordinated pair may also be able to exclude larger solitary individuals of other species. Simultaneous pair defense may also allow males to monitor their mate's contact with other territorial males, thus reducing the possibility of cuckoldry. In Section B, duetting is discussed in terms of communication between mates.

B. Prolonged Pair Bonds: Communication between Mates

For the simple reason that mates are vocalizing at the same time, many authors have concluded that duets represent communication between mates. Thorpe (1961, 1967, 1972) proposed that duets function primarily to maintain pair bonds, and that the process of learning the duet is an integral part of pair-bond formation (Thorpe and North, 1965, 1966). He explained the predominantly tropical distribution of duetting species by suggesting that duets replace visual pair-bonding displays in species living in dense tropical vegetation where visual communication is hindered. However, it is now known that many species from many different taxonomic groups usually duet when in visual contact (Wickler, 1980; Todt and Fiebelkorn, 1980; S. M. Farabaugh, unpublished data), and that duets are often accompanied by elaborate visual displays (Kunkel, 1974; Baptista, 1978; Wickler, 1980; Todt and Fiebelkorn, 1980; S. M. Farabaugh, unpublished data).

Although duets probably did not evolve because visual communication was hindered, they may have some pair-bonding function. In fact, because most duetting species have prolonged monogamous pair bonds, duets may function to develop and maintain these long-term bonds. Two questions remain to be answered: why are prolonged monogamous bonds advantageous, and how does duetting maintain these bonds?

1. Why Are Prolonged Monogamous Pair Bonds Advantageous?

Kunkel (1974) proposed that prolonged monogamous pair bonds and other tropical mating systems function to eliminate the danger of asynchrony in the sexual cycles of potential mates in the humid tropics, where breeding seasons last for most or all of the year. He suggested that duetting is a ritualized appeasement of aggression and allows mates to remain together for extended periods of time. It is doubtful, however, that gonadal asynchrony is a problem for most tropical birds. First, breeding seasons of equatorial forest birds are not necessarily continuous or even ill-defined (Ward, 1969). This impression has arisen because breeding data from wide areas of similar vegetation but with dissimilar local climates have been combined, a bias that may explain some records of continuous breeding in African birds (Bates, 1908; Chapin, 1932–1954). Even in equatorial evergreen forests, a distinct breeding season is the rule (Baker *et al.*, 1940; Moreau, 1936, 1950, 1964; Ward, 1969; Fogden, 1972). Second, though it is clear that changes in day length could not be used as a proximate cue for sexual synchronization in tropical species (Moreau, 1936; Ward, 1969), it is possible that a combination of external cues (e.g., vegetational changes, rainfall, cloud cover) or internal cues (e.g., level of protein reserve in flight muscle) are available for sexual synchronization (Ward, 1969; Fogden, 1972). Further, Jones and

Ward (1976) claimed that the protein reserve theory could apply to many bird species, including those for which photoperiod or other environmental stimuli had been invoked as proximate factors controlling breeding.

Immelmann (1961, 1963) agrees that proximate cues for breeding are available for most tropical birds, but he suggests that when the timing of such cues, for example the onset of wet season, can vary by weeks or months from year to year, other behaviors such as duetting, group singing, and joint nest-building can enable birds to respond quickly to these proximate cues as soon as they are available. It is doubtful that there is a greater need for these birds to respond quickly to proximate cues than for any breeding bird. Quick response would be selected for, whether or not the timing of proximate cues varies, if early breeding results in more successfully fledged young because the breeding season is short or breeding success declines as the season progresses—due to higher nest predation, depletion of food, etc. There is little evidence that duetting actually promotes early breeding, though it is possible that birds which have already formed pairs before the breeding season begins may breed earlier. If the latter is true, selection may favor early pairing, but this would not necessarily promote year-round pair bonds. Immelmann (1963) has shown that central Australian birds which breed at any time of the year, depending on the erratic pattern of rainfall, and which must breed quickly because favorable conditions for breeding do not last long, are unusual because they show physiological and behavioral responses to the rainfall itself rather than to an increase of food caused by the rains. Even in these species there is no strong evidence that duetting is involved in this rapid physiological response; nor are these species thought to have prolonged pair bonds. In general, lack of unvarying proximate cues for breeding, such as changes in day length, should not necessarily promote pair bonds outside of the breeding season. However, there are other factors which would favor such bonds.

The necessity for joint parental care of young throughout the nonbreeding season would favor the development of year-round pair bonds whether or not the birds remain on a permanent territory. Common Crows remain with their parents for many months after fledging and associate with parents and new young as yearlings (Good, 1952). E. D. Brown (1979) has found that captive crow families have family- and individual-specific song elements, and these elements are used in duets between family members. Balda and Balda (1978) have found that Piñon Jays (*Gymnorhinus cyanocephala*) can recognize their young from the previous year on the basis of vocalizations alone. J. L. Brown (1975) suggests that for corvids in general, song may be important for individual recognition in the large winter flocks. It is possible that crow families are able to maintain family integrity in the winter flocks using duet song. Similarly, duets and triumph ceremonies may facilitate recognition and cohesion of families of geese and swans during migration. Banko (1960; also cited in Johnsgard, 1975) observed that mutual triumph ceremonies occurred regularly among Trumpeter Swans (*Cygnus cygnus*) in winter flocks, and Johnsgard (1975) reported that triumph displays involving more than two swans in captive flocks represent participation by the pair's offspring from the previous year.

Strong pair bonds may aid in defense of territory and mated status. Territorial rivals may be less likely to intrude if it is difficult to win a partner from an established pair because of a strong existing pair bond (Wickler, 1980). For this reason, if selection favors territorial maintenance by females, pair-bonding may also be advantageous to these females. If territories are permanent, year-round pair bonds would also be expected. Duets may function as advertisement of the presence and status of a mated pair. Even if duetting only indicated that a strong pair bond existed rather than creating or strengthening such bonds, duets would be advantageous because solitary birds, upon hearing a duet, would be less likely to intrude. However, these functions are not mutually exclusive.

2. *How Can Duets Function to Develop Pair Bonds?*

Wickler (1980) proposes that each male must invest time and energy in learning a potential mate's songs. Birds which stay together from one breeding season to the next gain reproductive success without further investment, while those that change mates must invest again (Dawkins, 1976; Maynard Smith, 1977). Wickler suggests that investment in learning may prevent desertion and that such investment is nontransferable; that is, if a rival takes over a territory it cannot benefit from this investment as it would if the original bird had invested in hoarded food or in building a nest. There are some examples of duet development which involve a learning period. Thorpe and North (1966) found that two captive Boubou Shrikes (*Laniarius ferrugineus*) formed a pair bond through the development of pair-specific duets over a period of several weeks. Brown (1979) found that when an individual Common Crow was introduced into a strange group of captive crows, it began to imitate song elements specific to that group and that improved imitation was correlated with decreased aggression and increased social contact. Finally, Wickler and Seibt (1980) described a duetting species, the African Forest Weaver (*Ploceus bicolor*), in which mates sang identical songs in unison. They suggest that pair-bonding would be the only possible function of such a duet, and its development should involve considerable learning. Crow duets, with their low sound intensity and high structural complexity, may also function only in social bonding.

The three previous examples lend some support to Wickler's hypothesis that learning is involved in bond formation even though all three are captive studies with small sample sizes. One might think that any elaborate, precisely timed duet must also involve a period of learning during bond formation. However, territorial rivals are often the members of a floater population which may live near or, in the case of Rufous-collared Sparrows (*Zonotrichia capensis*) in Costa

Rica, on the pair's territory (Carrick, 1963; Brown, 1969; Smith, 1978); therefore, rivals could have the opportunity to learn song repertoires of territorial birds. During my studies of duetting in Buff-breasted Wrens, I once watched a male which had been displaced from his territory attempt to set up another one nearby. At least three unmated females entered his new territory and performed precise duets with him moments after they had arrived. In this case, if there was any investment in learning it occurred before the pair was formed. It is also possible that performance of antiphonal duets in this species does not require learning the mate's songs.

In summary, prolonged pair bonds may be advantageous in species which have extended parental care of young. The existence of a strong pair bond may also discourage conspecific rivals for mates from entering the territory. Duets are advantageous because they may provide a means of maintaining pair and family cohesion during migration and winter flocking, and they may advertise the existence of a strong pair bond. Further, learning duets may be a fundamental part of pair-bond formation in some species.

C. Interspecific Communication

Some authors have suggested that duets may function in two interspecific contexts: as a mechanism for reproductive isolation of congeners and as an adaptive response to predation. However, it is possible that these are merely incidental effects rather than functions of duetting.

Diamond and Terborgh (1968) suggested that duetting in the Mountain Cuckoo-shrike (*Coracina montana*) may be involved in the maintenance of reproductive isolation from its lowland-dwelling congener, the Gray-headed Cuckoo-shrike (*C. shisticeps*). Display and song of *C. shisticeps* are similar to those of *C. montana* except that the female does not reply; hence, there is no duet. On the other hand, the Golden Cuckoo-shrike (*Campochaera sloetii*), the sole member of its genus, has duets very similar to *C. montana*. Based on these observations and the fact that many other species with altitudinally distinct congeners maintain reproductive isolation by various means without resorting to duets, Diamond and Terborgh conclude that maintenance of reproductive isolation is an additional or incidental role of duetting in these species.

Harcus (1977) proposed that increased protection from predation is one of the important selective factors involved in the evolution of vocal duets in many species. He based this hypothesis on his observations that the presence of a walking or running observer elicited duets from nearby pairs, that duetting declines during the breeding season when duets may advertise the location of the nest, and that adaptations which decrease the effects of predation are of considerable advantage for species which are long-lived, and have permanent pair bonds and year-round territories. These observations lend little support to the contention that protection from predation is achieved by duetting, much less, that this is its primary function. Harcus reported that disturbance is only one of many contexts in which duets are performed and that such disturbances also elicited solitary vocalizations. He also reported that the decline in duets during the breeding season was accompanied by a rise in the number of solitary vocalizations, indicating that, in duetters and nonduetters alike, birds are silent on the nest. This silence would reduce the probability that a vocalization by a bird off the nest would be answered by its mate. Finally, adaptations which reduce predation may be advantageous to any individual, not just a long-lived one. For instance, duets, especially antiphonal duets, could possibly confuse a predator, thus providing some protection. But in general, protection from predation, like reproductive isolation, is probably an incidental effect of duetting and not a major function.

V. MULTIPLE FUNCTIONS OF DUETS AND DUET STRUCTURE

It is widely agreed that male song serves two major functions: advertisement of territory and attraction of females (Thorpe, 1961; Armstrong, 1963; Falls, 1969; Lemon, 1977). In some species, one or the other of these functions may be the primary one (Wasserman, 1977a,b). In other species, one song type may be used in territorial contexts while another type is used in the presence of potential mates (Morse, 1970). Duets also seem to serve multiple functions, such as coordinated maintenance of territory and pair-bonding. In some duetting species, one function may be dominant. On Barro Colorado Island in Panama, pairs of Checker-throated Antwrens which forage together throughout the day duet only during territorial encounters with adjacent pairs (J. Gradwohl and R. Greenberg, unpublished data). On the other hand, Common Crow duets are used mainly for communication between family members (Brown, 1979). Baptista (1978) found that Melodious Grassquits (*Tiaris canora*) use one duet type in territorial interactions and a different duet type in sexual contexts. However, the majority of duetters use the same duet types during counterduetting bouts at territorial borders, before copulation, during nest-building, and when the pair is reunited after being separated while foraging. Even species which are not thought to be territorial may use duets during aggressive encounters with conspecifics; Orange-chinned Parakeets (*Brotogeris jugularis*) duet during pair chases directed at competitors or predators, as well as in sexual context (Power, 1966).

Near the beginning of this chapter four structural variables which could be used to describe and classify duets were presented: type of sound used by each mate, arrangement of male and female elements within duet bouts, relative participation of mates in vocalizations used in the duets, and temporal precision.

Now that different functions of duets have been discussed, the possible functional significance of such variation in duet structure can also be considered. In other words, do structural characteristics of the duet correlate with the specific functions of duetting, such as pair-bonding and territoriality?

The function of a duet will affect the type of sound used by each participant. Individual- and pair-specific duet repertoires may be favored if learning the mate's repertoire is an integral part of pair-bond formation. Such repertoires may also be favored in species with territorial duets because they allow recognition of neighbors and strangers over a distance. Wiley and Wiley (1977) have shown that Stripe-backed Wrens (*Campylorhynchus nuchalis*) can recognize neighbors' duets. The response of these wrens to playbacks of neighbors' and strangers' duets is similar to the response of White-throated Sparrows (*Zonotrichia albicollis*) to playbacks of neighbors' and strangers' songs (Brooks and Falls, 1975), for pairs respond more strongly to duets of strangers. Sexual specificity of mates' duet contributions can function in pair-bonding by attracting new mates following the loss of a mate. Sexual specificity in territoriality may allow birds to recognize the sex of an intruder, whereas the duets of sexually dimorphic birds may not need to be sexually specific. Birds whose territorial duets are used mainly during direct confrontations with neighbors and other possible intruders may have simpler vocalizations compared to birds that are not in visual contact with neighbors when they are duetting, because, in this case, recognition is not based on voice alone. For example, the duets of Checker-throated Antwrens, given only during territorial confrontations, consist of overlapping bouts of call notes (J. Gradwohl and R. Greenberg, unpublished data). Such duets may also be of lower amplitude. Low-amplitude duets may also be found in species whose duets are used mainly for short-distance communication between mates and family members. The duets of Common Crows, which are mainly in social bonding, include complex low-amplitude vocalizations (Brown, 1979).

The second characteristic, the arrangement of male and female elements within overlapping bouts, allows duets to be grouped into two main classes: antiphonal, where elements are alternated without overlap; and simultaneous, where elements are either partially or totally overlapping. Antiphony might be viewed as mates answering one another's calls, but it could also be a way for two birds to vocalize at the same time without interference. Various studies have shown that territorial males avoid singing when a neighboring conspecific is singing (Cody and Brown, 1969; Ficken *et al.*, 1974; Wasserman, 1977c), thus minimizing the loss of information contained in their songs. Transmission of individual identity without interference may be important for recognition of neighboring pairs and floaters.

Wickler and Seibt (1980) suggest that singing of identical songs specific to the pair with perfect simultaneity can function only in pair-bounding, because such duets would require learning and practice and because other birds would not be able to detect that two birds, rather than one, are singing. However, such duets could also have a territorial function if two singers are usually visible to neighbors when duetting. E. S. Morton (unpublished data) reports that the duets of the Zapata Finch (*Torreornis inexpectata*) are completely simultaneous, but because these finches maintain territories in open habitat, territorial pairs are highly visible when duetting. Other types of simultaneous singing could possibly represent intentional interference. For example, male song overlap of female song may lessen the possibility that the female song will be heard by unmated males, and thus, fewer males will be attracted to the territory. On the other hand, simultaneity does not necessarily involve interference. In some simultaneous duetting species, female song is of higher frequency than male song; possibly two vocalizations at different frequencies may not interfere as much as two at the same frequency (Fig. 4). Further, at the beginning and end of simultaneous duet bouts, at least one song of each bird may occur without overlap (Fig. 5).

Mates may not participate equally in vocalizations used in the duet; in a species with duet song, one mate may sing with and without the mate. Unequal participation may indicate that the song of the more frequent singer may have more functions than its mate's song; perhaps its song functions in pair-bonding and territorial defense while its mate's song functions only in pair-bonding. It may also be possible that both birds' songs appear in the same contexts but that one bird participates more in certain contexts. For example, if territorial intrusion is less likely by females than by males (e.g., there are fewer female than male nonterritorial floaters), selection may favor greater participation in territorial defense by males than by females. Equal participation by both mates in vocalizations used in the duet could indicate that these vocalizations are used only as communication between mates or that both mates participate equally in all contexts in which these vocalizations are used.

It has been implied that extreme temporal precision of some duets, especially antiphonal duets, indicates that they must function in pair-bonding, because such precision could be achieved only through learning and practice. This is not necessarily true. In some species, precise timing is based on the internal rhythmicity of each singer, rather than response to the mate's vocalizations. Such is the case for Vieillot's Barbet (*Lybius vieilloti*), D'Arnaud's Barbet (*Trachyphonus darnaudii*), Red-and-yellow Barbet (*T. erythrocephalus;* Payne and Skinner, 1970), Barbary Shrike (*Laniarius barbarus;* Payne, 1970), and various duetting rails (Huxley and Wilkinson, 1979). In other species, birds may really be answering each other. For example, when Buff-breasted Wrens move apart during a duet bout, each bird's song rate slows; apparently each bird begins its song only after hearing its mate's song (S. M. Farabaugh, unpublished data). Yet even if a bird is responding to its mate's song and not to an internally derived rhythym, it does not mean that learning was required to achieve precision. According to W. M. Schleidt (unpublished data) a turkey (*Meleagris gallopavo*)

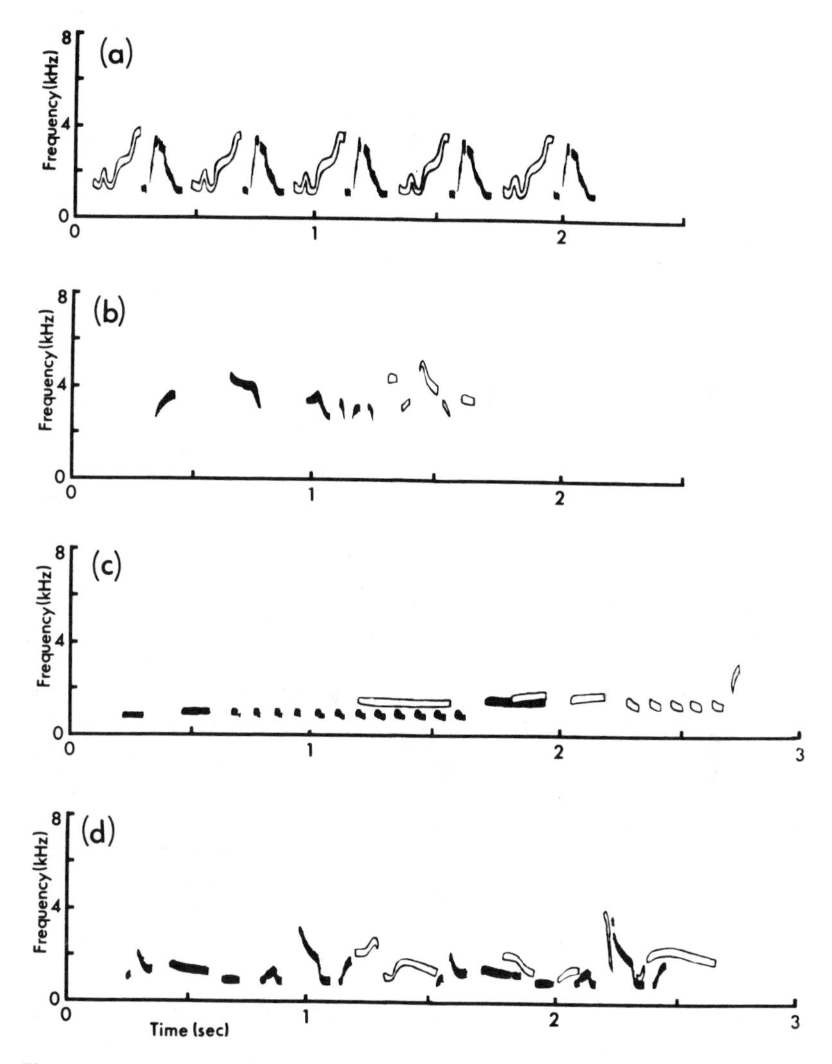

Fig. 4. Male (solid tracings) and female (open tracings) song in antiphonal and simultaneous duetting in *Thryothorus* species. In antiphonal species, male and female songs have the same frequency range [e.g., (a) Buff-breasted Wrens (*T. leucotis*) and (b) Rufous-breasted Wrens (*T. rutilus*)]. The frequency ranges of male and female songs overlap partially in species which sing simultaneously with the female at a slightly higher frequency [e.g., (c) Rufous-and-white Wrens (*T. rufalbus*) and (d) Black-bellied Wrens (*T. fasciatoventris*)].

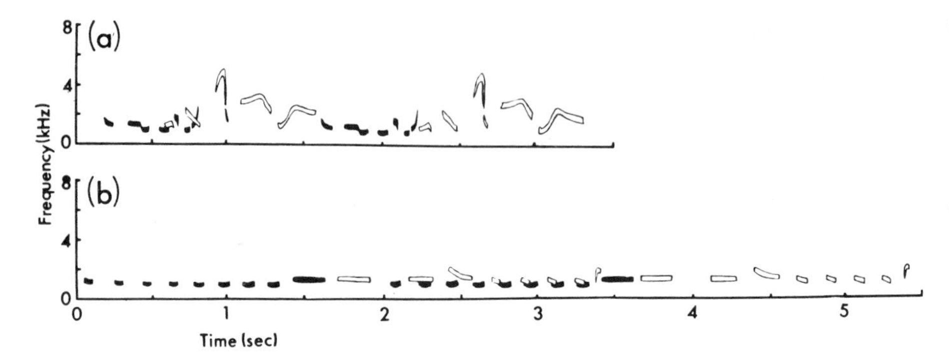

Fig. 5. Structure of duet bouts in simultaneous singers. Songs differ slightly in frequency, and notes of males (solid tracings) and females (open tracings) can occur without overlap during a bout of several songs. (a) Last male and last female song do not overlap in a four-song bout of a pair of Black-bellied Wrens (*T. fasciatoventris*). (b) First male and last female song occur without overlap in a four-song bout of a pair of Rufous-and-white Wrens (*T. rufalbus*).

housed alone in an anechoic chamber will produce gobbles at a variable rate, but when given a sound stimulus consisting of a playback of either gobbles or pure tone signals, the turkey will respond antiphonally with great precision to this repeated stimulus. There is no evidence that the temporal precision of the turkey's response improves over time due to learning or any other process. Precision, in some species, may indeed show improvement over time, but the mere observation that a particular duet is precise does not indicate whether learned improvement has occurred or that this was a part of pair-bond formation.

In conclusion, specific structural characteristics of duets can be related to the two main functions of duetting: pair-bonding and territoriality. However, these characteristics alone cannot indicate whether a particular duet has one or both of these functions because both could select for many of the same duet structural characteristics. Examination of both structural characteristics and the contexts in which the duet is used is necessary for determining the function or functions of any particular duet. Further, since the proposed role of duets in pair-bond formation involves learning, some characteristic of the duet should be observed to improve with time, and this improvement should correlate with decreased aggression between the potential mates.

VI. CONCLUSIONS

In this chapter, I have attempted to develop an objective definition of duetting and have examined its ecological and social significance.

Duetting is described as overlapping bouts of vocal or nonvocal sounds given by the members of a mated pair or extended family group. Overlapping bouts can be considered to be duets if the percentage overlap is high. Organization of a mate's elements within bouts may be such that the coefficient of variation of intervals between these elements is low or percentage alternation of them is high, or both. Techniques for measurement of these three variables, percentage overlap, coefficient of variation of intervals, and percentage alternation, are described. It is suggested that once these variables are measured for many species, critical limits for these variables can be set to define duetting. Alternatively, duets could be sorted from nonduets using discriminant analysis for two classes.

The phenomenon of duetting has polyphyletic origins and is distributed widely in birds. Comparisons of unrelated duetting species and contrasts of a few duetting species with their nonduetting kin reveal that duetting is correlated with year-round territoriality, prolonged monogamous pair bonds, and occurrence in the tropics. Sexual monomorphism is no more common among duetting species than among all bird species. Year-round territoriality and prolonged monogamous bonds appear to be the most important characteristics, because the higher proportion of tropical rather than temperate breeding duetting species may be related to the fact that these two characteristics are more common in the tropics, where most birds are resident throughout the year.

The role of duetting in territoriality and pair-bonding was examined, and it was found that duets can function as joint, territory-claiming, acoustic communication between neighboring pairs, as well as communication between mates. Female participation in defense of territory should be favored if benefits gained by excluding intruders outweigh the cost of such exclusion. Female defense of territory is particularly profitable in species which defend territories year-round. Coordinated defense by both mates allows for an organized defense against other pairs while reducing the possibility of misdirected aggression toward the mate. Simultaneous vocalizations may also allow males to monitor their mate's contact with other territorial males, thus reducing the possibility of cuckoldry. In terms of pair-bonding, duets are advantageous because they may provide a means of maintaining pair and family cohesion during migration and winter flocking. Duets may also advertise the existence of a strong pair bond, thus discouraging unmated conspecifics from entering the territory in search of mates. In addition, learning duets may be a fundamental part of pair-bond formation in some species. Finally, the role of duetting in reproductive isolation and protection from predation is thought to be additional or incidental.

Like male birdsong, duetting has two main functions: territoriality and pair-bonding. Any one duet may have one or both functions. The possible effects of these functions on four variable structural characteristics of duets were discussed, and it was concluded that since both functions could select for similar structural characteristics, examination of both the structure of the duet and the context in which it is used is necessary before speculations can be made about the function or functions of a particular duet.

ACKNOWLEDGMENTS

The ideas in this chapter have developed during the course of my study of the comparative duetting behavior of *Thryothorus* wrens in Panama and Venezuela. I am especially grateful to my advisor, Eugene S. Morton, who is mainly responsible for my present interest in communication, wrens, and the tropics. I would like to thank those people who provided many useful comments and ideas about duetting in general and this manuscript in particular: Ellie Brown, Ed Buchler, Judy Gradwohl, Russ Greenberg, Michael Greenfield, Nora Helgeson, Gene Morton, Stan Rand, Mike Ryan, Wolfgang Schleidt, and Andrea Worthington. For their help on the statistical definition of duetting, I thank Larry Douglass, Julie Riedel, Estelle Russek, and Wolfgang Schleidt. For critically reading this manuscript I am particularly grateful to Ellie Brown and also to Gene Morton, Steve Childs, Julie Riedel, and Jack Putz. I also thank Richard and Colette Farabaugh, Luigi de Limon, Georganne Neubauer, and Bernie Schoch for their technical assistance. Finally I thank my editors, Don Kroodsma and Ted Miller, and my parents, Michael and Ruth Farabaugh for their encouragement.

This research has been funded by the George A. Harris Foundation, the Smithsonian Tropical Research Institute, the National Zoological Park Office of Zoological Research's Venezuelan Project, the National Geographic Society, and the Friends of the National Zoo.

REFERENCES

Aldous, S. E. (1944). Winter habits of crows in Oklahoma. *J. Wildl. Manage.* **8,** 290–295.

Altmann, S. A., Wagner, S. S., and Lenington, S. (1977). Two models for the evolution of polygyny. *Behav. Ecol. Sociobiol.* **2,** 397–410.

Armstrong, E. A. (1955). "The Wren." Collins, New York.

Armstrong, E. A. (1963). "A Study of Bird Song." Oxford Univ. Press, London and New York.

Austin, O. L., Jr. (1960). "Birds of the World." Golden Press, New York.

Baker, J. R., Marshall, A. J., and Harrison, T. H. (1940). The seasons in a tropical rain forest (New Hebrides). Part 5: Birds (*Pachycephala*). *J. Linn. Soc. Zool.* **41,** 50–70.

Balda, R. P., and Balda, J. H. (1978). The care of young Piñon Jays (*Gymnorhinus cyanocephalus*) and their integration into the flock. *J. Ornithol.* **119,** 146–171.

Banko, W. (1960). The Trumpeter Swan: Its history, habits, and population in the United States. *U.S. Fish Wildl. Serv., North Am. Fauna* No. 63.

Baptista, L. F. (1978). Territorial, courtship, and duet songs of the Cuban Grassquit (*Tiaris canora*). *J. Ornithol.* **119,** 91–101.

Bates, G. L. (1908). Observations regarding the breeding seasons of the birds of Southern Kamerun. *Ibis* **9,** 558–570.

Birkhead, T. R. (1979). Mate guarding in the Magpie, *Pica pica*. *Anim. Behav.* **27,** 866–874.

Brooks, R. J., and Falls, J. B. (1975). Individual recognition by song in White-throated Sparrows. I. Discrimination of songs of neighbors and strangers. *Can. J. Zool.* **53,** 879–888.

Brown, E. D. (1979). The song of the Common Crow, *Corvus brachyrhynchos*. Master's Thesis, Univ. of Maryland, College Park.

Brown, J. L. (1964). The evolution of diversity in avian territorial systems. *Wilson Bull.* **76,** 160–169.

Brown, J. L. (1969). Territorial behavior and population regulation in birds. *Wilson Bull.* **81,** 293–329.

Brown, J. L. (1975). "The Evolution of Behavior." Norton, New York.

Carrick, R. (1963). Ecological significance of territory in the Australian Magpie, *Gymnorhina tibicen*. *Proc. Int. Ornithol. Congr.* **13,** 740–753.

Chamberlain, D. R., and Cornwell, G. W. (1971). Selected vocalizations of the Common Crow. *Auk* **88,** 613–634.

Chapin, J. P. (1932–1954). The birds of the Belgian Congo (4 vols.). *Bull. Am. Mus. Nat. Hist.* **65, 75, 75A, 75B.**

Chapman, F. (1929). "My Tropical Air Castle." Appleton, New York.

Cody, M. L., and Brown, J. H. (1969). Song asynchrony in neighboring bird species. *Nature (London)* **222,** 778–780.

Collias, N. F., and Jahn, L. (1959). Social behavior and breeding success in Canada Goose (*Branta canadensis*) confined under semi-natural conditions. *Auk* **76,** 478–509.

Davis, L. I. (1972). "A Field Guide to the Birds of Mexico and Central America." Univ. of Texas Press, Austin.

Dawkins, R. (1976). "The Selfish Gene." Oxford Univ. Press, London and New York.

Dewsbury, D. A. (1978). "Comparative Animal Behavior." McGraw-Hill, New York.

Diamond, J. M., and Terborgh, J. W. (1968). Dual singing by New Guinea birds. *Auk* **85,** 62–82.

Erickson, M. M. (1938). Territory, annual cycle and numbers in a population of Wrentits (*Chamaea fasciata*). *Univ. Calif., Berkeley, Publ. Zool.* **42,** 247–333.

Falls, J. B. (1969). Functions of territorial song in the White-throated Sparrow. *In* "Bird Vocalizations" (R. A. Hinde, ed.), pp. 207–232. Cambridge Univ. Press, London and New York.

Ficken, R. W., Ficken, M. S., and Hailman, J. P. (1974). Temporal pattern shifts to avoid acoustic interference in singing birds. *Science* **183,** 762–763.

Fogden, M. P. L. (1972). The seasonality and population dynamics of equatorial forest birds in Sarawak, *Ibis* **114,** 307–343.

Golani, I. (1976). Homeostatic motor processes in mammalian interactions: A choreography of display. *In* "Perspectives in Ethology," (P. P. G. Bateson and P. H. Klopfer, eds.), Vol. 2, pp. 69–134. Plenum, New York.

Good, E. E. (1952). The life history of the American Crow—*Corvus brachyrhynchus* Brehm. Ph.D. Thesis, Ohio State Univ., Columbus.

Gradwohl, J., and Greenberg, R. (1980). The formation of Antwren flocks on Barro Colorado Island, Panama. *Auk* **97,** 385–395.

Griscom, L. (1945). "Modern Bird Study," Harvard Univ. Press, Cambridge, Massachusetts.

Harcus, J. L. (1977). The functions of vocal duetting in some African birds, *Z. Tierpsychol.* **43,** 23–45.

Haverschmidt, F. (1947). Duetting in birds. *Ibis* **89,** 357–358.

Haverschmidt, F. (1953). Notes on the life history of the Black-crested Antshrike in Surinam. *Wilson Bull.* **65,** 242–251.

Hooker, T., and Hooker, B. I. (1969). Duetting. *In* "Bird Vocalizations" (R. A. Hinde, ed.), pp. 185–205. Cambridge Univ. Press, London and New York.

Huxley, C. R., and Wilkinson, R. (1977). Vocalizations of the Aldabra White-throated Rail, *Dryolimnas cuvieri aldabranus*. *Proc. R. Soc. London, Ser. B* **197,** 315–331.

Huxley, C. R., and Wilkinson, R. (1979). Duetting and vocal recognition by Aldabra White-throated Rails, *Dryolimnus cuvieri aldabranus*. *Ibis* **121,** 266–273.

Huxley, J. S. (1919). Some points on the habits of the Little Grebe, with some notes on the occurrence of vocal duets in birds. *Br. Birds* **13,** 155–158.

Immelmann, K. (1961). Beitrage zur Biologie und Ethologie australischen Honigfresser (Meliphagidae). *J. Ornithol.* **102,** 164–207.

Immelmann, K. (1963). Tierische Jahresperiodik in ökologischer Sicht. *Zool. J. Syst.* **91,** 91–200.

Janzen, D. H. (1973). Sweep samples of tropical foliage insects: Effects of season, vegetation types, elevation, time of day, and insularity. *Ecology* **54,** 687–708.

Janzen, D. H., and Schoener, T. W. (1968). Differences in insect abundance diversity between wetter and drier sites during a tropical dry season. *Ecology* **49,** 96–110.

Johnsgard, P. A. (1965). "Handbook of Waterfowl Behavior." Cornell Univ. Press (Comstock), Ithaca, New York.

Johnsgard, P. A. (1975). "Waterfowl of North America." Indiana Univ. Press, Bloomington.

Jones, P. J., and Ward, P. (1976). The level of reserve protein as a proximate factor controlling the timing of breeding and clutch-size in the Red-billed Quelea, *Quelea quelea*. *Ibis* **118,** 547–574.

Keith, S., Benson, C. W., and Irwin, M. P. S. (1970). The genus *Sarothura* (Aves, Rallidae). *Bull. Am. Mus. Nat. Hist.* **143,** 1–84.

Kilham, L. (1958). Pair formation, mutual preening and nest hole selection of the Red-bellied Woodpecker. *Auk* **75,** 318–329.

Kilham, L. (1959). Mutual tapping of the Red-headed Woodpecker. *Auk* **76,** 235–236.

Kilham, L. (1960). Courtship and territorial behavior of Hairy Woodpeckers. *Auk* **77,** 259–270.

Kunkel, P. (1974). Mating systems of tropical birds. *Z. Tierpsychol.* **34,** 265–307.

Lemon, R. E. (1968). The relation between organization and function of song in Cardinals. *Behaviour* **32,** 158–178.

Lemon, R. E. (1977). Bird song: An acoustic flag. *BioScience* **27,** 402–408.

Machlis, L. (1977). An analysis of the temporal patterning of pecking in chicks. *Behaviour* **63,** 1–70.

McKinney, F. (1975). The evolution of duck displays. *In* "Function and Evolution in Behavior—Essays in Honour of Professor Niko Tinbergen, F. R. S." (G. Baerends, C. Beer, and A. Manning, eds.), pp. 331–337, Oxford Univ. Press, London and New York.

McKinney, F., Siegfried, W. R., Ball, I. J., and Frost, P. P. G. (1978). Behavioral specializations for river life in the African Black Duck. *Z. Tierpsychol.* **48,** 349–400.

Marshall, J. T. (1960). Inter-relations of Albert and Brown towhees. *Condor* **62,** 49–64.

Maynard Smith, J. (1977). Parental investment: A prospective analysis. *Anim. Behav.* **25,** 1–9.

Mayr, E. (1963). "Animal Species and Evolution." Belknap Press, Cambridge, Massachusetts.

Miller, E. H., and Baker, A. J. (1980). Displays of the Magellanic Oystercatcher (*Haematopus leudopodys*). *Wilson Bull.* **92,** 149–168.

Moreau, R. E. (1936). Breeding seasons of birds in East African evergreen forest. *Proc. Zool. Soc. London* pp. 631–653.

Moreau, R. E. (1950). The breeding seasons of African birds. I. Land birds. *Ibis* **92,** 223–267.

Moreau, R. E. (1964). Breeding seasons. *In* "A New Dictionary of Birds" (A. Landsborough-Thompson, ed.), pp. 106–108. Nelson, London.

Morse, D. H. (1970). Territorial and courtship song in birds. *Nature (London)* **226,** 659–661.

Morse, D. H. (1980). "Behavioral Mechanisms in Ecology." Harvard Univ. Press, Cambridge, Massachusetts.

Morton, E. S., and Farabaugh, S. M. (1980). The ecological background of vocal sounds used at close range. *Proc. Int. Ornithol. Congr.* **17,** 737–747.

Munn, C. A., and Terborgh, J. W. (1979). Multi-species territoriality in neotropical foraging flocks. *Condor* **81,** 338–347.

Orians, G. H. (1969). On the evolution of mating systems in birds and mammals. *Am. Nat.* **103,** 589–603.

Parker, G. A. (1974). Courtship persistence and female guarding as male time investment strategies. *Behaviour* **68,** 157–184.

Payne, R. B. (1970). Temporal pattern of duetting in the Barbary Shrike, *Laniarius barbarus*. *Ibis* **112,** 106–108.

Payne, R. B. (1971). Duetting and chorus singing in African birds. *Ostrich, Suppl.* No. 9, 125–146.

Payne, R. B., and Skinner, N. J. (1970). Temporal patterns of duetting in African Barbets. *Ibis* **112,** 173–183.

Powell, G. V. N. (1980). Migrant participation in neotropical mixed species flocks. *In* "Migrant Birds in the Neotropics: Ecology, Behavior, Distribution, and Conservation" (A. Keast and E. S. Morton, eds.), pp. 477–483. Smithson. Inst. Press, Washington, D.C.

Power, D. M. (1966). Antiphonal duetting and evidence for auditory reaction time in the Orange-chinned Parakeet. *Auk* **83,** 314–319.

Ridgely, R. S. (1976). "A Guide to the Birds of Panama." Princeton Univ. Press, Princeton, New Jersey.

Robbins, C. S., Bruun, B., and Zim, H. S. (1966). "Birds of North America: A Guide to Field Identification." Golden Press, New York.

Rudegeair, T. J. (1975). The reproductive behavior and ecology of the White Ibis (*Eudocimus albus*). Ph.D. Thesis, Univ. of Florida, Gainesville.

Schleidt, W. M. (1974). How fixed is a fixed action pattern? *Z. Tierpsychol.* **36,** 184–211.

Seibt, U., and Wickler, W. (1977). Duettieren als Revier-Anzeige bei Vögeln. *Z. Tierpsychol.* **43,** 80–87.

Skutch, A. F. (1940). Social and sleeping habits of Central American wrens. *Auk* **57,** 293–312.

Skutch, A. F. (1947). Life history of the Marbled Wood-Quail. *Condor* **49,** 217–232.

Skutch, A. F. (1964). Life history of the Blue-diademed Motmot, *Momotus momota*. *Ibis* **106,** 321–332.

Smith, S. M. (1978). The "underworld" in a territorial sparrow: Adaptive strategy for floaters. *Am. Nat.* **112,** 571–582.

Sokal, R. R., and Rohlf, F. J. (1973). "Introduction to Biostatistics." Freeman, San Francisco, California.

Stokes, A. W., and Williams, H. W. (1968). Antiphonal calling in quail. *Auk* **85,** 83–89.

Thorpe, W. H. (1961). "Bird-Song; the Biology of Vocal Communication and Expression in Birds." Cambridge Univ. Press, London and New York.

Thorpe, W. H. (1963). Antiphonal singing in birds as evidence for avian auditory reaction time. *Nature (London)* **197,** 774–776.

Thorpe, W. H. (1967). Vocal imitation and antiphonal song and its implications. *Proc. Int. Ornithol. Congr.* **14,** 245–264.

Thorpe, W. H. (1972). Duetting and antiphonal singing in birds: Its extent and significance. *Behaviour, Suppl.* No. 18.

Thorpe, W. H., and North, M. E. W. (1965).Origin and significance of the power of vocal imitation: With special reference to the antiphonal singing of birds. *Nature (London)*, **208,** 219–222.

Thorpe, W. H., and North, M. E. W. (1966). Vocal imitation in the tropical Bou-bou Shrike, *Laniarius aethiopicus major*, as a means of establishing and maintaining social bonds. *Ibis* **108,** 432–435.

Todt, D. (1975). Effect of territorial conditions on the maintenance of pair contact in duetting birds. *Experientia* **31,** 648–649.

Todt, D., and Fiebelkorn, A. (1980). Display, timing, and function of wing movements accompanying duets of *Cichladusa guttata*. *Behaviour* **72,** 82–106.

Van Tyne, J., and Berger, A. J. (1959). "Fundamentals of Ornithology." Wiley, New York.

Verner, J. (1964). Evolution of polygamy in the Long-billed Marsh Wren. *Evolution* **18,** 252–261.

von Helverson, D. (1980). Structure and function of antiphonal duets in birds. *Proc. Int. Ornithol. Congr.* **17,** 682–688.

Ward, P. (1969). The annual cycle of the Yellow-vented Bulbul, *Pycnonotus goiavier*, in a humid equatorial environment. *J. Zool.* **157,** 25–45.

Wasserman, F. E. (1977a). Territorial behavior of White-throated Sparrows (*Zonotrichia albicollis*) with special emphasis on song. Ph.D. Thesis, Univ. of Maryland, College Park.

Wasserman, F. E. (1977b). Mate attraction function of song in White-throated Sparrow (*Zonotrichia albicollis*). *Condor* **79,** 125–127.

Wasserman, F. E. (1977c). Intraspecific acoustical interference in the White-throated Sparrow (*Zonotrichia albicollis*). *Anim. Behav.* **25,** 949–952.

Wickler, W. (1976). Duetting songs in birds: Biological significance of stationary and nonstationary processes. *J. Theor. Biol.* **61,** 493–497.

Wickler, W. (1980). Vocal duetting and the pair bond: I. Coyness and partner commitment. A hypothesis. *Z. Tierpsychol.* **52,** 201–209.

Wickler, W., and Seibt, U. (1980). Vocal duetting and the pair bond. II. Unison duetting in the African Forest Weaver, *Symplectes bicolor*. *Z. Tierpsychol.* **52,** 217–226.

Wiley, R. H., and Wiley, M. S. (1977). Recognition of neighbor's duets by Stripe-backed Wrens, *Campylorhynchus nuchalis*. *Behaviour* **62,** 10–34.

Wittenberger, J. F. (1976). The ecological factors selecting for polygyny in altricial birds. *Am. Nat.* **110,** 779–799.

Wolda, H. (1978a). Fluctuations in abundance of tropical insects. *Am. Nat.* **112,** 1017–1045.

Wolda, H. (1978b). Seasonal fluctuations in rainfall, food and abundance of tropical insects. *J. Anim. Ecol.* **47,** 369–381.

Zahavi, A. (1979). Why shouting? *Am. Nat.* **113,** 155–156.

Zar, J. H. (1974). "Biostatistical Analysis." Prentice-Hall, Englewood Cliffs, New Jersey.

5

Song Repertoires: Problems in Their Definition and Use

DONALD E. KROODSMA

I. INTRODUCTION

Superficially, the sounds of each participant in a dawn chorus seem functionally equivalent, broadcasting information about each singer's breeding status, territorial occupancy, etc. But what an incredible diversity of means are used to accomplish the same end, that of contributing as many offspring as possible to future generations. The performances of the participants differ on a multitude of dimensions, far more than seem essential for mere species, population, or individual recognition (see Chapters 7 and 8, Volume I; Chapter 7, Volume II). Thus, male Brown Thrashers (*Toxostoma rufum*) sing up to several thousand different song types (Kroodsma and Parker, 1977) whereas male White-throated

ACOUSTIC COMMUNICATION IN BIRDS
VOLUME 2

ISBN 0-12-426802-1

Sparrows (*Zonotrichia albicollis*) utter only one song type (Borror and Gunn, 1965). Songs of male Winter Wrens (*Troglodytes troglodytes*) may last 10 sec and consist of hundreds of notes, while a simple two-noted "tsi-lick" or "fitz-bew" is sufficient for a Henslow's Sparrow (*Ammodramus henslowii*) or Willow Flycatcher (*Empidonax traillii*), respectively. Why are successive songs of male Fox Sparrows (*Zonotrichia iliaca*) very different (Martin, 1977), yet the closely related Song Sparrow (*Zonotrichia melodia*) repeats a particular song over and over before switching? What rules govern which song is used next? And why do some species have different songs for different contexts (e.g., the Chestnut-sided Warbler, *Dendroica pensylvanica;* Lein, 1978) or during countersinging match themes from a large repertoire (e.g., the Long-billed Marsh Wren, *Cistothorus palustris;* Verner, 1975; Kroodsma, 1979). Peculiarities of life history and notions about how natural selection works are consistent with some of these species differences (e.g., Kroodsma, 1977; Catchpole, 1980; see also Chapter 9, Volume 1), but in general the variety of singing behaviors and song organizations remains delightfully perplexing.

It is a simple matter to recognize the existence of differences in vocal behaviors, yet it is very difficult to quantify those differences adequately. A satisfactory approach in identifying song types or estimating repertoire sizes for one species may be inappropriate for another. Vocal behaviors among species can differ so much that the comparative approach of seeking correlations between breeding biologies and singing styles must be undertaken with great care. Rules that seem to govern the vocal performances of one individual or species may be ignored by others. Yet, major advances in understanding the evolution of different song structures, repertoire sizes, sequencing behaviors, etc., must address these problems. In this chapter I first take a close look at the phenomenon of "vocal repertoires," examining (1) definitions of repertoires, (2) nonrandom singing and the problems of estimating repertoire size, and (3) the comparability problem (Krebs and Kroodsma, 1980). In the second section, I address several issues dealing with the organization and use of those song repertoires: (4) the degree of variability within and among song repertoires, (5) Hartshorne's (1956, 1973) "Monotony-Threshold," and (6) antihabituation as a candidate principle in organizing singing behaviors.

II. REPERTOIRE SIZE

A. What Is a Repertoire?

A repertoire is, by definition, "the list of plays, operas, pieces, or parts which . . . a performer is prepared to present" (Merriam-Webster, 1973). In dealing with repertoires of songs in birds, there is a primary problem of categorizing the units and a secondary problem of deciding what to call those units. The problem of terminology has been more of a menace than a handicap, but investigators continue to define old terms in new ways or establish new and unnecessary vocabularies. The recommendations of Shiovitz (1975) are the most thorough and reasoned to date and should be used where possible. The primary problem, that of defining the categories, varies considerably among species, as the examples below illustrate.

In the White-throated Sparrow, the basic pattern of male song consists of several introductory notes, then a series of "Peabody" triplets; the male may stop anywhere in the song. There is some variation among an individual's songs, even though each male sings only one basic pattern (Borror and Gunn, 1965). Such variation has often been regarded as "noise," but such seemingly minor variations on a single basic pattern are undoubtedly capable of conveying considerable information regarding the male's motivational states (e.g., Ficken and Ficken, 1973).

A series of ten consecutive songs from a Song Sparrow exhibits variation which is comparable to that within the single song type of a White-throated Sparrow. After ten or so songs, however, another very different song type is introduced, and later still another, and another, until the male eventually but unmistakably returns to the first type. There is variation among successive renditions of each song type, but far less than that found among song types (Mulligan, 1966; Harris and Lemon, 1972; Eberhardt and Baptista, 1977). Partly because of the human desire to categorize and simplify, the attention of the bioacoustician is drawn to how many different song types a male has, not to the variation within the song types. Such variation in successive or nonsuccessive renditions of one song type is surely biologically significant and merits further study.

In species such as the Song Sparrow, each song type is repeated several times before another is introduced (i.e., AAA . . . BBB . . . CCC . . .); the male here provides a convenient measure by which to classify his songs. In those species that sing with immediate variety, however, males may offer no immediate clue as to the magnitude of variation within or between song type categories. In the Long-billed Marsh Wren, for example, successive songs are often strikingly different, and a male may sing 50 rather different songs in succession (Verner, 1975). Only after graphing hundreds of songs per male is it clear that males sing discrete song types and that a distinct repertoire of song types does exist (Verner, 1975). As in typical AAA . . . BBB . . . songsters, variation among different renditions of one song type is far less than the variation between song types (see Verner, 1975, Figs. 3–5, 7, and 8).

This criterion of variation within and among song-type categories is easily stated, of course, but sometimes rather difficult to put into practice. For example, a Long-billed Marsh Wren may sing what appears to be the same song type at two different frequencies. If the bird were consistent in doing this, most bioacousticians would classify these as two different song types. Occasionally, however, the male may utter a "hybrid" song; here, the frequency of the repeated syllables changes, beginning on the frequency typical of one of the "song types" and gradually shifting so that the final syllable is at the frequency of the second "song type." Defining song types becomes more arbitrary under such conditions. The lumping/splitting dilemma faced by all taxonomists arises: are these two distinct song types with an odd hybrid, or would further sampling reveal a continuum and therefore only one highly variable song type? On occasion, some help is available from the birds. Among western Long-billed Marsh Wrens, song sequences may be so orderly that like songs may be classified to some extent based on the location in the sequence (e.g., see Verner, 1975). Among eastern Long-billed Marsh Wrens, like songs often occur every second or third song (e.g., ABABCBDC . . .), and this song patterning, while not as unambiguous as the AAA . . . BBB . . . patterning, may sometimes aid in classification.

Birds which sing more continuously pose even more difficult problems, for here discrete songs may not even exist. Temporal discontinuities in the singing may suggest the units of analysis (Isaac and Marler, 1963), but whether the song units of continuous songsters such as the Brown Thrasher or Sedge Warbler (*Acrocephalus schoenobaenus*) are comparable to the songs of discontinuous songsters (e.g., Long-billed Marsh Wren, Song Sparrow) is a serious problem for the comparative biologist. (These problems and some possible solutions are addressed further under "The Comparability Problem," Section II, C.)

B. Quantitative Estimations of Repertoire Size

1. *Estimating the Number of Different Song Types*

Determining the number of song types, song units, etc., an individual uses is easily accomplished in many species with small repertoires, especially if males cycle through their repertoires rapidly and regularly. For example, Fox Sparrows and Swainson's Thrushes (*Catharus ustulatus*) may have three to five song types presented in highly regular sequences of immediate variety (e.g., ABCDABCDABCD . . . ; Martin, 1977; Dobson and Lemon, 1977). As repertoires increase in size, however, sequences may become less regular and the frequency of occurrence of different repertoire components often varies; *estimating* the total repertoire size then becomes necessary. Such estimations are made difficult primarily because birds usually emit irregular and nonrandom sequences of song types. In addition, the relative frequencies of occurrence of song types often differ. These difficulties can be illustrated by examining the use of an exponential curve to estimate repertoire sizes in the Mockingbird (*Mimus polyglottos*) and the Rock Wren (*Salpinctes obsoletus*).

In Mockingbirds Wildenthal (1965) plotted the number of new repertoire components in a recorded sample against the total number of components sampled. The resultant graph was approximated by an exponential curve of the form

$$n = N\,(1 - e^{-T/N}),$$

where n = the number of different repertoire components in the sample, T = the total number of repertoire components sampled, and N = the number of different repertoire components in the total repertoire.

Two conditions for the singing behavior of the Mockingbird to fit this curve are (1) that the number of song units in the repertoire remains constant during the sample period and (2) that the choice of the next song type is totally random. Over relatively short sample periods (hours), only the second condition poses a problem, for songbirds rarely emit random sequences of song types. Rather, the next song type is often dependent on the immediately previous one or even two song types (e.g., Lemon and Chatfield, 1971). Furthermore, songbirds tend to utter a larger fraction of their repertoire during a brief sample period than would be expected by chance alone (Verner, 1975; Dobson and Lemon, 1977), or, in other words, bouts of the same song type tend to be overdispersed in time (Beck, 1971). Finally, some song types are quite rare while others are very common (Mulligan, 1966). These deviations from random singing and the corresponding problems with the exponential curve can best be illustrated by a close examination of the singing behavior of a typical songbird.

From a male Rock Wren at Malheur National Wildlife Refuge near Burns, Oregon, I recorded 1200 songs and then plotted the cumulative number of distinct song types against the total number of songs sampled (see Kroodsma, 1975, Fig. 3). Eighty-three different songs occurred in this sample, and the asymptote of the curve, and therefore the estimated repertoire size, was between 85 and 90 song types. After every 100 recorded songs I estimated N, the total repertoire size, by adjusting the exponential to pass through the last point of actual data (Table I). As T becomes large relative to N, $e^{-T/N}$ becomes increasingly insignificant, and N increases with n (compare columns 3 and 4 in Table I). However, not all occurrences of each song type are independent of one another, for the male alternates several renditions of different song types before abandoning them for some longer period of time (e.g., ABCBDCBDEFDEFGEHGEHGEHGI . . . +44 songs . . . XBYBXBCBC . . . see Kroodsma, 1975, Fig. 1). The number of songs sampled (T) must be adjusted to reflect the number of *independent* occurrences of each song type. Recurrence numbers (number of other songs occurring between renditions of the

TABLE I

Use of the Exponential $n = N(1 - e^{-T/N})$ to Estimate the Repertoire Size (N) after Every 100 Songs for a Male Rock Wren[a]

T_{songs}[b]	T_{bouts}[c]	n[d]	N_{songs}[e]	N_{bouts}[f]
1	27	25	26	173
2	48	44	45	272
3	69	59	59	214
4	98	71	71	143
5	121	74	74	112
6	150	76	76	96
7	190	80	80	91
8	218	80	80	87
9	239	80	80	85
10	267	81	81	85
11	285	83	83	86
12	304	83	83	85

[a] Two methods are used, one based on the total number of songs sampled and the other on the estimated number of independent occurrences (i.e., bouts) of each song type.
[b] Hundreds of songs sampled.
[c] Number of bouts sampled.
[d] Number of distinct song types in the sample.
[e] Estimated repertoire size, based on number of songs sampled.
[f] Estimated repertoire size, based on number of bouts sampled.

same song type) of five or less were common, and then a song type usually would not recur for at least 50 songs. Thus, renditions of a song type that occur within about five (here I arbitrarily chose ten) recurrence numbers of one another could be considered as a single bout of that particular type (Kroodsma, 1975). If only "independent" occurrences of each song type are considered, the exponential at first overestimates but eventually improves in estimating N, the total repertoire size (Table I).

There are two opposing biases in these calculations, both a result of nonrandom singing patterns of this Rock Wren. First, sequences of song types are more regular than random. For example, given any of the 21 most commonly used songs of this male, the following song type could be predicted with 40% accuracy. As a result, this male presented during a burst of 50 or so song bouts (independent occurrences of a song type: see above) many more different song types than would be predicted by chance alone; hence, 45 different song types in a sample of 48 bouts produced an estimate of 272 (Table I), about three times the more reasonable extrapolation of approximately 90. Howard (1974) tried to circumvent this bias by randomly sampling songs from a lengthy recording of an individual Mockingbird; if that approach is used with this Rock Wren, however, a second bias appears, this one tending to produce a serious underestimate of N. This bias is produced by a disparity in frequencies of occurrence: some song types occur very frequently, while others are very rarely used. The 10 most frequently used song types account for 28.8% of the songs given, while the 10 most infrequently used song types account for only 1.4%. Thus, when 50, 100, 150, and 200 songs were drawn at random from the sample of 1200, cumulative estimates of N by the exponential were 56, 77, 67, and 70 song types, respectively. The median of four separate estimates based on 50 songs apiece was only 61, a serious underestimate of the more realistic estimate of 85 to 90 song types.

These two biases can best be perceived by considering three hypothetical individual songsters, each with 10 song types in the repertoire. Male 1 sings a highly regular sequence, always uttering all 10 songs in the sequence ABCDEFGHIJABC . . . ; all songs, of course, have an equal frequency of occurrence. Male 2 always chooses his next song at random, and all song types are equally likely to occur. Male 3 also chooses songs at random, but one of the 10 songs occurs 90.0% of the time and each of the other 9 only 1.1% of the time. If a sequence of 10 songs is sampled from each individual, Males 1, 2, and 3 will, on an average, use 10.0, 6.3, and 2.0 song types, respectively. If the exponential curve is fitted to these data and adjusted to pass through the final point on the curve (at $T = 10$ songs sampled), the estimated repertoire sizes for birds 1, 2, and 3 are infinity, 10.00, and 2.01, respectively (see Fig. 1).

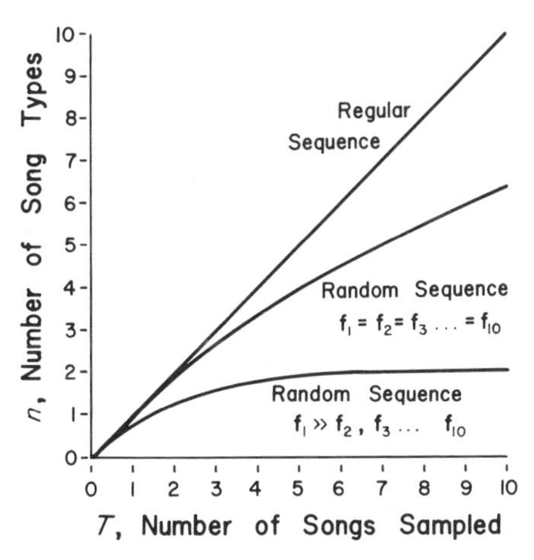

Fig. 1. An illustration of how regular or random sequencing of song types can influence estimates of repertoire sizes from limited recording samples. Each of three males has ten song types in his repertoire, but the expected number of different song types (n) in a sample (T) depends on the relative frequencies of occurrence and regularity (or randomness) of sequences. (See text for details.)

Thus, the only way in which the singing of this Rock Wren (and probably most songbirds) can be satisfactorily approximated by the exponential is if the two biases offset one another. Limited random samples of songs underestimate and limited highly regular continuous samples of songs overestimate the true repertoire size N. For this Rock Wren, it is clear from column 5 in Table I that the regularity of sequences and overdispersion of independent occurrences of song types weigh more heavily, causing an early overestimate of the actual repertoire size.

Alternative approaches for estimating repertoire sizes do exist. Fagen and Goldman (1977) applied the truncated lognormal Poisson, negative binomial, and Poisson distributions to the relative abundances of different behaviors in a sample; the goal was to estimate the number of behavioral acts that fall into abundance category 0, i.e., to estimate the number of behaviors in the repertoire which did not occur in the sample. The success of their approach was mixed, undoubtedly for the same reasons illustrated above: animals do not display in random sequences.

2. *Relative Abundances of Song Types*

The total repertoire size weights commonly and rarely used song types equally, but there are other measures that consider the relative frequencies of occurrence of the different song types. Bertram (1970), for example, introduced the "99% repertoire," the minimum number of song types that could account for 99% of the songs sampled. In the above Rock Wren example, 75 of the 83 song types (90%) accounted for 99% of the total songs, and 65 song types (78%) accounted for 95% of the total songs.

Fagen and Goldman (1977), following Good (1953), introduce the concept of sample coverage. This is the probability that in a new, independent sample of behavior, a randomly chosen act (here, song type) will belong to a type already represented in an initial sample I. The probability is estimated by $1 - N_1/I$, where N_1 is the number of behaviors that have occurred only once in a sample of size I. Again, using the Rock Wren example, after $I = 304$ bouts, $N_1 = 8$ song types had occurred only once: the probability that the next sampled bout would already have been sampled is $1 - 8/304 = 0.97$.

C. The Comparability Problem

Identifying the number of song types or song units in the repertoire of a particular individual can be relatively straightforward, but a comparative approach demands that repertoires of different populations and species be assessed on a single meaningful scale. The difficulty can be illustrated by an hypothetical example. Individuals 1, 2, and 3 each have a repertoire of three song components or building blocks, a, b, and c, each 0.25 sec long and consisting of a pure tone of 2, 4, and 8 kHz, respectively. Male 1 combines all his song units into his only song type, abc. Male 2 sings three song patterns, aaa, bbb, and ccc. Finally, Male 3 uses the three song units to create all combinations and permutations of the three song units: abc, acb, bca, and so on, for 27 different song types. Which male, then, has the largest repertoire size?

Clearly, "repertoire size" can be judged in at least two ways: by the number of song units and the number of unique song types composed of those units. All three males have an identical song unit repertoire, but the males have repertoires of 1, 3, and 27 song types, respectively.

This is a greatly simplified example, though, for all song units are of equal complexity and all songs consist of three song units. Suppose a fourth pure tone, also 0.25 sec long, is slurred from 4 to 2 kHz, or 8 to 2 kHz. This intuitively seems to be a more "complex" song unit, but can the difference in complexity be expressed quantitatively? And when more realistic frequency and amplitude modulations of typical bird vocalizations are considered, the potential difference in complexity of song units and song types makes comparison of the repertoire sizes of conspecifics, but especially heterospecifics, extremely difficult. This, quite simply, is the "comparability problem" (see Krebs and Kroodsma, 1980).

To illustrate further some of the problems and some possible solutions, I compare here repertoire sizes of Bewick's Wrens (*Thryomanes bewickii*) from six locations in the western United States (Corvallis, Oregon; Point Reyes, California; Santa Barbara, California; Santa Cruz Island, California; Madera Canyon, Arizona; and Grand Junction, Colorado). The Bewick's Wren typically sings with eventual variety (AAA . . . BBB . . .), and I recorded males at each location throughout one morning or until the male began his third cycle through his repertoire. The songs usually consist of two to eight phrases (terminology is that of Mulligan, 1966; see Kroodsma, 1974, Fig. 2), but song complexity varies geographically (Fig. 2).

Consider first the number of song types per male at each location (row 1 of Table II). This varies from 9 to 22 over the western United States, and further sampling would likely verify that males at Grand Junction, Colorado, and Santa Barbara, California, have the smallest and largest repertoires, respectively.

Other measures of complexity agree broadly with this trend. Each song consists of several phrases (phrases of repeated syllables alternated with phrases of unrepeated "note complexes"), and the *total phrase repertoire* (row 2 of Table II) is the number of songs multiplied by the number of phrases per song. The same phrases may occur in several different song types, though, so the repertoire of different *phrase types* (row 3) is somewhat smaller than the total phrases. Thus far, when considering either song types or song units, males at Santa Barbara and Santa Cruz Island tend to rank first and second in the overall size of the vocal repertoire, respectively.

However, not all component song units are of equal duration. The number of

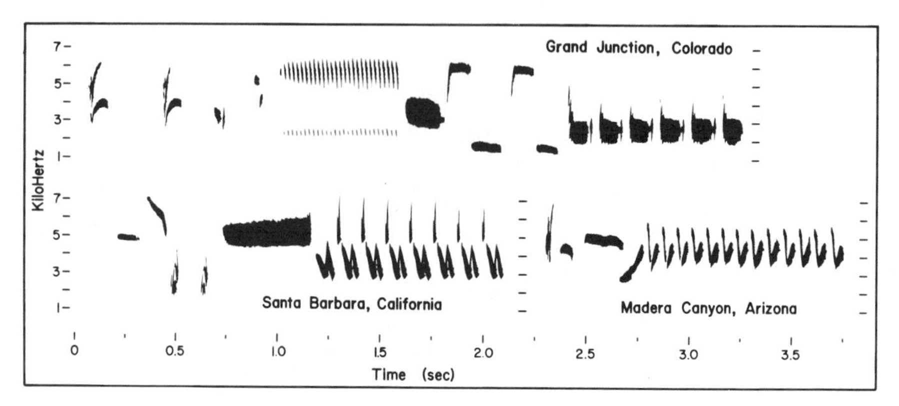

Fig. 2. Geographic variation in songs of the Bewick's Wren, as illustrated for three locations in the western United States. Analyzing bandwidth is 300 Hz. Horizontal and vertical scales are time (sec) and kHz, respectively.

song types multiplied by their average length gives *song repertoire duration* (row 4, Table II). This duration contains some redundancy, though, for the same phrase type may occur in different song types and the entire phrase of repeated syllables is included. The *corrected repertoire duration* (row 5) includes only once the duration of different song units used to construct the different song types in the repertoire. Interestingly, Santa Cruz Island and Grand Junction rank highest on this measure, Santa Barbara and Madera Canyon are intermediate, while Point Reyes and Corvallis males rank last.

Finally, phrases of equal duration may differ in amplitude and frequency modulations (see Fig. 2). In an attempt to include the frequency dimension in the "complexity rating," I have calculated the product of duration and frequency range (the maximum minus the minimum frequency) for each phrase in the males' repertoires of song types. The sum of the frequency–time products for each phrase in each song type for a given male is termed the *kHz-sec complexity* (row 6, Table II). By including only one example of each type of repeated syllable and only one example of each note complex in the different song types, a *corrected kHz-sec complexity* may be calculated (row 7).

No one index of repertoire size is entirely satisfactory by itself. For example, the number of song types is a standard measure of repertoire size, but in the present case it cannot reflect differences in song length, number of building blocks (= song units) per song, or complexity in the frequency dimension. Because these birds sing with eventual variety (i.e., AAA . . . BBB . . .), the number of song types is the most objective and simplest measure, for the singing male clearly presents each of his song types in turn. It might be desirable to strive for a single "complexity index" which reflects not only the song type repertoire but also the total number and the relative complexity of song building blocks.

TABLE II

A Multidimensional Comparison of Repertoire Size and Complexity for Bewick's Wrens from Six Geographic Locations in the Western United States

	Point Reyes, California	Santa Barbara, California	Santa Cruz Is., California	Madera Canyon, Arizona	Grand Junction, Colorado	Corvallis, Oregon
1. Song types[a]	14.5	20	17.5	17.5	10	16
	(2:12–17)	(4:16–22)	(4:17–20)	(4:16–19)	(4:9–11)	(31:13–20)
	5	1	2.5	2.5	6	4
2. Total phrases[a]	56	87.5	56	35.5	40.5	50
	(2:47–65)	(4:66–93)	(4:52–69)	(4:32–38)	(4:33–47)	(7:43–63)
	2.5	1	2.5	6	5	4
3. Phrase types[a]	45	69.25	48.25	35.5	38	40
	(2:36–54)	(4:49–78)	(4:46–61)	(4:32–38)	(4:32–45)	(7:30–53)
	3	1	2	6	4.5	4.5
4. Song repertoire duration[b]	31.2	40	34.6	27.8	29.2	27.4
	3	1	2	5	4	6
5. Corrected repertoire duration[b]	9.5	13.4	15.4	13.2	15.1	8.5
	5	3	1	4	2	6
6. kHz-sec complexity[b]	117.6	139.4	147.0	86.8	120.3	111.8
	4	2	1	6	3	5
7. Corrected kHz-sec complexity[b]	39.8	51.0	71.9	56.4	71.4	31.3
	5	4	1	3	2	6

[a] Median number of repertoire units/male, with sample size and range in parentheses (*n*: range), followed below by the ranking for each of the six geographic locations.
[b] The sample size in each column remains the same, but ranges are not available for rows 4–7.

Such an index would necessarily obscure the separate contributions to the index, though, but more importantly, there is at this time no simple way to weight the different indexes.

Birds do respond with renewed vigor when song types are changed during playback experiments (Krebs, 1976), but there is no evidence as to how different successive song types must be in order to maintain or elicit new interest. Thus, going back to the original example of the three males with the a, b, and c song units, we know neither the relative difficulty of producing all permutations nor the stimulus value of those different permutations. A clue to the relative difficulty of different singing behaviors could perhaps be addressed by relating the volume of brain song control centers to different styles of singing (see Chapter 3, Volume 1). Ideally, one would want to find conspecific males 1, 2, and 3 discussed above, or perhaps males in captivity could be trained to sing differently from one another. The stimulus value of a large repertoire has been demonstrated partially in female Canaries (*Serinus canaria;* Kroodsma, 1976) and male Great Tits (*Parus major;* Krebs, 1976), but additional field experiments are required where repertoires of song units are synthetically combined to create differing numbers of song types. An ideal species for a study of this type may be the Five-striped Sparrow (*Aimophila quinquestriata*), where males have sizeable repertoires of song units which are often recombined in different sequences to produce different song types (K. Groschupf, unpublished data).

III. ORGANIZATION AND USE OF SONG-TYPE REPERTOIRES

A. Stimulus Contrast: Variation within and among Repertoires

Regardless of its function, a repertoire of song types is effective only if listeners can detect differences among the song types. Therefore, a singer should insure some minimum level of contrast among song types within his repertoire. If a songster is striving to deceive listeners (as proposed in the Beau Geste hypothesis), his songs should not reveal his identity; within-repertoire variation should then be *at least as great* as that among repertoires (Krebs, 1976, 1977). However, if the mechanism in Krebs' Beau Geste hypothesis is habituation, the songster should strive for increased contrast between the songs in his repertoire; the within-repertoire variation should then be *greater than* the among-repertoire variation (Whitney, 1979). In other words, the variation within a repertoire should be as great as or greater than the variation of a repertoire drawn randomly from the song population.

Krebs (1976) first tested this variability hypothesis in Great Tits. He found that variability of phrase duration within repertoires was greater than that among repertoires; the results from an analysis of note types were less clear, yet suggested that songs within repertoires were at least as variable as songs of random repertoires. With songs of the Red-winged Blackbird (*Agelaius phoeniceus*), Yasukawa (1981) found no differences in within- and among-repertoire variability for four variables. Finally, Whitney (1979) described songs of the Varied Thrush (*Zoothera naevia*) and found that for dominant frequency and period of modulation "songs within repertoires are less similar than would be expected if they were drawn at random from all the songs in the population."

Krebs (1976, p. 225) realized that his results could have been "simply due to song sharing between birds," yet he discounted this possibility because a frequency distribution of phrase lengths did not "seem to indicate . . . a few discrete types." I disagree with this conclusion. *Any* vocal learning within a population will produce a convergence in the song parameters of neighboring males and therefore a relative divergence in the songs within a male's repertoire. The greater "between-repertoire" variance found by Krebs (1976) and Whitney (1979) may merely be an epiphenomenon of song learning. This effect of vocal learning can be demonstrated with two simulation studies where the extent of song sharing among neighboring males varies.

For simplicity, consider a population of only four males, where each male has four song types in his repertoire. The songs vary along only one dimension (e.g., frequency or modulation rate) and are described by values ranging from 1 to 99 which are chosen from a table of random numbers. The variability within a repertoire may be determined by a difference index,

$$\mathrm{DI} = \sum_{i=1}^{3} \sum_{j=i+1}^{4} |x_i - x_j|$$

where x_1, x_2, x_3, and x_4 are numbers representing the four song types in a repertoire. This index offers in a single number the degree of contrast achieved by all possible pair-wise comparisons of songs within a single repertoire.

Consider first the extreme form of song sharing where all four males have learned identical song type repertoires. What is the nature of the within- and among-repertoire variability under such conditions? To study this problem, I selected four "song types" from a random numbers table, and assigned that repertoire to all four males in the population (see Table III). The relative contrast (DI) within each male's repertoire is now 285 (calculation in Table III). To assess the "among-repertoire" variability, I assembled at random a total of 100 repertoires of four song types from the pool of 16 songs (actually only four song types, since the repertoires of the four males are identical); I then compared the DI of these 100 repertoires with that of the original repertoire of each of the four males (DI = 285). Only 11 ABCD repertoires, identical to those of the four

TABLE III

Repertoires of Identical Song Types and Calculation of the Difference Index (DI) in a Hypothetical Four-Male Population

		Male			
		1	2	3	4
Song type	A	39	39	39	39
	B	97	97	97	97
	C	2	2	2	2
	D	45	45	45	45

DI = |39−97|+|39−2|+ . . . . +|2−45|= 285

males, were selected; of the remaining 89 repertoires the DI of 60 repertoires was less than 285, 16 equaled 285, while only 13 repertoires had a DI greater than 285. If this extreme form of song sharing had no influence on among-repertoire variability, the DI of randomly selected repertoires would have been equally likely to be either greater than or less than 285. Sixty achieved less contrast, while only 13 achieved greater contrast, indicating that extreme song sharing within a population can by itself create an *apparent* enhancement of ''within-repertoire'' variability. (Of particular interest here is the fact that a repertoire consisting of BBCC would achieve the greatest DI, 380. The songs would have to be slightly different from one another in order to be recognized as different song types, of course, but the point is that there are different ways of achieving a great variance within a repertoire. Of utmost importance, then, may be the sequence with which the male uses his songs—this point will be discussed later.)

The apparently enhanced within-repertoire variability under conditions of extreme song sharing comes as no surprise, and Krebs (1976, p. 225) was careful to document that Great Tit songs did not come in a ''few discrete types.'' Hence, while this example is instructive, it is highly unnatural, because songbirds do not achieve perfect song sharing. Consider a more realistic example, then, where all four males share only one song type and the other three songs in each male's repertoire are determined at random. After assembling these repertoires for the four males, I calculated a median DI for the four males (see Table IV). To assess among-repertoire variability, I selected ten repertoires at random from the population of 16 songs and then compared their DI's to the original DI's of the four males (see Table IV). I repeated this entire process 20 times, creating 20 four-male populations, for each comparing 10 randomly selected repertoires, and so

TABLE IV

Repertoires of Song Types in a Hypothetical Four-Male Population[a]

		Male			
		1	2	3	4
Song type[a]	A	5	5	5	5
	B_{1-4}	31	21	11	88
	C_{1-4}	93	75	40	36
	D_{1-4}	22	1	46	56
	DI[b]	273	238	152	269

[a] Only one song type (A) is shared by all males.

[b] Median DI for four males = 254. DI for 10 repertoires selected at random compared to 254: 134, 135, 153, 158, 170, 179, 222 < 254 < 280, 290, 293.

on. Of the 200 repertoires, 55 contained at least two renditions of the shared song type; 42 of those were less variable, while only 13 were more variable than the median DI for the four males. When all 200 replicates are pooled, the 42:13 ratio weighs heavily, and the variability (as measured by the difference index) of songs within repertoires of the four males is greater than the variability of song repertoires drawn randomly from the population ($\chi = 4.2$, $p = 0.042$).

In the above example, the probability of obtaining two or more identical songs in the randomly drawn repertoire is 0.24. If there exist only two identical songs in the population of 16 and the other 14 are determined at random, the probability drops to 0.05. But regardless of the number of males and the number of songs in a population, vocal learning and the sharing of songs among the members of that population will to some extent create a convergence in the song parameters of population members and thus a relative divergence of parameters within the repertoire of a single male.

These simulations also reinforce the observations of how young songbirds enter and learn the songs in a given population. For example, a juvenile Bewick's Wren establishing a territory in the Willamette Valley of Oregon is often exposed to the song repertoires of at least five adult males. Each of these males sings, on the average, 16 different song types. Altogether, then, there are at least 80 song types from which a juvenile must choose his repertoire of only 16. The songs of each male are always distinguishable and unique in some way, but many of the song types of neighboring males are so similar that these 80 types reduce to perhaps only 20 basic song patterns (very similar song types of different males may be of the same basic pattern—the song pattern is a more subjective category than is the song type, for the former requires an evaluation of the relative similarity of song types across repertoires of different males, the latter only the relative similarity within repertoires). The 20 patterns are unmistakably different from one another and easily identified by the human ear (and therefore highly contrasted with one another), and generally each male sings only one rendition of each song pattern. Likewise, a juvenile male entering such a population does not learn songs at random from the pool of 80 song types. Learning primarily the songs of a single male would assure a nonrandom selection from the 80 songs, and much song learning undoubtedly occurs on a one-to-one basis like this. Additional learning from other males would require, however, that the juvenile categorize the 80 songs just as I did in classifying 20 song patterns; otherwise the incidence of multiple renditions of a single one of the 20 song patterns would be far higher within individual repertoires. Repertoires are limited in size, of course, and learning only one rendition of each general song pattern does yield a repertoire of relatively dissimilar songs (see Kroodsma, 1974, for data on which the above discussion is based).

Perhaps two levels of stimulus generalization are working simultaneously in song development. In the Bewick's Wren, there is clearly (1) a predisposition to learn only one rendition of each song pattern available, but perhaps there is also (2) a tendency to develop a rendition of especially dissimilar song patterns. Because the first level of generalization appears so dominant in many songbird populations, the second level is very difficult to detect. Song repertoires in populations with little or no song sharing among neighboring males should be studied to test adequately this "contrast hypothesis" of Krebs (1976), Whitney (1979), and Yasukawa (1981).

Because vocal learning appears almost ubiquitous among songbirds (see Appendix), the best test of this hypothesis may be to assess the organization of repertoires in acoustically isolated, hand-reared males. Vocal convergence via learning can then be eliminated, and within-repertoire variance may then be explained entirely by efforts on the part of the individual to produce different, highly contrasted songs.

B. "The Monotony-Threshold and the Correlation of Continuity with Immediate Variety" (Hartshorne, 1973, p. 119)

When a bird has a repertoire of different song types, there are several ways in which the repertoire can be presented during a singing performance. Hartshorne (1956, 1973) classified singing performances into categories of no (only one song type, AAA . . .), eventual (AAA . . . BBB . . .), and immediate (ABCD . . .) variety, where different letters indicate different song types. Hartshorne also estimated the actual amount of time that birds sang during a singing performance (i.e., continuity), and using "some hypothetical statistics," he (Hartshorne, 1973, p. 124) claimed to find among different species a rough correlation between the degree of variety (= versatility) and the continuity of that singing performance. Hartshorne postulated the existence of a "monotony-threshold," where variety and continuity were adjusted to prevent monotonous singing behaviors.

Hartshorne's nonrigorous approach was unacceptable to many biologists, and because he was more of a philosopher than a scientist, his ideas were embedded in discussions of aesthetics and nonevolutionary thought and therefore led to considerable resistance. If his hypothesis had only involved a "habituation-threshold" rather than a "monotony-threshold," perhaps his idea would have proven more acceptable.

Some of the criticism of Hartshorne's hypothesis is based on misinterpretation. He stated, for example, that "For most purposes . . . eventual versatility is . . . nonversatility. This qualification is important" (p. 120). Thus, Hartshorne clearly specified that versatility should be measured by the degree of contrast in successive songs, not by the total repertoire of song types, a mistake often made in assessing Hartshorne's contributions (see Dobson and Lemon, 1975; Kroodsma, 1978).

Hartshorne's hypothesis was concerned primarily with cross-species comparisons, but interspecific differences in singing behaviors present a number of problems. (1) A major problem involves the "comparability problem" discussed above. The discrete songs of discontinuous songsters are quite different from the ramblings of many continuous songsters and perhaps they should not be considered equivalent. (2) Hartshorne was concerned with presence or absence of variety in successive songs, but contrast is not an all or none phenomenon; e.g., Rose-breasted Grosbeaks (*Pheucticus ludovicianus*) may alter only small portions of successive songs (Lemon and Chatfield, 1973), while the entire song may change in other species (e.g., Verner, 1975). (3) Complexity *within* some longer songs of non-versatile songsters (e.g., Winter Wren; Kroodsma, 1980) may provide signal variety comparable to that of immediately versatile species. (4) Finally, sequencing of song types may not be a static characteristic of a species; individuals may use different singing styles depending on the immediate context (Kroodsma and Verner, 1978).

Nevertheless, I feel that Hartshorne's proposed correlations between continuity and versatility are real. Dobson and Lemon (1975) tabulated data for 39 species with a great diversity of singing behaviors; 15 of the species usually sing with either no or eventual variety, while 19 typically sing with immediate variety. Overall, there was no difference in continuity between the two groups (one-tailed Mann–Whitney U test, $p > 0.1$), but this is largely because nine *Vireo* species with very short songs ($\bar{X} = 0.33$ sec) and very low continuity ($\bar{X} = 15\%$ performance time) heavily bias other trends in the data. The non-*Vireo* data do support the continuity–versatility hypothesis, for the remaining species which use immediate variety sing with greater continuity compared to the less versatile songsters (median of 39 and 22%, respectively; $p < 0.001$, one-tailed Mann–Whitney U test).

Data from several wren species also support the continuity–versatility correlation. In Rock Wrens, Long-billed Marsh Wrens, and Short-billed Marsh Wrens (*Cistothorus platensis*), the silent interval between unlike songs is shorter than that between like songs (Kroodsma, 1975; Verner, 1975; Kroodsma and Verner, 1978). In the Short-billed Marsh Wren, three additional measures of versatility were introduced. Those were the song type and transition versatility (the number of different song types and transitions in a sequence of songs, respectively) and the total versatility (product of song type and transition versatility). Males did not adjust song lengths for different occasions, and both song rates and percentage performance time were highly correlated with all three measures of versatility (Kroodsma and Verner, 1978). Thus, data from within a species avoid many of the interspecific comparability problems and support the continuity–versatility hypothesis very strongly.

In Short-billed Marsh Wrens, males alter the continuity and versatility of their singing performance, presumably depending upon the context; one carefully studied male significantly increased both measures when countersinging with another male. Other wren species may increase continuity and versatility when courting and actively chasing a female (Kroodsma, 1977). Detailed studies are needed among a variety of species before the functional significance of this continuity–versatility relationship can be understood thoroughly, but one of the most viable hypotheses is that males are attempting to reduce habituation or, alternatively, increase stimulation by presenting a highly contrasting performance during critical interactions.

C. Antihabituation: A Candidate Rule for Song Organization?

If natural selection favors a highly contrasting vocal performance, a number of predictions can be made with regard to the style of singing:

(1) Successive song types, especially in immediately versatile singers, should be more different from one another than song-type pairs drawn randomly from the repertoire. Some limited evidence does support this prediction. Verner (1975) examined 15 parameters in Long-billed Marsh Wren songs, and found that two of the four most noticeable characteristics were especially different in successive songs. Similarly, favored song type pairs in Rock Wrens and Varied Thrushes appear especially dissimilar (Kroodsma, 1975; Whitney, 1979).

(2) The same or especially similar song types should be regularly spaced throughout a song performance. Beck (1971), Todt (1975), and others have shown that renditions of the same song type are more regularly than randomly spaced during a singing performance (see Section II,B). Dobson and Lemon (1977) did not address this point directly, but presented highly regular song sequences (e.g., ABCDEABCDE . . .) for several Swainson's Thrushes where especially similar songs did not follow one another.

(3) The songs within a repertoire should be especially dissimilar. Careful choices during song development can lead to high stimulus contrast, but the extent to which this occurs awaits a critical test (see Section III,A).

(4) During intense interactions, contrast between successive songs should increase. Under these conditions, wrens which normally sing with eventual variety sing fewer repetitions of a song type before advancing to the next type, or song patterning may switch to that of immediate variety, where successive songs are all of a different type. Also, wren species living in dense populations, where interactions are likely to be more frequent and intense, use immediate variety more than do wren species living in populations of reduced wren density (Kroodsma, 1977; Kroodsma and Verner, 1978). Thus, the relative contrast in a vocal signal is undoubtedly a direct indication of the motivational state of the vocalizer.

(5) Listeners should show renewed interest when a singer changes song types. Playback experiments tend to confirm this (Krebs, 1976; Yasukawa, 1981).

(6) Immediate repetition of a song type may be inhibited but eventually facilitated if other song types occur in the intervening time. In Rock Wrens, "AA" sequences actually require more time than do "ABA" sequences (Kroodsma, 1975).

Several attempts have been made to incorporate some or all of these features of observed or predicted singing behaviors into a descriptive model. Hinde (1958) first discussed the roles of inhibition and facilitation in singing of the Chaffinch (*Fringilla coelebs*) and Todt (1975) developed a model which incorporated a short-term inhibition ("component of throttling back") and periodic and regular recurrences of particular song types in a repertoire. Whitney (1979) has addressed most of these issues directly by developing a "software" model of how song is controlled in Varied Thrushes. In this species, males have three to seven different songs and typically sing with immediate variety. All songs must "compete" for the same output (Slater, 1978), and a given song type is sung only when its corresponding controlling unit reaches a threshold. The expression of a given song control unit results in immediate self-inhibition *and* inhibition of other song control units in relation to the similarity of songs controlled by those

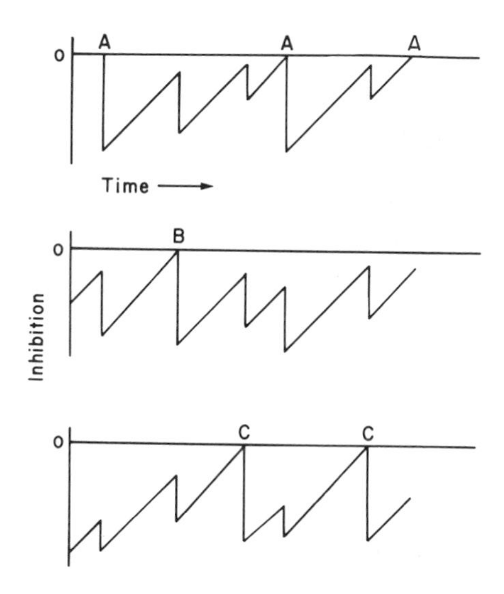

Fig. 3. A "software" model of the singing behavior of the Varied Thrush. Three song control units are depicted, each controlling one song type, either A, B, or C. When a given song control unit reaches threshold (i.e., an inhibition of zero), the song type controlled by that unit is uttered. Song control units inhibit not only themselves but also, to a lesser degree, other song units in relation to the similarity of the songs controlled by those units. Here, songs A and B are more similar to one another than are A and C or B and C. The song sequence produced here is ABCACA. (After Whitney, 1979.)

units (Fig. 3). Controlling units gradually recover until they once again reach threshold. By making further provisions for a low level of random variability and asymmetrical inhibition between song units, the model mimicked a number of the most distinctive features of Varied Thrush singing behavior.

Apparent exceptions to this antihabituation hypothesis remain perplexing, though; the Carolina Wren (*Thryothorus ludovicianus*), for example, may have a repertoire of 30 to 40 song types, yet repeat one song type 100 times before switching to a second type (see Chapter 6, Volume 1). By human standards, such a singing behavior is very monotonous. It would be reassuring if versatility (either song type or transition versatility; Kroodsma and Verner, 1978) and singing rates were correlated in the singing of individual Carolina Wrens, but such a habituation-prone performance is certainly contrary to that expected by the "monotony-threshold," or "antihabituation," hypothesis. Clearly, further refinements of this hypothesis together with a detailed examination of singing behaviors of wild birds will be necessary before it can be concluded that antihabituation is the major controlling factor in the organization and use of song repertoires.

IV. CONCLUDING REMARKS

In this chapter I have discussed what I perceive to be some of the current problems in the study of song repertoires in birds. Estimating repertoires of song types in birds can be a difficult and time-consuming task, but both careful estimates and definitions of the units of analysis are essential before it is possible to use the comparative approach in studying correlations between vocal repertoires and breeding biologies. Further *descriptive* work is also needed to document exactly how birds use their song types under different circumstances. There is no substitute for intensive recording of individuals, pairs, or communities of males which are using different song types or song patterning under different circumstances. Both the nature of the singing and the "intended" recipient must be documented in nature. Description and correlation are not enough, of course, but description is not a second-class approach to that of experimentation; rather, it is a fundamental prerequisite to any experimental approach.

Understanding how birds perceive and develop their different repertoire components can best be approached experimentally. If repertoires of dissimilar and therefore contrasting song types minimize habituation, then playback experiments to territorial males should readily demonstrate a greater response to contrasting than to nondistinctive song types. Using techniques developed by Searcy and Marler (1981), female songbirds can be used to assess the intersexual components of these vocal displays. In developmental studies in the laboratory, juveniles should select dissimilar song types from an array of model songs, and especially dissimilar songs should follow one another during a singing performance. Thus, there exists a diversity of experimental techniques for assessing the significance of both the existence and the manner of use of these song type repertoires.

ACKNOWLEDGMENTS

I thank J. R. Krebs and C. L. Whitney for comments on Section III, A. This chapter was written while I was supported by the National Science Foundation (BNS78-02753 and BNS80-40282).

REFERENCES

Beck, R. M. (1971). Sequence patterning of Wood Thrush song. *Am. Zool.* **11,** 66. (Abstr.)

Bertram, B. (1970). The vocal behaviour of the Indian Hill Mynah, *Gracula religiosa. Anim. Behav. Monogr.* **3,** 79–192.

Borror, D. J., and Gunn, W. W. H. (1965). Variation in White-throated Sparrow songs. *Auk* **82,** 26–47.

Catchpole, C. K. (1980). Sexual selection and the evolution of complex songs among European warblers of the genus *Acrocephalus. Behaviour* **74,** 149–166.

Dobson, C. W., and Lemon, R. E. (1975). Re-examination of monotony threshold hypothesis in bird song. *Nature (London)* **257,** 126–128.

Dobson, C. W., and Lemon, R. E. (1977). Markovian versus rhomboidal patterning in the song of Swainson's Thrush. *Behaviour* **62,** 277–297.

Eberhardt, D., and Baptista, L. F. (1977). Intraspecific and interspecific song mimesis in California Song Sparrows. *Bird-Banding* **48,** 193–205.

Fagen, R. M., and Goldman, R. N. (1977). Behavioural catalogue analysis methods. *Anim. Behav.* **25,** 261–274.

Ficken, M. S., and Ficken, R. W. (1973). Effect of number, kind and order of song elements on playback responses of the Golden-winged Warbler. *Behaviour* **46,** 114–128.

Good, I. J. (1953). The population frequencies of species and the estimation of population parameters. *Biometrika* **40,** 237–264.

Harris, M., and Lemon, R. E. (1972). Songs of Song Sparrows (*Melospiza melodia*): Individual variation and dialects. *Can. J. Zool.* **50,** 301–309.

Hartshorne, C. (1956). The monotony-threshold in singing birds. *Auk* **83,** 176–192.

Hartshorne, C. (1973). "Born to Sing. An Interpretation and World Survey of Bird Song." Indiana Univ. Press, Bloomington.

Hinde, R. A. (1958). Alternative motor patterns in Chaffinch song. *Anim. Behav.* **6,** 211–218.

Howard, R. D. (1974). The influence of sexual selection and interspecific communication on Mockingbird song (*Mimus polyglottos*). *Evolution* **28,** 428–438.

Isaac, D., and Marler, P. (1963). Ordering of sequences of singing behavior of Mistle Thrushes in relationship to timing. *Anim. Behav.* **30,** 344–374.

Krebs, J. R. (1976). Habituation and song repertoires in the Great Tit. *Behav. Ecol. Sociobiol.* **1,** 215–227.

Krebs, J. R. (1977). The significance of song repertoires: The Beau Geste hypothesis. *Anim. Behav.* **25,** 475–478.

Krebs, J. R., and Kroodsma, D. E. (1980). Repertoires and geographical variation in bird song. *In*

"Advances in the Study of Behavior" (J. S. Rosenblatt, R. A. Hinde, C. Beer, and M.-C. Busnel, eds.), Vol. 11, pp. 143–177. Academic Press, New York.

Kroodsma, D. E. (1974). Song learning, dialects, and dispersal in the Bewick's Wren. *Z. Tierpsychol.* **35,** 352–380.

Kroodsma, D. E. (1975). Song patterning in the Rock Wren. *Condor* **77,** 294–303.

Kroodsma, D. E. (1976). Reproductive development in a female songbird: Differential stimulation by quality of male song. *Science* **192,** 574–575.

Kroodsma, D. E. (1977). Correlates of song organization among North American wrens. *Am. Nat.* **111,** 995–1008.

Kroodsma, D. E. (1978). Continuity and versatility in bird song: Support for the monotony threshold hypothesis. *Nature (London)* **274,** 681–683.

Kroodsma, D. E. (1979). Vocal dueling among male Marsh Wrens: Evidence for ritualized expressions of dominance/subordinance. *Auk* **98,** 506–515.

Kroodsma, D. E. (1980). Winter Wren singing behavior: A pinnacle of song complexity. *Condor* **82,** 357–365.

Kroodsma, D. E., and Parker, L. D. (1977). Vocal virtuosity in the Brown Thrasher. *Auk* **94,** 783–785.

Kroodsma, D. E., and Verner, J. (1978). Complex singing behaviors among *Cistothorus* wrens. *Auk* **94,** 703–716.

Lein, M. R. (1978). Song variation in a population of Chestnut-sided Warblers (*Dendroica pensylvanica*): Its nature and suggested significance. *Can. J. Zool.* **56,** 1266–1283.

Lemon, R. E., and Chatfield, C. (1971). Organization of song in Cardinals. *Anim. Behav.* **19,** 1–17.

Lemon, R. E., and Chatfield, C. (1973). Organization of song of Rose-breasted Grosbeaks. *Anim. Behav.* **21,** 28–44.

Martin, E. J. (1977). Songs of the Fox Sparrow. I. Structure of song and its comparison with other Emberizidae. *Condor* **79,** 209–221.

Merriam-Webster (1973). "The New Merriam-Webster Pocket Dictionary." Simon & Schuster, New York.

Mulligan, J. A. (1966). Singing behavior and its development in the Song Sparrow, *Melospiza melodia. Univ. Calif., Berkeley, Publ. Zool.* **81,** 1–76.

Searcy, W. A., and Marler, P. (1981). A test for responsiveness to song structure and programming in female sparrows. *Science* **213,** 926–928.

Shiovitz, K. A. (1975). The process of species-specific song recognition by the Indigo Bunting, *Passerina cyanea,* and its relationship to the organization of avian acoustical behavior. *Behaviour* **55,** 128–179.

Slater, P. J. B. (1978). A simple model for competition between behaviour patterns. *Behaviour* **67,** 236–257.

Todt, D. (1975). Short term inhibition of outputs occurring in the vocal behaviour of Blackbirds (*Turdus merula m.* L.). *J. Comp. Physiol.* **98,** 289–306.

Verner, J. (1975). Complex song repertoire of male Long-billed Marsh Wrens in Eastern Washington. *Living Bird* pp. 263–300.

Whitney, C. L. (1979). The control of singing in Varied Thrushes. Ph.D. Thesis, Univ. of British Columbia, Vancouver.

Wildenthal, J. L. (1965). Structure in primary song of the Mockingbird (*Mimus polyglottos*). *Auk* **82,** 161–189.

Yasukawa, K. (1981). Song repertoires in the Red-winged Blackbird (*Agelaius phoeniceus*): A test of the Beau Geste hypothesis. *Anim. Behav.* **29,** 114–125.

6

Microgeographic and Macrogeographic Variation in the Acquired Vocalizations of Birds

PAUL C. MUNDINGER

I. INTRODUCTION

Morphologists have had centuries of experience describing and cataloging avian species by analyzing selected morphological traits. For the past few decades, ethologists have embarked on a similar path of describing and cataloging bird vocalizations. But in many respects the task facing the ethologist is more formidable than that facing the morphologist. In large part this is because behavioral variations are often extreme, and different species vary greatly in how much

ACOUSTIC COMMUNICATION IN BIRDS
VOLUME 2

ISBN 0-12-426802-1

their vocal patterns are learned, or in how their vocal behaviors vary regionally and locally. How can one organize and analyze diverse behavioral data of this kind and arrive at meaningful generalizations? The classic ethological approach has been to treat behavioral characters like morphological ones. This has been a useful strategy. The expectation is that after vocal traits have been described, categorized, and compared, we can then deduce generalizations from the comparisons. However, it is worth examining this morphological parallel in more detail.

A standard morphological procedure is to measure a trait such as wing length, then analyze the mensural data statistically. The sample variance is attributed to differences in genotype, in environmental effects, and in gene–environment interactions. The environment, acting as a component of natural selection and/or through its effect on development, is clearly an organizing factor, and many morphological traits vary in concordance with environmental factors, e.g., clinally.

Clinal variation in bird vocalizations might be expected to be common as well. But in contrast to morphological traits many bird vocalizations include a new and important source of variation—tradition—the effects of which can disrupt or obscure clinal tendencies. In contrast to *Homo sapiens,* for whom flattened heads, scarification, and foot-binding are examples of traditional modifications of morphological traits, avian morphological characters presumably are not affected by tradition. However, many avian vocal characters certainly are affected by socially learned traditions since many birds acquire their vocal patterns by imitating conspecifics. So this important factor, tradition, must be accounted for in the analysis of geographic variation of vocal behavior.

There are at least two ways of treating the effects of vocal traditions. One is to try to eliminate the effects by focusing on the basic structure of vocalizations. *Basic structure* refers to the fundamental, underlying, species-typical features of a vocalization, uninfluenced by tradition (Mundinger, 1979). One way to conceptualize basic structure is to examine a sample of songs uttered by birds reared individually in social isolation. A description of a sample of isolated song patterns is comparable to morphological structure since tradition has been eliminated as a factor. But description and analysis of geographic variation using such techniques are arduous at best. Eggs or nestlings collected from various regions would have to be reared in acoustical isolation. Description and analysis of their song patterns might then reveal clinal patterns of variation in such fundamental features as frequency measures, duration, tonal characteristics, or the pulsing of sound energy. I am unaware of such analyses so clinal variation will not be an important aspect of this review. But it is important to point out that the effect of social learning may be extremely limited or absent for some species, and in such instances songs of wild birds might represent basic structure, and thus clinal variation may be revealed. This appears to be the case for some species of *Myiarchus* flycatchers. Lanyon (1960, 1978) has documented instances of clinal variation in the carrier frequency used in the songs and calls of the Mexican Crested Flycatcher (*Myiarchus tyrannulus*) and the Olivaceous Flycatcher (*M. tuberculifer*).

A second way to cope with tradition is to focus on it and on its effects on variation. The advantage of this approach is the presence of a good data base: many species learn their specific call and song patterns; many local populations are characterized by population-specific features in their acquired vocal patterns (dialects); and the fundamental mechanisms of vocal learning are fairly well understood for a number of species (Marler and Mundinger, 1971; see also Chapter 1, this volume). And since vocal traditions have been reported from a wide variety of species, this second approach may provide sufficient comparative material from which to derive some basic generalizations. With that thought in mind, the special focus of this article is the analysis of micro- and macrogeographic variation in learned call and song patterns. Microgeographic variation refers to differences among neighboring populations whose close geographic proximity makes direct interbreeding and social interaction a potential consideration. Macrogeographic variation refers to differences between populations that are separated by distances (or other barriers) sufficiently great as to inhibit these populations from directly mixing either their genes or their vocal traditions on a regular basis.

II. MICROGEOGRAPHIC VARIATION

A. The Dialect Problem

The concept of a vocal dialect is both a potentially useful analytic tool and a problem. On the one hand, a description of microgeographic variation of a socially shared song or call pattern might logically employ the dialect concept since dialects were originally conceived of as situations where vocal features are shared by members of a local population of birds (Sick, 1939; Poulsen, 1951; Marler, 1952; Marler and Tamura, 1962). But on the other hand, the dialect concept poses serious problems. It is a concept that is not accepted by all workers, several of whom have serious doubts about the reality, value, and significance of dialects (e.g., Smith, 1972). Also, there is no uniformly used and accepted definition. Some workers equate song or call sharing with dialects, whereas others do not. Therefore, the alleged existence or absence of dialects cannot always be accepted at face value.

It is possible that the dialect problem is largely the result of not having a uniform definition, one that can be applied to any species by any investigator. At present there are several alternative definitions. Early papers on the subject (Marler, 1952; Marler and Tamura, 1962) focused on the phenomenon of shared song features or song patterns that seemed to characterize a local population. But several investigators have since shown that although individual birds may share partial or entire song patterns, this does not necessarily indicate dialects, at least as these investigators conceive of them (Thompson, 1970; Romanowski, 1978; Bitterbaum and Baptista, 1979; Kreutzer, 1979). Thielcke (1965, 1969) defined dialects as vocal variants with a mosaic distribution. This "mosaic" definition is a general one and has been widely used, but most studies of microgeographic variation are not sufficiently extensive in geography to determine whether mosaic patterns are present or not. So in many such instances investigators have employed *ad hoc* definitions which might not be generalized to other species. Thus the dialects of the White-crowned Sparrow (*Zonotrichia leucophrys;* Marler and Tamura, 1962) and Rufous-collared Sparrow (*Z. capensis;* Nottebohm, 1969) are defined on the basis of the terminal trill in their songs. Cardinal (*Cardinalis cardinalis;* Lemon, 1965, 1966), Winter Wren (*Troglodytes troglodytes;* Kreutzer, 1974), House Finch (*Carpodacus mexicanus;* Mundinger, 1975), and Bobolink (*Dolichonyx oryzivorus;* Avery and Oring, 1977) dialects are based on sharing of certain syllable repertoires or syllable sequences; the dialects of the Splendid Sunbird (*Nectarinia coccinigastra;* Grimes, 1974) were originally determined on the basis of shared temporal features in the songs.

Another attempt at a general definition focuses on the presence of dialect boundaries (Mulligan, 1975), and some investigators have made an effort to identify or describe dialect boundaries between contiguous populations (Sick, 1939; also cited in Thielcke, 1969; Bjerke, 1974; Kreutzer, 1974; Baker, 1975; Baptista, 1975; Mundinger, 1975; McGregor, 1980). But, as they are reported in the literature, the behaviors of few wild populations of birds actually fit this particular dialect concept—namely, that dialects be temporally stable where contiguous populations meet, and the integrity of the respective dialects be maintained with a boundary between them. A race of the White-crowned Sparrow (*Z. l. nuttalli;* Marler and Tamura, 1962, 1964; Baker, 1975; Baptista, 1975), some eastern populations of the House Finch (Mundinger, 1975), the Redwing (*Turdus iliacus;* Bjerke, 1974), and the Corn Bunting (*Emberiza calandra;* McGregor, 1980) seem to fit this concept, but many species do not show such clear-cut boundaries between populations. This is especially evident in birds, such as wrens, characterized by complex song repertoires of many different song types per male (e.g., Kreutzer, 1974; Kroodsma, 1974). Each song type may have its own unique pattern of geographic distribution. Where can a dialect boundary be drawn if all song types are considered simultaneously?

B. A Modified Linguistic Definition of Dialect

1. *Description*

The word "dialect" was borrowed from linguistics, and it is instructive to determine how linguists identify speech dialects. Although the descriptive study of speech dialects *per se* is no longer a major focus in modern linguistics, the field of dialect geography (the mapping of speech dialect distributions and boundaries) is still an important component of descriptive linguistics. Once speech dialects have been mapped and identified they can then be more productively analyzed.

The classic methods of dialect geography are outlined in standard texts (e.g., Bloomfield, 1933; Kurath, 1972). These reveal differences between the linguist's definition of speech dialect and the ornithologist's definition of song dialect. Boundaries between dialects are important in both instances, but the linguist's concept of boundary is different from the ornithologist's. Borrowed words and individual differences in vocabulary, both common occurrences in man, are often disruptive of clear-cut, easily recognized, boundaries.

The boundaries of human speech dialects are somewhat quantitative in that they are based on a number of different isoglosses (also called heteroglosses)* which are lines on a map that separate related speech variants. Isoglosses are the key to mapping and identifying speech dialect boundaries. A dialect boundary is simply a bundle of isoglosses, and wherever the map shows a number of different isoglosses running close together and in parallel a dialect boundary is indicated. If the isoglosses meander independently, or fan out into a loose network, then no dialect boundary is present. Large bundles identify major dialect boundaries, and small bundles minor ones.

Figure 1 illustrates several major dialect boundaries in Italy. Two major bundles mark the boundaries of three major Italian dialects. The northern boundary runs along a major mountain range. Over time, this geographical barrier has limited the social contact between northern and central peoples, leading to the dialect diversification observed today.

The methods of dialect geography can be fruitfully applied to avian song traditions. For example, sonagrams can readily be used to identify vocal variants, whose distributions can be mapped with avian isoglosses. Methods are straightforward. An adequate sample of birds is recorded, and the birds' positions are mapped. Sonagrams are visually inspected for similarities and differences in the fine details of syllable patterns. This often identifies two kinds of variants, pronunciation differences and vocabulary differences. Similar pronunciations of a given syllable type are mapped by an isogloss called an isophone;

*A short glossary of linguistic terms and symbols is provided on p. 200.

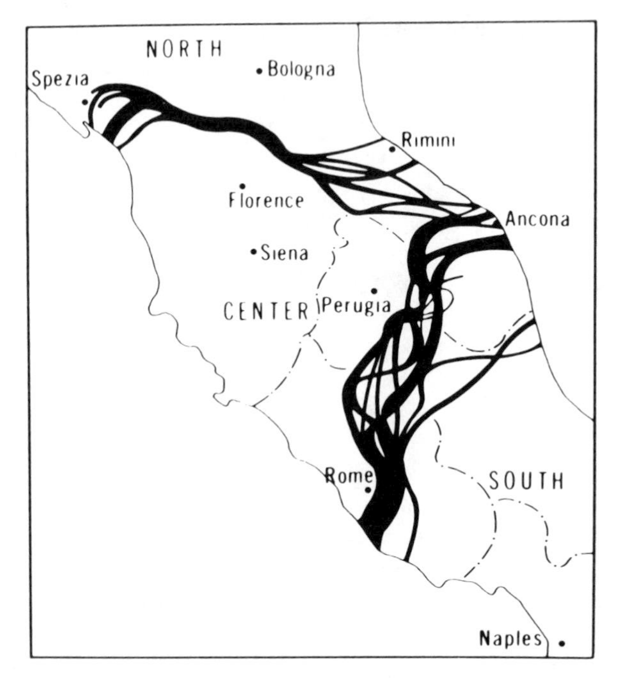

Fig. 1. Human speech dialects. Isogloss bundles, based on 7 northern and 11 southern features, delineating the three major dialect areas of Italy. (From Kurath, 1972.)

different pronunciations of the same syllable (e.g., compare the forms of syllable *u* in Figs. 2A–C) are mapped by different isophones. Isophones are most important in microgeographic analyses where pronunciation differences predominate. Vocabulary differences (e.g., compare syllables *r–z,* which are generally common to the vocabularies represented in Figs. 2A–C, to the totally different set of syllables illustrated in Fig. 2D) are marked by isoglosses called isolexes. Isolexes will be discussed more fully in the section on macrogeographic variation.

When the geographic distributions of many different phonetic or lexical (vocabulary) variants are plotted, the isoglosses may run close together and in parallel in some areas. These are avian isogloss bundles, and they mark song dialect boundaries. For example, eastern populations of the House Finch have local song dialects (Mundinger, 1975) (Fig. 3A). Recently, the songs used in that study were reanalyzed using isoglosses to map the distribution of pronunciation and vocabulary variants (Fig. 3B). The two analyses identified the same dialects and boundaries, but the isogloss analysis provided additional, important information. Thus Fig. 3B reveals that some boundaries are major ones, while others are minor. Also some dialects are related to one another in that they share the same syllable vocabularies and can even share some pronunciations, whereas other

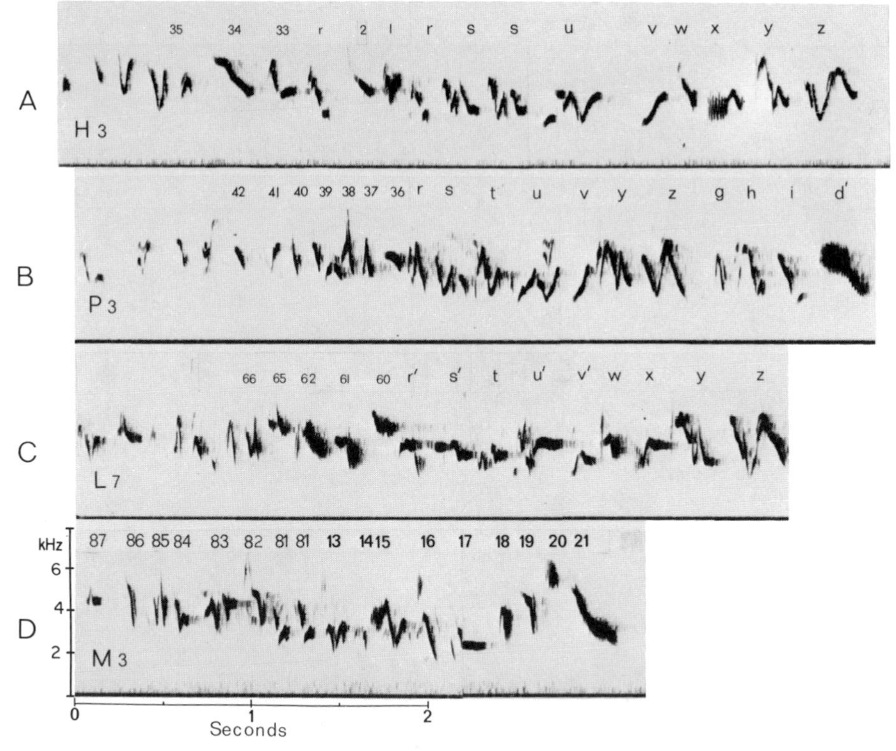

Fig. 2. Representative song patterns of four different House Finch populations. (A) Harrison dialect pattern; (B) Peningo dialect pattern; (C) Larchmont dialect pattern; (D) pattern from a song belonging to a song institution (see text) found in Mamaroneck, New York, and unrelated to the song institution represented by A, B, and C. (From Mundinger, 1975.)

regions contain birds singing a totally different syllable vocabulary (see cross-hatched area in Fig. 3B). Finally, members of a given local population of males may not all be characterized by the same song pattern(s). Some minor individual variation almost always exists. This is reflected by the meanderings of some isoglosses in the interior of some dialect areas. None of this information was present in the original analysis (Fig. 3A).

When interpreting Fig. 3B it is important to realize that the isogloss lines do not map bird positions, nor song positions, but rather they map the distribution of many different syllable variants. In essence one is looking at geographic clustering within a pool of song syllable types. The bundles mark subsets of different syllable types that cluster together and presumably function together in communication.

In sum, methods of human dialect geography can be applied, somewhat modi-

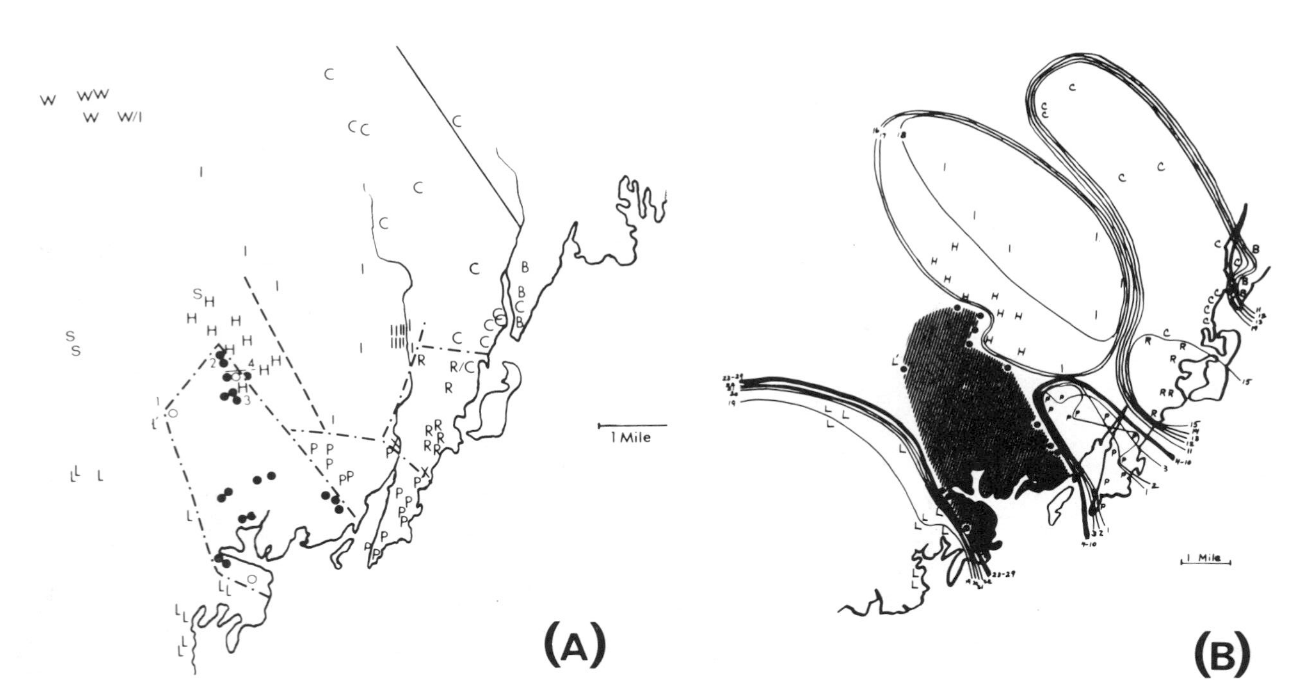

Fig. 3. House Finch song dialects. (A) Dialects as determined by comparing song patterns. Each letter represents a male singing a characteristic set of song patterns it shares with neighboring males; solid circles represent males from the Mamaroneck population which has a unique syllable repertoire (from Mundinger, 1975). (B) Dialects determined by isogloss analysis (see text). Crosshatched area represents the Mamaroneck institution; the letters represent birds belonging to a set of related dialects from a second song institution. Thin lines are individual isoglosses, thick lines are bundles of six isoglosses.

fied, to avian song data. The resultant analysis yields a picture of geographic variation that is comparable to, but is more detailed than, results obtained from previously used methods. And the method is compatible with the following general definition: a *song dialect* is a *variant song tradition* shared by members of a local population of birds, with a dialect *boundary* delineating it from other variant song traditions. Note that this definition does not require the sharing of entire song pattern(s), but only that traditionally learned song components characteristically appear in the songs of a local population. This is very much like Marler's (1952) early definition. What is added is the concept of a boundary, which can be determined by employing the methods of dialect geography. Finally, note two other important features: (1) Song dialects are populations of behavior patterns (i.e., song syllables); they are not populations of birds. (2) Birdsong dialects, like speech dialects, are learned traditions in contrast to the "dialects" of honeybees (*Apis mellifera*) which are genetically programmed behavioral variants (Lindauer, 1971).

2. *General Application*

The isogloss method is readily generalizable to any species with a vocal learning capacity. However, certain preconditions must be met. Maps giving the location of individual birds, sonagrams of songs (or the equivalent) to provide individual syllable repertoires, and adequate sampling techniques are all necessary. Sampling is important because a small sample of geographically dispersed males, or an uneven sampling of song or syllable types, could give spurious and misleading results.

To assess its applicability, the isogloss method is applied to the songs of four species previously analyzed with alternative methods: the Little Hermit (*Phaethornis longuemareus;* Wiley, 1971); the House Finch (Mundinger, 1975); Bewick's Wren (*Thryomanes bewickii;* Kroodsma, 1974); and the Hill Myna (*Gracula religiosa;* Bertram, 1970). This is a phylogenetically diverse group representing a non-passerine and three distantly related passerine families. Behaviorally, this sample also represents a broad spectrum of singing behavior: the myna shares call syllables extensively, but these vocal patterns are not identified as songs; the Little Hermit has a repertoire of one song per male; the House Finch has several themes (2–6) per male; Bewick's Wren has a large song repertoire (16 themes per male). Song dialects were reported for the Little Hermit and the House Finch but not for the Hill Myna or Bewick's Wren, for which Kroodsma (1974) found syllable and song theme sharing but no clear evidence of dialects.

Figure 4 illustrates the pattern of microgeographic variation in Little Hermit song syllable patterns. The bird positions are represented by uppercase letters A–Y (following Wiley, 1971). All but bird "Q" sang songs composed of the same four or five syllable types, designated *a–e* in Fig. 5. Wiley's spectrograms revealed several forms of the same syllable types: I identified five forms of

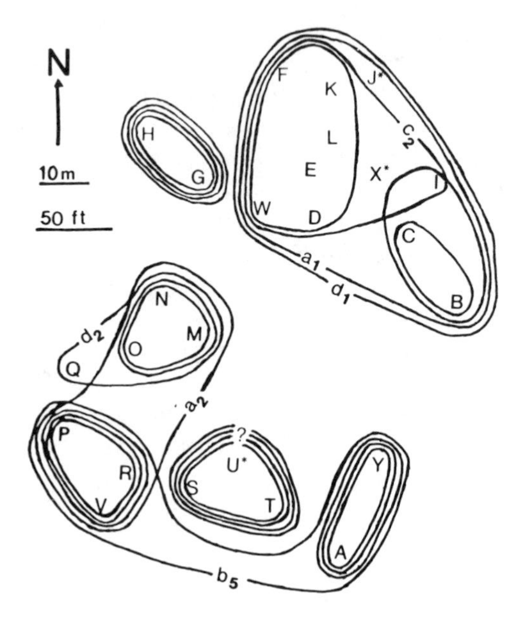

Fig. 4. Isogloss analysis of Little Hermit (Trochilidae) songs. Capital letters map bird positions, and an asterisk indicates no song was recorded from that individual. Lower case letters with subscripts identify the isoglosses for some of the widespread syllable forms. (Adapted from Wiley, 1971.)

syllable a—a_1, a_2, a_3, a_4, a_5; syllable c appeared in seven alternative forms, etc. (see Fig. 5). Isoglosses are drawn to illustrate the geographic distribution of the different syllable variants. All birds sharing a_1 are delineated by one line, and the groups of birds sharing b_2, c_2, and d_1 are each delineated by other lines. I was unable to differentiate consistently forms of e. The analysis reveals that the population studied by Wiley is grouped into a number of major and minor song dialects (see Fig. 4). For example, birds B, C, D, E, F, I, K, L, and W are separated from other birds by a major dialect boundary of four isoglosses. (Birds J and X are within these boundaries but their songs were unrecorded.) This group is then further subdivided by a minor boundary, and only a small cluster of neighboring males share the same song pattern, e.g., birds F, W, K, L, E, D share the song $a_1b_2c_2d_1e$. Some disjunct groups are linked by sharing very similar or almost identical syllables, e.g., birds P, R, V and A, Y (Fig. 4). This grouping of these behavior patterns by isogloss bundles is consistent with Wiley's original analysis. However, isogloss analysis reveals such additional information as dialect boundaries, major versus minor boundaries, and quite different spatial patterns for the different variants of the syllable types constituting the song of the Little Hermit.

Behaviorally, the House Finch represents a more complex situation. Individual males have repertoires of about 25–40+ different syllables, which are organized into two to six themes. Mundinger (1975, 1980) reports that eastern populations near New York City have dialects. Bitterbaum and Baptista (1979) report that western local populations, in and near Los Angeles, California, share syllables as well, but lack dialects. Figure 3B shows that isogloss bundles demark the same set of local dialects that Mundinger (1975) previously identified using other methods, but that isoglosses add other, important details. The data in Bitterbaum and Baptista (1979) do not permit isogloss analysis, due to an insufficient sample of published sonagrams, but there is no reason to necessarily expect to find isogloss bundles. Conditions there, such as high mobility and an abundance of alternative models for learning, could produce a meandering meshwork pattern of isoglosses instead of identifiable bundles.

In another isogloss analysis of an eastern House Finch population, this one in the vicinity of Jones Beach, Long Island, isoglosses map the geographic range of syllable variants found in just one song theme (males there sing two or three themes). The map pattern of geographic variation shows 13 different Theme I sister dialects arranged in a patchwork mosaic, as opposed to a clinal pattern (Fig. 6). But the map fails to reveal that some of these sister dialects resemble one another more than they do others. Thus dialect 9 has many syllables with a fine structure (details in "shape" on a sonagram) strikingly like those in dialects 7 and 8. All 13 dialects are similar, however, since they all share a common syllable vocabulary. This general similarity is presumably due to a cultural evolutionary history that can be traced back to a common ancestor song pattern (Mundinger, 1980). The striking similarities of dialects 7–9 presumably reflect either the very recent historical separation of these dialects, or minimal cultural evolutionary divergence. The word "mosaic" does not reflect this historical component of geographic variation in acquired song patterns. Therefore, the descriptive phrase "historically variegated" is applied to describe a mosaic-like pattern of geographic variation involving *historically related* song dialects. This phrase also distinguishes this type of pattern both from a clinal distribution and from a patchwork of historically unrelated (i.e., independently derived) dialects.

Kroodsma (1974) studied a population of 108 territorial male Bewick's Wrens in Oregon, mapping their positions as in Fig. 7. He reported: (1) an average song repertoire of 16 different song patterns (themes) per male; (2) that the structure of some song patterns remained constant over several kilometers (e.g., six of the themes were invariant within his study area); (3) that other patterns appeared confined to small patches of habitat (e.g., six themes were quite variable, and one set of variations was restricted to the northeast portion of his study area); (4) that dialect areas based on a few songs seemed easy to establish, but the "dialect" concept was more complex if the dialect areas were evaluated by the degree of song change in all the song patterns taken simultaneously. In other

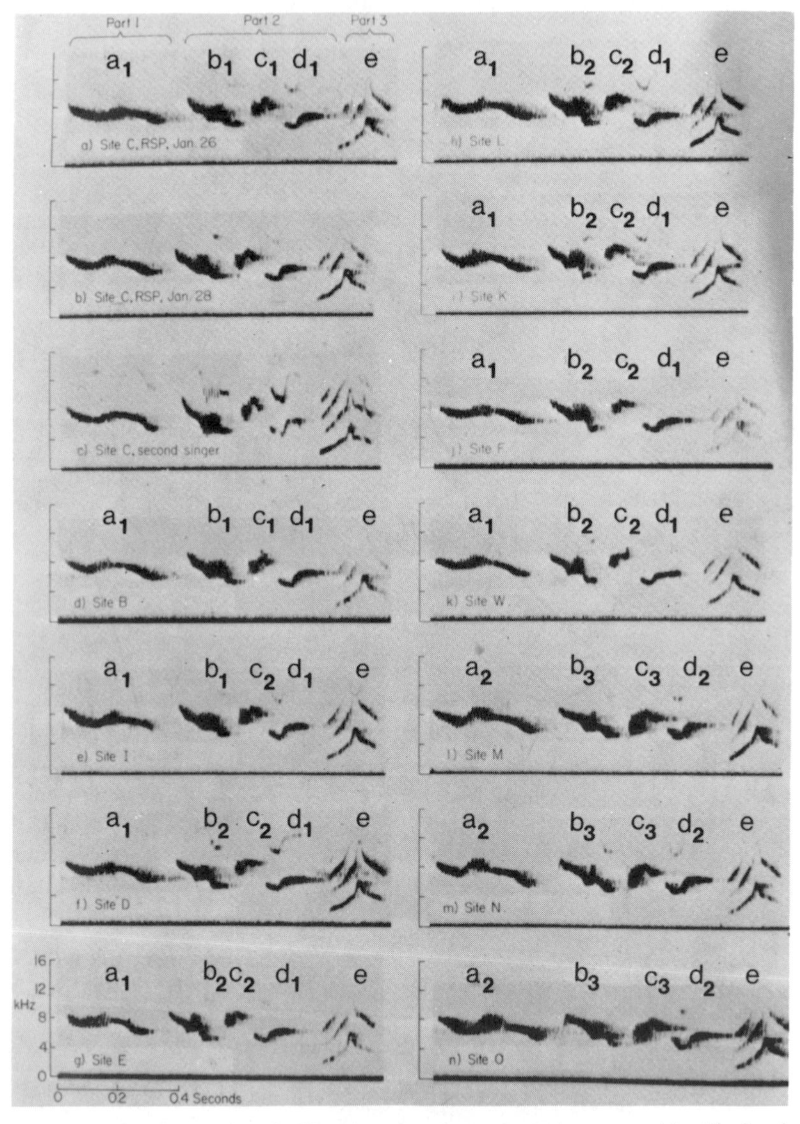

Fig. 5. Sonagrams of Little Hermit songs. Lower case letters *a–e* identify five basic syllable types. Subscript numbers identify the phonetic variant of each of the five basic syllable types. (Adapted from Wiley, 1971.)

words, the large song repertoires precluded an objective identification of a dialect boundary. The limited set of published sonagrams prevents a thorough

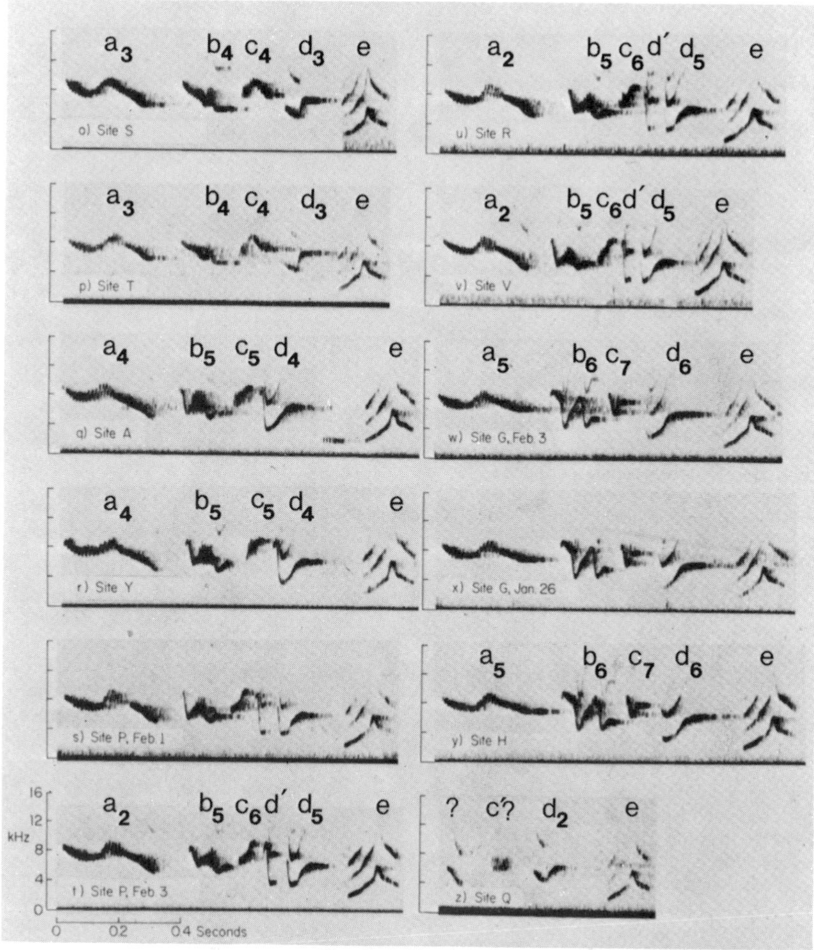

Fig. 5. *(Continued)*

isogloss analysis, but a table (Kroodsma, 1974, Table 3) lists those birds in the northeastern subpopulation sharing one or more of the six variable song patterns found there. Each of these themes was identified by one or more distinctive syllable types. Therefore, isoglosses mapping the distribution of these song patterns are each equivalent to one or more isophones. The distribution of these isoglosses (Fig. 7) reveals a fairly well defined bundle enveloping several birds in the northeastern part of the study area (i.e., birds 202, 275, 182, 32, 199, 180, 181, and 60). Although many different song themes were considered simultaneously, a boundary delimiting a variant song tradition is still apparent. Therefore, at least some local populations of Bewick's Wren in western Oregon exhibit song dialects.

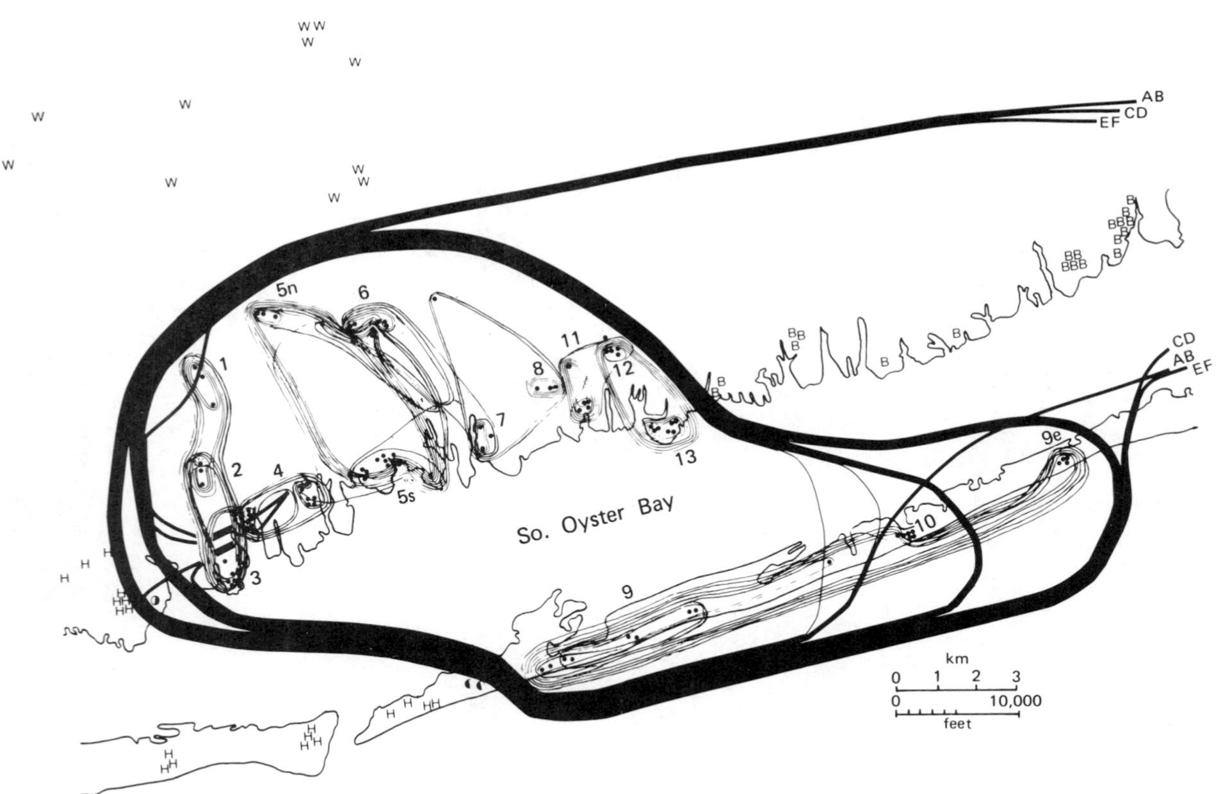

Fig. 6. Isogloss analysis of House Finch song patterns. Thick lines are isolexes (see Appendix) delineating the Merrick song institution. Thin lines within the institution boundary are isophones (see Appendix) which delineate 13 historically related dialects. Solid dots represent Merrick bird positions; the letters H, W, and B represent the positions of birds belonging to three other song institutions. (From Mundinger, 1980.)

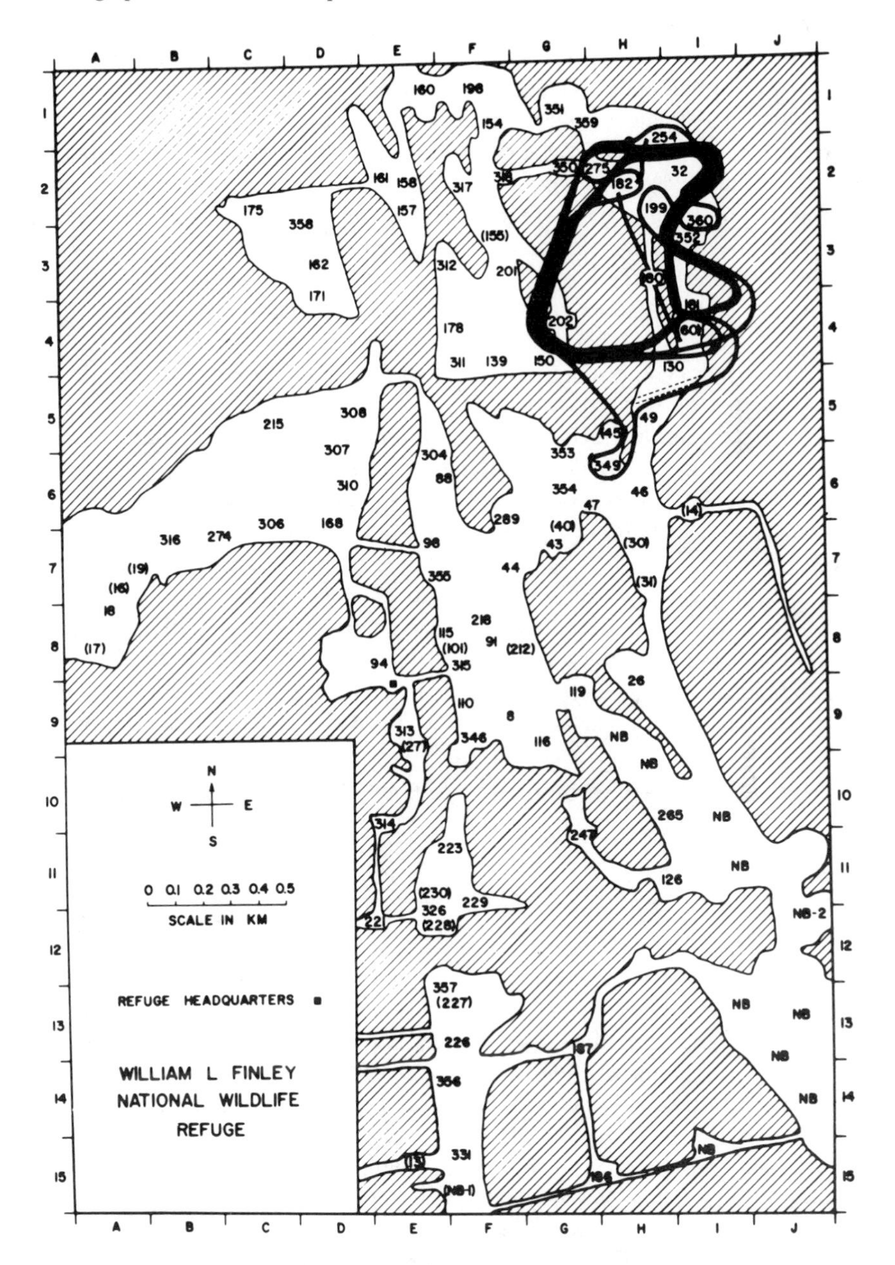

Fig. 7. Isogloss analysis of six song themes unique to the NE portion of a population of Bewick's Wrens studied by Kroodsma (1974). Numbers represent individual males. (Modified from Kroodsma, 1974.)

An isogloss analysis can also lead to the conclusion that no dialects exist even though vocal imitation is evident. Bertram (1970) studied vocal imitation in the Hill Myna. He found neighboring birds held one to several call patterns in common, males imitating neighboring males and females imitating females. Figure 8A illustrates the geographic pattern of call sharing among one member of each pair, presumably males. Figure 8B is the pattern for (presumed) females. There are no isogloss bundles; instead the pattern can be described as a series of variable and overlapping links, forming a chain. Although the links reveal syllable sharing among a number of males or females nesting in close proximity, the local population is not characterized by a common set of shared syllables.*

Examining still another study, no isogloss lines could be drawn for the Berkeley population of the Dark-eyed Junco (*Junco hyemalis oreganus*) since no sampled male exhibited syllable sharing (Konishi, 1964). Theoretically, "isoglosses" could be drawn, but would appear as a set of scattered points rather than lines. This junco can modify its song by learning, in this instance learning by improvisation, which masks a limited tendency for vocal imitation (Marler *et al.*, 1962).

In sum, the reanalysis of previously published data shows that the modified linguistic definition of song dialect can be applied to describe the patterns of geographic variation in the songs and calls of several different species. In some cases, dialect boundaries are delineated, in other instances syllable sharing is organized into patterns that lack discrete boundaries. The method of isogloss analysis appears to be independent of the complexity of singing behavior, and it seems to be capable of objectively differentiating a dialect boundary if one exists.

C. Generalizations

The effective data base now available for learned vocal traditions can be used to uncover some basic generalizations. A good portion of these data is summarized in Table I.

Table I lists those species for which there is some evidence of syllable sharing (or lack of it), or modification of vocal pattern by learning, or both. Most examples involve songs, but for some species the vocalizations studied were calls.

Evidence for syllable sharing (Table I, column 2) is usually based on reports of the original investigator, often supported by sonagrams. Any exceptions, such as inferences I have made from published sonagrams or other data, are bracketed in parentheses in Table I. This indicates that the original investigator had not

*A rather narrow concept of bundle is employed in this example; a less rigid one might alter the interpretation. The objective analysis of bundles is a task for the future.

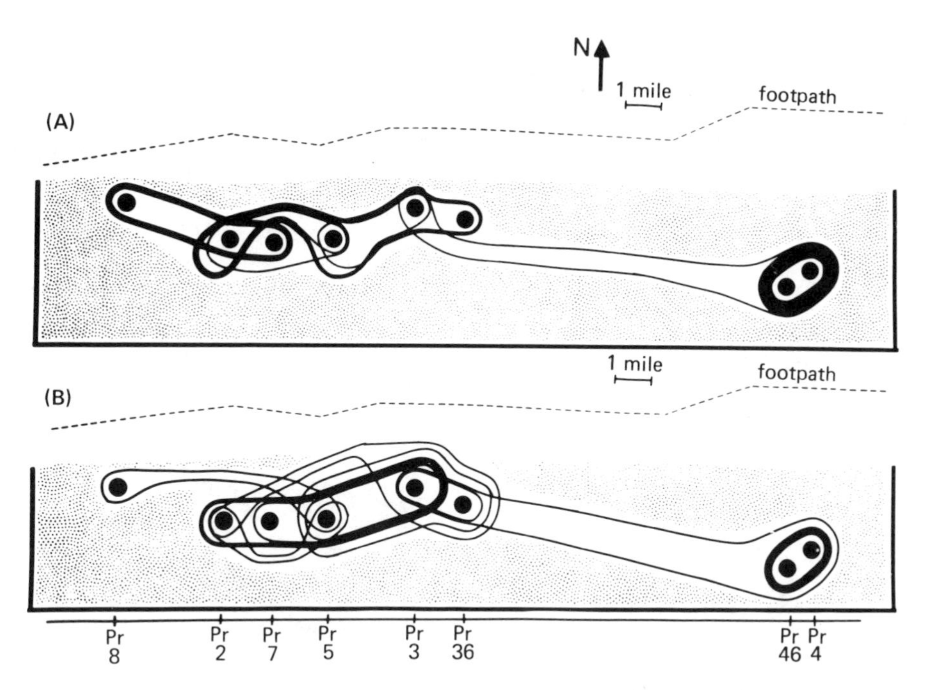

Fig. 8. Isogloss analysis of Hill Myna shared call patterns. Thickness of lines is proportional to the number of different isoglosses and shared calls, i.e., one isogloss per call pattern. Shaded area is forested hills; unshaded is open plains. (A) Presumed males; (B) presumed females. (Modified from Bertram, 1970.)

commented on the issue. Column 3 indicates studies in which song (or call) dialects were reported as present or absent, or where this information could be inferred (inferences are bracketed). However, many different definitions of song dialect are represented in the studies summarized in column 3, so criteria vary. Occasionally the original data in some of the studies could be reanalyzed with isoglosses permitting application of the modified linguistic definition and relatively uniform criteria (column 4). Finally, the last two columns indicate basic features of singing behavior: column 5 gives song repertoire size; and column 6 reveals if there is evidence for song learning other than the criterion of syllable sharing. This evidence was usually either from studies of song or call ontogeny, or reports of mimesis. (References to the studies summarized in Table I are found in its Appendix on pp. 175–176.)

Several generalizations follow from the data surveyed in Table I:

(1) Table I represents a taxonomically diverse assemblage. The 124 species surveyed represent 28 families, including four non-passerine and four sub-oscine

TABLE I

Sound Pattern Sharing in Avian Local Populations[a]

Species number[b] and species	Syllable sharing	Dialect	Bundles	Repertoire	Evidence of learning
Nonpasserines					
Tetraonidae					
1. *Tympanuchus cupido* (Greater Prairie Chicken)	+				+
2. *Phasianellus* (Sharp-tailed Grouse)	+				+
Phasianidae					
3. *Phasianus colchicus* (Ring-necked Pheasant)	(+?)			1	
Psittacidae					
4. *Amazona amazonica* (Orange-winged Parrot)	+	+			+
Trochilidae					
5. *Phaethornis longuemareus* (Little Hermit)	+	+	+	1	
6. *Calypte anna* (Anna's Hummingbird)	(+)				
Passerines (sub-oscines)					
Cotingidae					
7. *Procnias tricarunculata* (Three-wattled Bellbird)	(+)	+		3	
Tyrannidae					
8. *Myiozetes inornatus* (White-bearded Flycatcher)	+				

9. *Empidonax virescens* (Acadian Flycatcher)		(+)	−		1	
Menuridae						
10. *Menura novaehollandiae* (Superb Lyrebird)		+	(+)		Indeterminate	+
Atrichornithidae						
11. *Atrichornis clamosus* (Noisy Scrub-Bird)		+	(+)		Indeterminate	+
Passerines (Oscines)						
Alaudidae						
12. *Mirafra rufocinnamomea*[c] (Flappet Lark)		+	+	+	1	
13. *Galerida cristata* (Crested Lark)		+				+
Hirundinidae						
14. *Petrochelidon pyrrhonota* (Cliff Swallow)		(−?)	(−)			
Laniidae						
15. *Laniarius ferrugineus* (Boubou Shrike)	a.	+			Several	+
	b.	+	+			(+)
16. *L. erythrogaster* (Black-headed Gonolek)		+				+
17. *L. funebris* (Slate-colored Boubou Shrike)		+			Several	(+)
Troglodytidae						
18. *Salpinctes obsoletus* (Rock Wren)		+			69–119	
19. *Cistothorus platensis* (Short-billed Marsh Wren)		+	−		80–148	+

(continued)

TABLE I (*Continued*)

Species number[b] and species	Syllable sharing	Dialect	Bundles	Repertoire	Evidence of learning
20. *Cistothorus palustris* (Long-billed Marsh Wren)	+	+		107–114	
21. *Thryomanes bewickii* (Bewick's Wren)	+	(+)	+	16	+
22. *Thryothorus ludovicianus* (Carolina Wren)	+			1	
23. *T. felix* (Happy Wren)	(+)			2	
24. *T. sinaloa* (Bar-vented Wren)	(+)			2	
25. *Troglodytes troglodytes* (Winter Wren)	+	+		1–32+	
26. *Troglodytes aedon* (House Wren)	+				+
Mimidae					
27. *Dumetella carolinensis* (Gray Catbird)	+	+		Indeterminate	
28. *Mimus polyglottos* (Mockingbird)	+			Indeterminate	+
Muscicapidae (Turdinae)					
29. *Turdus merula* (Blackbird)	+				+
30. *T. iliacus* (Redwing)	+	+		1	+
31. *T. viscivorus* (Mistle Thrush)	+	(+?)		Indeterminate	

32. *T. rufopalliatus* (Rufous-backed Robin)		+	(−)	26	
33. *T. migratorius* (American Robin)	a.	−		Indeterminate	+
	b.	+			
Muscicapidae (Sylviinae)					
34. *Acrocephalus palustris* (Marsh Warbler)		+	+	Indeterminate	+
35. *Sylvia communis* (Common Whitethroat)		+	+		+
36. *S. cantillans* (Subalpine Warbler)			+		
37. *S. melanocephala* (Sardinian Warbler)		+	+		
38. *Phylloscopus collybita* (Chiffchaff)		+?		9–10	
39. *P. bonelli* (Bonelli's Warbler)		(+?)	+	5	
40. *P. tytleri* (Tytler's Leaf Warbler)		(+)		1–6	
41. *P. affinis* (Tichell's Leaf Warbler)		(−)	(−)		
42. *P. pulcher* (Orange-barred Leaf Warbler)		+			
43. *P. inornatus* (Yellow-browed Leaf Warbler)		+			
44. *P. maculipennis* (Gray-faced Leaf Warbler)		+		12	
45. *P. magnirostris* (Large-billed Leaf Warbler)		+		3	
46. *P. occipitalis* (Western Leaf Warbler)		+	(?)	5–7	

(*continued*)

TABLE I (*Continued*)

Species number[b] and species	Syllable sharing	Dialect	Bundles	Repertoire	Evidence of learning
47. *P. reguloides* (Blyth's Leaf Warbler)	(+)	+?		4–6	
48. *Regulus regulus* (Goldcrest)	+	+		18	
49. *R. ignicapillus* (Firecrest)	+	+		1–5	
50. *Cisticola hunteri* (Hunter's Warbler)	+			Several	
Paridae					
51. *Parus palustris* (Marsh Tit)	+	+		?–5	+
52. *P. montanus* (Willow Tit)	+	–		?–3	
53. *P. atricapillus* (Black-capped Chickadee)	(+)	(+)		1	
54. *P. carolinensis* (Carolina Chickadee)	+	+		1–4	
55. *P. ater* (Coal Tit)	+			6	+
56. *P. major* (Great Tit)	+	+		4–7	
57. *P. caeruleus* (Blue Tit)	+			1–3	
58. *P. inornatus* (Plain Titmouse)	+	+		10	
59. *P. bicolor atricristatus* (Tufted Titmouse)	+			10	
Certhiidae					
60. *Certhia familiaris* (Brown Creeper)	+	+		1	+

61. *C. brachydactyla* (Short-toed Tree Creeper)		+	+		1–(2)	+
Nectariniidae						
62. *Nectarinia coccinigastra* (Splendid Sunbird)		+	+	+	1	
Emberizidae (Emberizinae)						
63. *Emberiza calandra* (Corn Bunting)		+	+		2–3	
64. *E. citrinella* (Yellowhammer)		+	+		1–3	
65. *E. buchanani* (Stone-ortolan)		+	(+)		1	
66. *E. hortulana* (Ortolan Bunting)		+	+		1,(2–3)	
67. *E. cirlus* (Cirl Bunting)		+	–	–	1–4	+
68. *E. schoeniclus* (Reed Bunting)			–		Several	
69. *Zonotrichia iliaca* (Fox Sparrow)		+	(+)		2–7	
70. *Z. melodia* (Song Sparrow)	a.	+	–			+
	b.	+	+			
71a. *Z. leucophrys nuttalli* (White-crowned Sparrow)		+	+	+	1	+
71b. *Z. l. pugetensis* (White-crowned Sparrow)		+	+	+	1	
71c. *Z. l. oriantha* (White-crowned Sparrow)		+	+	+	1	
71d. *Z. l. gambelii* (White-crowned Sparrow)		+	–		1	

(*continued*)

TABLE I (*Continued*)

Species number[b] and species	Syllable sharing	Dialect	Bundles	Repertoire	Evidence of learning
72. *Z. capensis* (Rufous-collared Sparrow)	+	+	+	1	+
73. *Z. albicollis* (White-throated Sparrow)	−, (+)	−		1–(2)	+
74a. *Junco hyemalis* (Dark-eyed Junco)	+	+		1–7	
74b. *J. h. oreganus* (Dark-eyed Junco)	−	(−)		1–2	+
75. *J. phaeonotus* (Yellow-eyed Junco)	−	(−)		1–2	
76. *Ammodramus sandwichensis beldingi* (Savannah Sparrow)	+	+		1	
77. *Spizella passerina* (Chipping Sparrow)	+			1	+
78. *S. pusilla* (Field Sparrow)	(+)	−?		1	
79. *Pooecetes gramineus* (Vesper Sparrow)	+	+		Indeterminate	+
80. *Amphispiza bilineata* (Black-throated Sparrow)	+			7–9	
81. *A. belli* (Sage Sparrow)	+	(+)		1	
82. *Pipilo chlorurus* (Green-tailed Towhee)	+			Several	

83. *P. erythrophthalmus* (Rufous-sided Towhee)	a.	+	+	(−)	?–22	+
	b.	+	−		4	
84. *P. fuscus* (Brown Towhee)		−	(−)			
Emberizidae (Cardinalinae)						
85. *Pheucticus ludovicianus* (Rose-breasted Grosbeak)		+			Indeterminate	
86. *Cardinalis cardinalis* (Cardinal)		+	+		8–11	+
87. *C. sinuatus* (Pyrrhuloxia)		+	+		13–15	
88. *Passerina cyanea* (Indigo Bunting)		+	−		1	+
89. *P. amoena* (Lazuli Bunting)		+				+
90. *P. ciris* (Painted Bunting)		+			4+	
Parulidae						
91. *Vermivora chrysoptera* (Golden-winged Warbler)		+	(+)		2	
92. *V. pinus* (Blue-winged Warbler)		+	(+)		2	+
93. *Dendroica pensylvania* (Chestnut-sided Warbler)		+			10–12+	
94. *Geothlypis trichas* (Common Yellowthroat)		+			1	

(continued)

TABLE I (*Continued*)

Species number[b] and species	Syllable sharing	Dialect	Bundles	Repertoire	Evidence of learning
Vireonidae					
95. *Vireo griseus* (White-eyed Vireo)	+	(−)		5–20	+
96. *V. flavifrons* (Yellow-throated Vireo)	+			8	
97. *V. solitarius* (Solitary Vireo)	+	(+)		15	
98. *V. olivaceus* (Red-eyed Vireo)	(−)			35–46	
Icteridae					
99. *Cacicus cela* (Yellow-rumped Cacique)	+	+		2–4	+
100. *Sturnella magna* (Eastern Meadowlark)	+	(+)		6–9	
101. *S. neglecta* (Western Meadowlark)	+	(+)			
102. *Quiscalus quiscula* (Common Grackle)	−	−		1	
103. *Molothrus ater* (Brown-headed Cowbird)	+	+			+
104. *Dolichonyx oryzivorus* (Bobolink)	+	+		2	
Fringillidae (Fringillinae)					
105. *Fringilla coelebs* (Chaffinch)	+	+		1–5	+

Fringillidae (Carduelinae)						
106. *Carduelis chloris* (Greenfinch)		+	(+)		Indeterminate	+
107. *C. spinus* (Siskin)		+				+
108. *C. pinus* (Pine Siskin)		+				+
109. *C. tristis* (American Goldfinch)		+				+
110. *Acanthis flavirostris* (Twite)		+				+
111. *Carpodacus cassinii* (Cassin's Finch)		+				
112. *C. mexicanus* (House Finch)	a.	+	+	+	2–6	+
	b.	+	–			
113. *Loxia leucoptera* (White-winged Crossbill)		+				+?
114. *Pyrrhula pyrrhula* (Bullfinch)		+				+
Drepanididae						
115. *Himatione sanguinea* (Apapane)		+	(+)		1?	
Estrildidae						
116. *Lonchura cucullata* (Bronze Mannikin)		+	+			+
Ploceidae						
117. *Vidua fischeri* (Fischer's Whydah)		+				
118. *V. chalybeata* (Senegal Indigo-bird)		+	+		12 n[d] 6 m	+
119. *V. purpurascens* (Dusky Indigo-bird)		+	(+)		12 n 6 m	+

(continued)

TABLE I (*Continued*)

Species number[b] and species	Syllable sharing	Dialect	Bundles	Repertoire	Evidence of learning
120. *V. funerea* (Dusky Indigo-bird)	+			12 n 6 m	+
121. *V. wilsoni* (Pale-winged Indigo-bird)	+			12 n 6 m	+
Sturnidae					
122. *Gracula religiosa* (Hill Myna)	+		−		+
Callaeidae					
123. *Creadion carunculatus* (Saddleback)	+	+	+	1	+
Corvidae					
124. *Cyanocitta cristata* (Blue Jay)	+	+			

[a] See Section II,C of text for explanation.
[b] See References to Table I, p. 175.
[c] Refers to a nonvocal, wingflap "song."
[d] These four *Vidua* spp. (118–121) have about 12 nonmimetic (n) and 6 mimetic (m) songs.

References to Table I

Species number	References
Non-passerines	
1,2	Sparling (1979)
3	Heinz and Gysel (1970)
4	Nottebohm (1976)
5	Snow (1968); Wiley (1971)
6	Mirsky (1976)
Passeriforms (suboscines)	
7	Snow (1977)
8	Thomas (1979)
9	Payne and Budde (1979)
10	Robinson (1974, 1975)
11	Smith and Robinson (1976)
Passeriforms (oscines)	
12	Payne (1973a); Seibt (1975); Bertram (1977)
13	Tretzel (1965)
14	Samuel (1971)
15a,16	Hooker and Hooker (1969); Thorpe (1972)
15b	Harcus (1977)
17	Wickler (1972); Thorpe (1972)
18	Kroodsma (1975)
19	Kroodsma and Verner (1978)
20	Verner (1975)
21	Kroodsma (1974)
22	Borror (1956)
23,24	Grant (1966)
25	Kreutzer (1972, 1974); Kroodsma (1980)
26	Kroodsma (1973)
27	Thompson and Jane (1969)
28	Wildenthal (1965); Hatch (1967); Howard (1974)
29	Messmer and Messmer (1956); Thielcke and Thielcke (1960); Tretzel (1967)
30	Bjerke (1974); Espmark (1981)
31	Isaac and Marler (1963)
32	Grabowski (1979)
33a	Konishi (1965)
33b	Bensen (1975)
34	Lemaire (1974, 1975); Dowsett-Lemaire (1979)
35	Bergmann (1976a)
36,37	Bergmann (1976b)
38	Thielcke and Linsenmair (1963)
39	Bremond (1976)
40–47	Martens (1980)
48,49	Becker (1974, 1977)
50	Todt (1970)
51	Romanowski (1978); Becker (1978)
52	Thönen (1962); Romanowski (1978)
53	Bagg (1958); Dixon and Stefanski (1970); Ficken *et al.* (1978)
54	Ward (1966); Smith (1972)
55	Thielcke (1973)
56	Gompertz (1961); Thielcke (1969)
57	Becker *et al.* (1980b)
58	Dixon (1969)
59	Lemon (1968)
60,61	Thielcke (1969, 1972)
62	Grimes (1974); Payne (1978)
63	McGregor (1980)
64	Kaiser (1965); Hansen (1978); Møller (1977)
65	Conrads and Conrads (1971); Conrads (1976)
66	Martens (1979)
67	Kreutzer (1979)
68	Ewing (1977)
69	Martin (1977, 1979)
70a	Mulligan (1963, 1966)
70b	Harris and Lemon (1972); Eberhardt and Baptista (1977)
71a	Marler and Tamura (1962, 1964); Baker (1974); Baptista (1975); Baker and Mewaldt (1978)

(*continued*)

References to Table I (*Continued*)

Species number	References	Species number	References
71b	Baptista (1977); Heinemann (1981)	91,92	Gill and Murray (1972); Russel (1976)
71c	DeWolfe and DeWolfe (1962); Orejuela and Morton (1975); Baker (1975); Lein (1979); Baptista and King (1980)	93	Lein (1978)
		94	Borror (1967)
		95	Adkisson and Conner (1978); Bradley (1981)
		96	James (1976); Smith *et al.* (1978)
71d	DeWolfe *et al.* (1974)		
72	Nottebohm (1969); King (1972); Egli (1971)	97	James (1976); Martindale (1980a)
73	Borror and Gunn (1965); Lemon and Harris (1974)	98	Lemon (1971b); James (1976)
		99	Feekes (1977)
74a	Williams and MacRoberts (1977, 1978)	100,101	Lanyon (1957, 1958, 1966)
		102	Wiley (1976)
74b	Konishi (1964); Marler *et al.* (1962)	103	Rothstein and Fugle (1980); King *et al.* (1980)
75	Marler and Isaac (1961)	104	Avery and Oring (1977)
76	Bradley (1977)	105	Sick (1939); Poulsen (1951); Marler (1952); Thorpe (1958); Conrads (1966); Metzmacher and Mairy (1972); Slater and Ince (1979); Pickstock *et al.* (1980)
77	Tasker (1955); Borror (1959)		
78	Goldman (1973); Heckenlively (1976)		
79	Kroodsma (1972)		
80	Heckenlively (1970)		
81	Rich (1981)		
82	Burr (1974)	106	Güttinger (1974, 1976)
83a	Kroodsma (1971)	107–110	Mundinger (1970, 1979)
83b	Borror (1975); Ewert (1978, 1979)	111	Samson (1978)
		112a	Mundinger (1975, 1980)
84	Marler and Isaac (1960)	112b	Bitterbaum and Baptista (1979)
85	Lemon and Chatfield (1973)		
86	Lemon (1965, 1966, 1967, 1971a); Lemon and Scott (1966)	113	Mundinger (1979)
		114	Nicolai (1959); Wilkinson and Howse (1975)
87	Lemon and Herzog (1969)	115	Ward (1964)
88	Thompson (1968, 1970); Rice and Thompson (1968); Shiovitz and Thompson (1970); Emlen (1971)	116	Güttinger and Achermann (1972)
		117	Nicolai (1973)
		118–121	Payne (1973b)
89	Emlen *et al.* (1975)	122	Bertram (1970)
90	Thompson (1968); Forsythe (1974)	123	Jenkins (1977)
		124	Kramer and Thompson (1979)

families. (This is a statement about the scope of research activity rather than a generalization per se.)

(2) A song- or call-learning capacity is taxonomically widespread. There is good evidence, other than syllable sharing, of vocal learning for 53 species representing 20 families (column 6). And syllable sharing, which *may* be due to learning, has been reported in many other passerine and non-passerine species (column 2, and item 3 below). Moreover, vocal learning may well be a primitive passerine condition, given its widespread occurrence within Passeriformes, a distribution which involves at least four sub-oscine families. Several phylogenetically remote non-passerine families also exhibit vocal learning.

(3) Syllable sharing is very widespread and common in nature, and was identified in 116 of 124 species surveyed. Excluding two species for which there are no data on sharing and four species for which the evidence for sharing is either equivocal (No. 33, *Turdus migratorius,** No. 73, *Zonotrichia albicollis*) or questionable (No. 1, *Phasianus colchicus*, No. 7 *Procnias tricarunculata*), then 112 of 118 species (95%) exhibit some degree of syllable sharing in at least some of their local populations.

(4) Complete absence of syllable sharing is therefore uncommon. Some of the evidence for nonsharing is also weak or equivocal (e.g., No. 14, *Petrochelidon pyrrhonota*, No. 33, *T. migratorius*). Excluding these two, only five species and one subspecies (No. 47, *Phylloscopus affinis*, No. 75, *Junco phaeonotus*, No. 83, *Pipilo fuscus*, No. 97, *Vireo olivaceus*, No. 101, *Quiscalus quiscala*, and No. 74, *Junco hyemalis oreganus*) appear to have communication systems in which syllable sharing is rare or absent (see column 2).

(5) Most species exhibiting syllable sharing have also been identified as having a song dialect pattern of microgeographic variation. Looking only at those species with shared syllables, there were 80 for which the original investigator(s) either made positive or negative statements about dialects or where this could be inferred (Table I, column 3). Of these 80 species, 64 (80%) had song or call dialects reported or inferable. These data include five species for which dialects have been variously reported present and absent by different investigators (see Table I: Nos. 70a, b, 71a–d, 74a, b, 83a, b, 112 a, b—see References to Table I).

(6) Song dialects are a taxonomically widespread phenomenon. The 64 reported or inferred examples of song or call dialects represent 2 non-passerine families, 3 sub-oscine families, and 18 oscine families (column 3). Although column 3 represents diverse viewpoints, definitions, and interpretations with regard to dialects, the same general picture appears when uniform criteria are applied. The published data for 11 species permit isogloss analysis. In 7 out of

*The number preceding each of these species is the species number listed in Table I.

these 11 instances (64%) I identified isogloss bundles. The seven species include a non-passerine and a broad representation of passerines (Table I, column 4).

D. Conclusions and Discussion

Several important conclusions follow from the generalizations. First, a great many songbirds, and a number of non-passerines as well, are capable of vocal learning. This learning capacity is usually expressed as vocal imitation, with other kinds of vocal modification (e.g., improvisation) infrequent. With regard to microgeographic variation, this vocal imitation reduces interindividual variation, although a careful analysis of song patterns is almost always likely to reveal some individualistic features (*ideolects*). Second, vocal dialects are common and taxonomically widespread. Therefore, song dialects may be the rule and not the exception. Third, a modified linguistic definition of song dialect, one that includes dialect boundaries, can be applied to the vocal behavior of many different species.

An important related issue is the temporal stability of dialects. On the one hand, the song patterns characterizing a social group can apparently change in a short period of time, within a breeding season in the Yellow-rumped Cacique *Cacicus cela* (Feekes, 1977), or on occasion between successive breeding seasons as in the Senegal Indigo-bird (*Vidua chalybeata;* Payne and Payne, 1977). On the other hand, song dialects can be stable for decades or longer. A distinctive Chaffinch (*Fringilla coelebs*) dialect, the Egge dialect, has been associated with a given population inhabiting a forest in northwest Germany for at least 20 years (Conrads, 1966). More recently, Ince *et al.* (1980) noted three Chaffinch themes still present in an English wood after a period of 18 years, although other themes there changed over that same period of time. The Berkeley dialect of the White-crowned Sparrow (*Z. l. nuttalli*) was still identifiable after 8–12 years, the period between Marler and Tamura's (1962) original descriptions in 1959–1960 and Baptista's (1975) later study done in the period 1968–1971. Baptista and King (1980) report the same song patterns present over a period of 10 years in the Mt. Lassen population of the montane White-crowned Sparrow (*Z. l. oriantha*), and the distinctive song patterns that characterize the Indian Village dialect of the House Finch (Mundinger, 1975) were present in that locale throughout the 10-year period 1970–1980 (P. C. Mundinger, unpublished data). Such temporally stable dialects are conservative vocal traditions passed on to succeeding generations by vocal imitation. Whether temporally stable or changeable, traditions passed on by imitative learning are cultures (Mundinger, 1980). This leads to an important conclusion with notable consequences—for many species the patterns of variation in acquired vocal traditions are the product of cultural evolution.

The presence of culture and cultural evolution can modify functional interpretations of dialects. It is therefore important to briefly review what we know about the impact of culture and cultural evolution on birdsong patterns.

The word "pattern" can be interpreted in two ways. It can refer to song (or call) pattern. It can also refer to spatial or geographic pattern. In both instances the effect of cultural factors must be considered for some species, though not necessarily all. With regard to song patterns, cultural components of the songs of Cardinals (Lemon, 1975), White-crowned Sparrows (Baptista, 1975), Saddlebacks (*Creadion carunculatus;* Jenkins, 1977), and House Finches (Mundinger, 1980) have been documented and discussed. Jenkins (1977) worked with a color-banded population, following the movements and behaviors of individual male Saddlebacks over a period of years. He uncovered clear instances of cultural mutations, e.g., a young male that miscopied the song pattern of an older male neighbor (the model) and subsequently passed this altered tradition on. Lemon (1975) described the process of "drift" in Cardinal song and showed how that process can also be a source of cultural mutation by generating new forms of a given syllable type. Mundinger (1980) discusses and gives examples of several major causal mechanisms operating in cultural evolution: psychological selection, meme flow, and memetic drift. And Baptista's observations (Baptista, 1974; Baptista and Wells, 1975) provide examples of meme flow (called cultural borrowing by anthropologists) involving cultural contact between sedentary and migratory populations of the White-crowned Sparrow. In sum, several independent studies have shown that cultural evolutionary processes are responsible for the variation we see in the fine details of song syllable patterns.

Turning to geography, song dialects have long been associated with a mosaic geographic patterning. An early description of a mosaic distribution (Thielcke, 1965, pp. 552–553) identifies it in terms of the same elements (syllables) which occur in different localities separated by populations with other forms of those elements. I use the term mosaic somewhat differently, namely as referring to a patchwork distribution of independent, irregularly shaped areas. Both concepts are evident at the macrogeographic level and are discussed later.

This survey of microgeographic variation has revealed spatial patterns other than strict mosaics. One isogloss analysis of House Finch songs yielded a set of irregularly shaped areas, but the song patterns in that dialect system were not independent since they were linked by a common cultural evolutionary heritage (Fig. 6) (Mundinger, 1980). Also, the same detailed form of given syllable types (i.e., Thielcke's elements) did not recur in areas separated by other forms of those syllables. So this is not a classic mosaic distribution. Therefore, the term *historically variegated* is applied to this pattern to emphasize both the historical links within the system of related song dialects, and the unpredictable outline and position of each dialect area. How common such patterns are in nature remains to be documented, but it is worth noting that the Cardinal (Lemon, 1965, 1966, 1975), Redwing (Bjerke, 1974; Epsmark, 1981), and White-crowned Sparrow (Baptista, 1975) appear to have similar systems of related song dialects.

In addition we know of still other kinds of microgeographic patterning, such as syllable sharing with no dialect boundaries, as in the Hill Myna, and instances of limited or no syllable sharing such as that exhibited by the Dark-eyed Junco (*J. h. oreganus*). Presumably, cultural evolution is a factor for species with historically variegated patterns of microgeographic variation, and perhaps also for species with syllable sharing even though bundles are not identified. But culture may have little or no effect on species like the Dark-eyed Junco that do not share syllables since these do not pass on their vocal patterns as imitated traditions.

III. MACROGEOGRAPHIC VARIATION

With macrogeographic variation the perspective shifts from localized patterns of similarities and differences in the fine details of acoustic structure to coarser-grained comparisons in which fundamental similarities in song or syllable pattern may extend over an extensive area. At this level the opportunity for deducing general principles is more limited since fewer studies are available for comparison and methods of data collection, analysis, and interpretation vary considerably.

A. Calls

Compared to songs, regional variation in call pattern has not been well studied. Four reports are briefly outlined.

Local variation in the "rain call" of the Chaffinch was studied very early on by Sick (1939), who found sharing of the same call syllable pattern within local breeding populations in Germany. There the rain call seems to function in a song-like capacity (Thielcke, 1969). Subsequently, regional variation was sampled across central and southern Europe and several distinctive call patterns were found. When mapped, the geographic distribution of these distinctive rain call patterns revealed a classic mosaic pattern, i.e., calls of a given fundamental pattern appear in different regions which are separated by areas in which calls of a very different pattern are found (see Thielcke, 1969, Fig. 3).

Marler and Mundinger (1975) analyzed geographic variation in the flight call pattern of the Twite (*Acanthis flavirostris*.). The fine structure of this call is modifiable by learning. At the local level members of a breeding pair share a given pattern, each pair in a local breeding population having its own distinctive pattern (this is an instance of call sharing, but no dialects). The regional sample included birds from the western coasts of Norway, Scotland, and Ireland, and represented two subspecies, *A. f. flavirostris* (Norway) and *A. f. pipilans* (Scotland, Ireland). The same fundamental call-note pattern, a sharply down-slurred syllable, was found throughout the sampled region. Although there were no striking regional differences in this qualitative character, there were regional differences in quantitative features such as temporal and frequency measures.

The Norwegian population had the shortest syllables with broadest frequency ranges; Irish populations uttered the longest notes with the narrowest frequency ranges; call note patterns in the Scottish sample were intermediate.

Island populations of both the Subalpine Warbler (*Sylvia cantillans*) and the Sardinian Warbler (*S. melanocephala*) differ from their counterpart populations on the European continent with regard to alarm-call patterns (Bergmann, 1976b). For the Subalpine Warbler the difference is primarily syntactic, i.e., the island population on Sardinia repeats the basic call note as a trill, whereas mainland populations do not trill the similarly patterned call note (syllable) they emit. Calls of populations of the Sardinian Warbler on Tenerife differ from those in European populations in such quantitative features as frequency (of the "short" form of the alarm) as well as qualitative differences in the overall pattern of the long form of the alarm call.

Goldstein (1978) sampled the *Hoy* call of female Bobwhite (*Colinus virginiana*) in four widely separated locales representing the limits of the species range: Massachusetts, Florida, Texas, and Nebraska. Whether female quail can modify their *Hoy* call pattern by learning is unknown, so this study may not represent regional variation in an acquired pattern of call behavior. But the work is notable in that it represents an early example of using objective, multivariate statistical methods in the analysis of quantifiable parameters (temporal and frequency measures). Goldstein found regional heterogeneity, and cautiously concluded that the geographic variation of the *Hoy* call was not clinal but was best described as a mosaic, so the term dialect was applied to the different regional forms. However, Thielcke's (1965, 1969) mosaic definition of dialect refers to qualitative traits, i.e., syllable (element) patterns. A mosaic distribution of quantitative traits may not match, or even be strictly comparable to, a mosaic distribution of call patterns.

In sum, some species appear to have the same basic call pattern throughout their sampled range (e.g., the Twite, but this species is unsampled in the Himalayas); other species may have quite different patterns of the same call type in different regions (Chaffinch). Regional variation can be detected in qualitative traits (syllable form, syntax) and quantitative traits (frequency, temporal measures). The dialect concept ought be considered and examined carefully before applying it to describe regional variation in call patterns.

B. Songs

Twenty-eight different species appear in this survey of regional studies (Table II). These represent most of the macrogeographic studies I found in the literature. A criterion for inclusion in the survey was that the species sample include populations separated by a distance of at least 50–100 km. Summary descriptions for each species, which represent the data base for making comparisons, are to appear in a more thorough treatment of variation in bird vocalizations (Mundinger, unpublished data). For the purpose of this comparison the survey data are very briefly summarized in Table II.

It is obviously inadequate to present a rich and varied set of data in the simplified format of a table. The perspectives and methods of analysis of the various investigators varied widely, as did the size of the areas sampled, sampling densities, terminology, and the applications of that terminology. The behavior of the species studied varied most of all. Little of this diversity is evident in Table II. Instead, the interpretations and biases of the tabulator become important, intervening between the original data and the reader. To offset this to some degree, a description of the table's format follows.

Description of Table II

Each column is explained in serial order:

a. Species/References. If two quite different studies were made on the same species, or if different subspecies were the focus of different studies, then the references in Table II are subdivided (i.e., a and b). There was no subdivision if more than one study was done but the methods and perspectives were similar, or if one study sampled a number of different subspecies [e.g., Common Yellowthroat (*Geothlypis trichas*), Rufous-sided Towhee (*Pipilo erythrophthalmus*)].

b. Analytic Methods. Methods of analysis varied considerably. The most common is referred to as *Analysis of Pattern* (*P*), in which the investigator focuses on a qualitative analysis of syllable or song patterns. Often summary quantitative measures are included (e.g., max/min frequencies, durations, etc.), but such data are usually not given statistical treatment. Quantitative Analysis (Q) refers to studies in which mensural data are the primary focus, although limited qualitative data (sonagrams) may be present as well. The mensural data are given univariate statistical analysis. Multivariate Analysis (MV) represents a recent development. Typically, this approach applies the methods and philosophy of numerical taxonomy. The methodology, as applied to bird vocalizations, is discussed in detail by Sparling and Williams (1978) and is briefly summarized and updated by Martindale (1980b). Institution Analysis (I) is another recent analytic method. It is a form of isogloss analysis. But, instead of mapping phonetic variants, isolexes are used to map the spatial distributions of qualitatively different song syllable lexicons (vocabularies). Fundamentally, this is a qualitative analysis since the analyst must decide whether a new syllable pattern is an extreme phonetic variant of a previously known syllable type, or is a new syllable type. Mundinger (1980) describes the method and defines the institution, which is a population of culturally acquired behavior patterns (e.g., a population of syllable types) with its own historical identity, and is the product of cultural evolution.

c. Sample Sizes (N). The number of birds recorded, or songs analyzed, in conjunction with the dimensions of the sampled region, gives a very rough estimate of sampling densities. However, a given sample of birds is rarely evenly distributed across a sampled area. Instead, the birds tend to be clustered in a few widely scattered localities visited by the investigator(s). Inadequate sampling often results in a distorted picture of song variation in space. Sample sizes were unavailable for some studies.

d. Region. For more adequate regional descriptions, see the original studies. The summary description in Table II barely suggests the outlines of the regional surveys. The only distance measure given is the approximate distance (in km) between the two most widely spaced sites. This is an oversimplication and is given only for the purpose of indicating the geographic scope of the study. No distance measure appears for the Greenfinch (*Carduelis chloris*) and one Great Tit study, as those works focused on a few recordings (some apparently made for other purposes) from several continents.

e. Regional Lexicon. The primary set of data for comparison is the regional lexicon column. It also represents a source of bias as it does not present original data but rather my interpretations. It is, therefore, important to underscore the criteria used in the interpretations, and to be aware of some personal biases. Explanations of syllable variation (+ if present) and of categories I–III, a typological classification for the kinds of syllable variation encountered, outline the criteria.

i. Syllable variation. Ideally, the comparison involves lexical variants only. However, it was occasionally very difficult to differentiate a potentially new lexical variant from an extreme phonetic variant of a familiar syllable type. This was often the case with species such as the Short-toed Tree Creeper (*Certhia brachydactyla*) or Carolina Chickadee (*Parus carolinensis*), with few syllables in individual repertoires and limited syllable diversity at the species level.

ii. Categories I–III. Category I refers to species that have qualitatively different syllable repertoires in different geographical regions. Category II represents species that are characterized by the same basic syllable repertoire throughout the sampled region. Category III describes one species, the White-throated Sparrow (*Zonotrichia albicollis*), which exhibits an enigmatic spatial distribution pattern featuring both syllable heterogeneity and limited syllable sharing at both the regional and local levels.

iii. Tabulator bias. (1) I began this survey with the hypothesis that all

TABLE II

Comparison of Species Sampled Regionally[a]

Species	References	Analytic method	*N*		Region	Regional lexicon			Different syntaxes
			Birds	Songs		I	II	III	
1. Winter Wren (*Troglodytes troglodytes*)	a. Kreutzer (1974)	P	36+	200	Scotland, England, France, Switzerland, Hungary, Russia, Morocco (2600 km)	(+)			
	b. Kroodsma (1980)	P	4+		New York, Maine, Oregon (4000+ km)	+			+
2. Rock Wren (*Salpinctes obsoletus*)	Kroodsma (1975)	Q	12	>428	Oregon, California, Nevada, South Dakota, Texas (1600 km)		+		
3. Mockingbird (*Mimus polyglottos*)	Howard (1974)	P	14		2 localities in Texas (480 km)	+			
4. Redwing (*Turdus iliacus*)	Bjerke (1974)	P			Near Oslo, Norway (85 km^2)	+			
5. Chiffchaff (*Phylloscopus collybita*)	Thielcke and Linsenmair (1963)	P	47	741	Scandinavia to Morocco (2200+ km)	+			+
6. Marsh Warbler (*Acrocephalus palustris*)	Lemaire (1975)	P	4	Many	Belgium, 3 sites (225 km)	+			
7. Goldcrest (*Regulus regulus*)	Becker (1977)	P	161	2,380	Denmark, Germany, Holland, England, Belgium, France, Spain (2200 km)	+			
8. Firecrest (*R. ignicapillus*)	Becker (1977)	P	150	770	Belgium, West Germany, France, Spain, Morocco (2300+ km)	+			
9. Willow Tit (*Parus montanus*)	a. Thönen (1962)	P	(209 sites)		Switzerland, Germany, France, Italy, Finland (900 km)	+			
	b. Romanowski (1978)		36						
10. Carolina Chickadee (*P. carolinensis*)	Ward (1966)	P	118	227	New Jersey, Pennsylvania, Virginia, Maryland, Florida (1500+ km)	+			+
11. Great Tit (*P. major*)	a. Gompertz (1968); Thielcke (1969)	P	13	24	Northern Europe, Afghanistan, India, Japan	+			
	b. Hunter and Krebs (1979)	Q	507	759	England, Norway, Sweden, Poland, Spain, Greece, Morocco, Iran (5,000 km)	+			
12. Blue Tit (*P. caeruleus*)	Becker *et al.* (1980b)	P	53	273	West Germany, Yugoslavia, Spain, Morocco, Tenerife (3,500 km)	+			+

13. Short-toed Tree Creeper (*Certhia brachydactyla*)		Thielcke (1965)	P	156	1,447	Northern Germany to Southern Spain (2100+)		+		
14. Splendid Sunbird (*Nectarinia coccinigastra*)		Payne (1978)	MV	~46 (16 map sites)		Ghana: coastal and interior (400 km)				
15. Cirl Bunting (*Emberiza cirlus*)		Kreutzer (1979)	P	89	3200	France: Paris and Perigeux regions (400 km)		+		
16. Fox Sparrow (*Zonotrichia iliaca*)		Martin (1979)	P and MV	133		Utah and Wyoming (57 km)		+[b]		
17. Rufous-collared Sparrow (*Z. capensis*)		Nottebohm (1969)	P	523		Argentina (800+ km)	+			
		King (1972)	P	55	150		+			
18. White-crowned Sparrow (*Z. leucophrys nuttalli*)	a.	Baptista (1975); Baker (1974, 1975)	P	414	2420	Northern California (50 km)	(+)[b,c]			
(*Z. l. pugetensis*)	b.	Baptista (1977)	P	270	1668	Northern Pacific coast (800 km)	(+)[c]			
(*Z. l. oriantha*)	c.	Orejuela and Morton (1975);	P	142	1000	Western U.S.: Sierra and Rocky Mountains (2000 km)	+			
		Baker (1975); Baptista and King (1980)		~60						
19. White-throated Sparrow (*Z. albicollis*)		Borror and Gunn (1965)	P	433	433+	8 U.S. states, 6 Canadian provinces (4,500+ km)			+	
		Lemon and Harris (1974)	P	85	91	New Brunswick			+	
20. Savannah Sparrow (*Ammodramus sandwichensis beldingi*)		Bradley (1977)	P	280		Coastal California (725 km)		+		
21. Rufous-sided Towhee (*Pipilo erythrophthalmus*)		Borror (1975)	P	492	10,000	19 U.S. States, 1 Canadian province (4,000+ km)	+			+

(continued)

TABLE II (*Continued*)

Species	References	Analytic method	N: Birds	N: Songs	Region	Regional lexicon: I	II	III	Different syntaxes
22. Cardinal (*Cardinalis cardinalis*)	Lemon (1966)	P	224		Ontario, New York, Pennsylvania, Ohio, Texas, Mexico, Honduras (4000+ km)		+		
23. Indigo Bunting (*Passerina cyanea*)	a. Shiovitz and Thompson (1970) b. Emlen (1971)	P	213		Michigan, New York, Kentucky (683 km)		+		
24. Common Yellowthroat (*Geothlypis trichas*)	Borror (1967)	P	411	472	30 U.S. states; 6 Canadian provinces (~4400 km)		+		
25. Solitary Vireo (*Vireo solitarius*)	a. Borror (1973)	P		424	New England, Arizona, Montana (4,000 km)	+			
	b. Martindale (1980a)	MV	22	29[d] (3,500 syll)	New England, Tennessee Virginia (500+ km)	+			
26. Bobolink (*Dolichonyx oryzivorus*)	Avery and Oring (1977)	P	62	620+	Minnesota, North Dakota (240+ km)	+			
27. Greenfinch (*Carduelis chloris*)	Güttinger (1976)	P		24+[d]	Denmark, England, Germany France, Spain, Corsica, New Zealand	+			
28. House Finch (*Carpodacus mexicanus*)	Mundinger. (1980)	P,I	~900[e]	~9,000[e]	New York: Long Island, New York City area (180 km)	+			+

[a] See Section III,B of text for explanation.
[b] Limited geographic sample.
[c] Limited lexical variation.
[d] Long bouts of continuous singing.
[e] From P. C. Mundinger (unpublished data).

species might have song institutions; that hypothesis has to be reexamined in light of the evidence uncovered. (2) I tend to be a lumper rather than a splitter, and tend to emphasize species similarities, or syllable similarities, rather than their differences. This is reflected in the small number of categories used to describe, in terms of spatial distribution, regional variation in the song-syllable lexicons of 28 species. (3) The comparison is based on syllable patterns, i.e, on a unit below the level of an entire song, and on qualitative instead of quantitative characters. This is partly because most of the data are qualitative, but also because of a personal bias favoring qualitative analysis and the syllable.

f. Different Syntax. Syllable order is surprisingly regular in some species, e.g., the Short-toed Tree Creeper. Many other species have a considerable number of permutations on a basically similar syntactic pattern, e.g., the three *Zonotrichia* species (*leucophrys, capensis, albicollis*). Still other species exhibit remarkably dissimilar syllable orderings, e.g., the Spanish population versus the Central European population of the Chiffchaff (*Phylloscopus collybita*) song patterns (Thielcke, 1969; Thielcke and Linsenmair, 1963), or eastern and western Winter Wrens (Kroodsma, 1980). This column identifies those species that have remarkably dissimilar syntactical arrangements in different regions.

C. Generalizations and Discussion

Although Table II is not a complete review of the literature, it demonstrates that regional variation in songs has been studied in at least 28 species. This group represents 11 higher oscine families: Troglodytidae, Mimidae, Muscicapidae (Turdinae, Sylviinae), Paridae, Certhiidae, Nectariniidae, Emberizidae (Emberizinae, Cardinalinae), Parulidae, Vireonidae, Icteridae, and Fringillidae. Any generalizations based on these data would seem to be representative of the higher oscines.

The studies also represent fundamental differences with regard to descriptive methods and analytic approach. The unit of analysis varied greatly. Occasionally it was a specific physical feature (e.g., frequency or duration); more often it was one of several possible structural units, ranging from the element (note, simple syllables), to the syllable (element, element groups), through the phrase or song unit, up to an entire song pattern (stroph). Thus, diversity of approach can be taken as one generalization, although it refers to how geographic variation is studied and says little about geographic variation itself, except that any subsequent general statements based on this diverse material are best thought of as first-order approximations. These will likely be modified later, when more comparable material becomes available. Also, these first generation generalizations are going to be very simple and basic. Five are outlined below:

First, the macrogeographic perspective tends to emphasize species-typical features and constraints. Kreutzer's (1974) study of variation in Winter Wren song focused on documenting species-typical features, and many other studies refer to various species commonalities brought to light by surveying many different populations. Comparisons of geographically widespread syllable types, such as those found in the Cardinal (Lemon, 1966) and Indigo Bunting (Shiovitz and Thompson, 1970), to similar types developed by untutored birds in song ontogeny experiments (Lemon and Scott, 1966; Rice and Thompson, 1968) provide additional documentation for the biological basis of some aspects of song variation. In contrast, cultural evolutionary processes, which are often evident at the microgeographic level, are not as readily apparent. However, cultural evolution can also be a significant factor in macrogeographic variation. For example, the song institution is not only a unit of regional variation, it is also both a culturally defined unit and a product of cultural evolution. Nevertheless, when macrogeographic variation is examined, then species-typical features are often very evident and interpretations tend to focus on the mechanisms of biological evolution.

Second, the macrogeographic perspective often shifts the focus of attention from phonetic variants to variation expressed in the form of different kinds of syllables in the species vocabulary. Some investigators with quantitative interests (e.g., Hunter and Krebs, 1979) focused on differences in maximum or minimum frequencies, etc., and not syllable patterns per se. But most investigators looked at qualitative variations in pattern (Table II, column 3) and how the different syllable types or song types were spatially distributed.

Third, it is possible to group the spatial distribution of 27 passerine syllable vocabularies into three broad, typological, descriptive categories (see Table II, column 6; the Splendid Sunbird is not in this comparison as there are no sonagrams representing its regional sample):

Category I

Different regions are characterized by qualitatively different syllable vocabularies. This description fits 18 of 27 species (67%). But there are substantial differences among their spatial patterns, so this category can be subdivided further.

a. Category Ia. The regional vocabularies are song institutions, or have a number of characteristics that suggest song institutions. This description seems to apply to seven species surveyed: the House Finch, White-crowned Sparrow, Bobolink, Chiffchaff, Redwing, Willow Tit, and Carolina Chickadee. In very few instances have song institutions been identified, however. Figure 9 illustrates one case, the mapping of 29 different House Finch song institutions in the New York metropolitan region. Many, and perhaps all, of these song institutions are composed of systems of culturally related sister dialects. Figure 6 illustrates

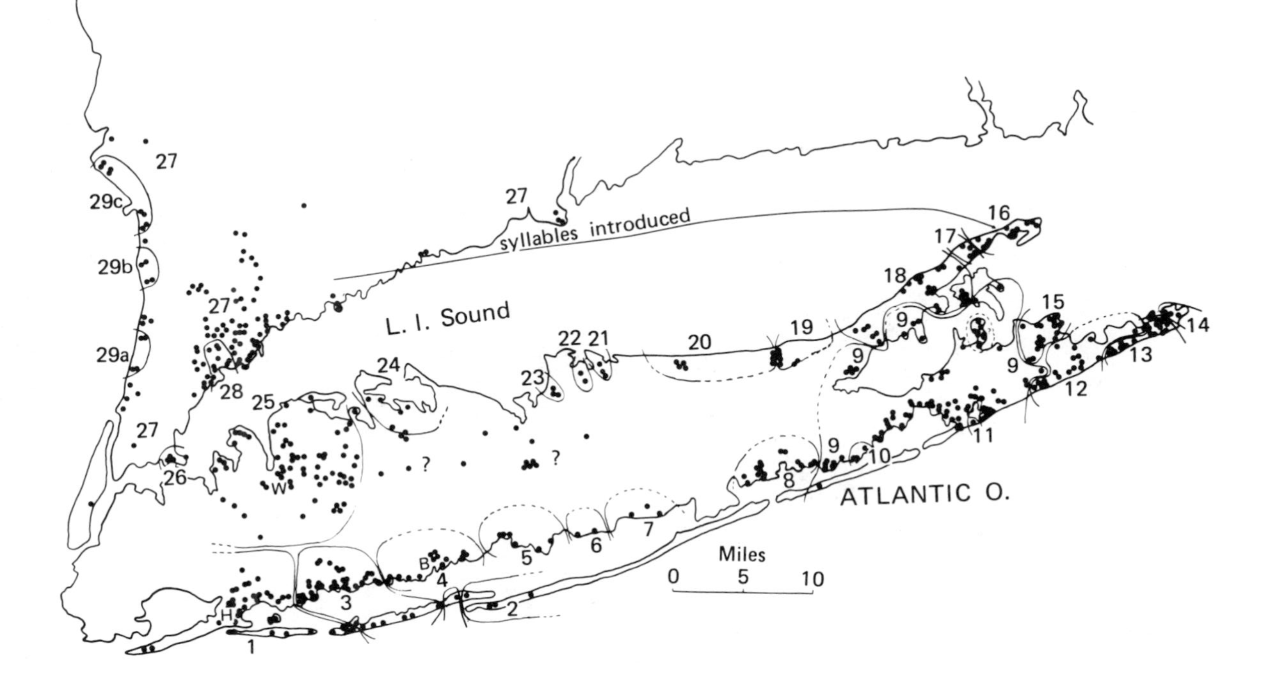

Fig. 9. House Finch song institutions on Long Island, New York, and in metropolitan New York. Black symbols are bird positions (one to several birds per position); solid lines surrounding black symbols are song institution boundaries (see text for details); dotted lines are boundary estimates. There are 29 institutions. Letters H, B, and W in institutions No. 1, No. 4, and No. 25 mark the sites of three of the original breeding colonies of this introduced bird. (From Mundinger, 1980. Reprinted by permission; copyright 1980 by Elsevier North Holland, Inc.)

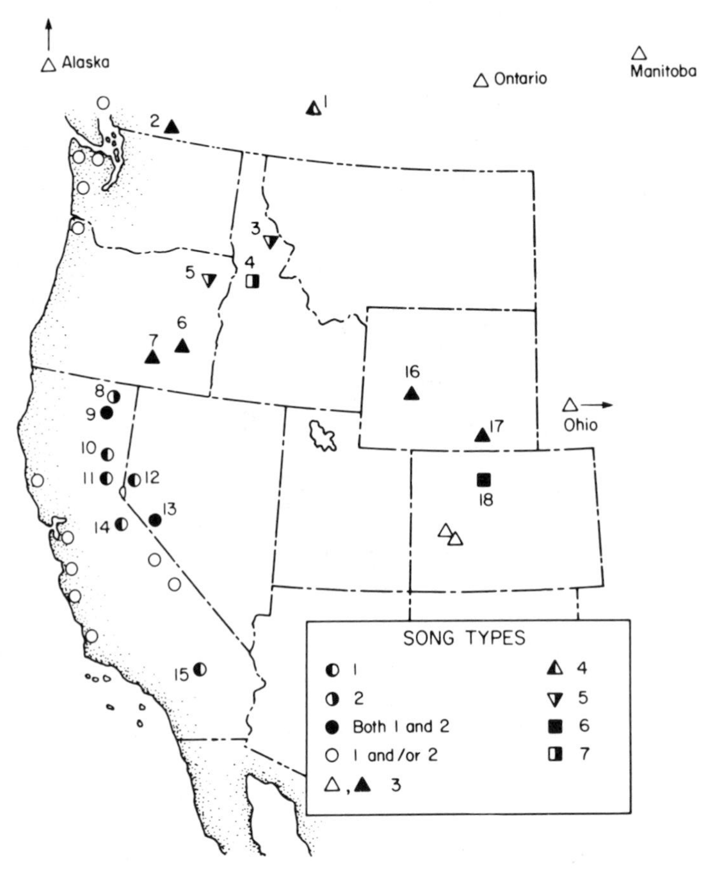

Fig. 10. Localities of populations of White-crowned Sparrows where songs were recorded by Baptista and King (black or half-black symbols) or by other investigators (clear symbols). In box: symbols for the seven different song types identified in the study, which identify seven different dialect regions. (From Baptista and King, 1980.)

such a system for institution No. 3 of Fig. 9. Most House Finch institutions in metropolitan New York are unitary, i.e., they are composed of contiguous sister dialects. But they can also have disjunct distributions, e.g., see institution No. 29a–c in Fig. 9. The set of 29 irregularly shaped, unpredictably positioned, and (largely) historically unrelated song institutions represents a patchwork mosaic. The disjunct institution represents a mosaic distribution *sensu* Thielcke. The thoroughly sampled White-crowned Sparrow has a rich data base that suggests this species may also have song institutions. Figure 10 illustrates the regional grouping of montane White-crowned Sparrow songs into dialect regions as determined by Baptista and King's (1980) comparative analysis of song formulae. Some of these dialect regions are like institutions, but others (e.g., area 3 in Fig. 10) are not. For example, I examined published sonagrams and identified the following sets of populations as representing four different institutions: (1) populations 2 and 5; (2) populations 6 and 7; (3) population 8; and (4) populations 9–15 (see Fig. 10). The geographically widespread songs of song type 3, however, were composed of a heterogeneous array of different syllable lexicons, which is inconsistent with the institution concept (i.e., song type 3 may represent several additional institutions). Studies of the other five species do not provide as rich a data base. Nevertheless, published descriptions of song variation of the Redwing, Bobolink, Chiffchaff, Willow Tit, and Carolina Chickadee all give some indication that these species might have institutions also.

b. Category Ib. Regional syllable repertoires in close proximity differ, but on a much broader regional scale the same basic kinds of syllables begin to reappear, e.g., as in the Greenfinch.

c. Category Ic. The syllable repertoire gradually changes over distance, e.g., as in the Mockingbird (*Mimus polyglottos;* but this study was based on a limited sample).

d. Category Id. Other, i.e., the data are unclear as to how the variation is organized spatially, other than that there are regional qualitative differences in syllable lexicon. Ten species were classified in this category: the Rufous-collared Sparrow, Rufous-sided Towhee, Solitary Vireo (*Vireo solitarius*), Marsh Warbler (*Acrocephalus palustris*), Goldcrest (*Regulus regulus*), Firecrest (*R. ignicapillus*), Winter Wren, Blue Tit (*Parus caeruleus*), and Great Tit. The Goldcrest and Firecrest may be borderline as there is also some reason for placing them in Category II. That would reduce the prevalence of Category I to 59%.

Category II

All sampled regions are characterized by basically the same syllable lexicon. Such a description fits 9 of the 27 species in the survey (33%), but there are basic differences among them that can be represented by the following subdivisions.

a. Category IIa. A basic, common denominator syllable vocabulary characterizes the entire sampled range. I identified five species with this type of regional variation—the Cardinal, the Savannah Sparrow (*Ammodramus sandwichensis beldingi*), the Short-toed Tree Creeper, the Common Yellowthroat, and perhaps the Fox Sparrow (*Zonotrichia iliaca;* based on a geographically limited sample, however).

b. Category IIb. There is some neighborhood heterogeneity (i.e., neighboring individuals do not necessarily share a common syllable repertoire, either in part or in whole), but regionally the same syllable types reappear regularly when the entire local population samples are considered. Such a description applies to two species, the Cirl Bunting (*Emberiza cirlus*) and the Indigo Bunting (*Passerina cyanea*).

c. Category IIc. Overlap of a substantial portion of the song or syllable repertoire with that of a reference individual, e.g., Rock Wren (*Salpinctes obsoletus*).

Category III

Regional and local heterohomogeneity: applies to one species, the White-throated Sparrow. This *Zonotrichia* species reveals an enigmatic pattern of individualistic song and syllable structure at both the local and regional levels. This is presumably the result of song learning or developmental mechanisms promoting individuality. Yet there are also instances of syllable sharing, both at regional and local levels. This sharing is not nearly as prevalent as is the individual distinctiveness (Borror and Gunn, 1965; Falls, 1969; Lemon and Harris, 1974). Furthermore, this sharing is not necessarily due to imitative learning, as independent development is also a likely cause.

Fourth, regional variation in syllable order (syntax) has not been a major focus of attention, but several studies reveal certain species with significant regional differences. Strikingly different syntactical organizations have been reported for six species in this regional comparison (Table II, column 7). The syllable repertoire size, syllable organization, song length, song repertoire size, etc., of Spanish Chiffchaffs differ in many fundamental ways from the songs of Central European Chiffchaffs (Thielcke, 1969; Thielcke and Linsenmair, 1963). Kroodsma (1980) found major differences in song organization and song repertoire size between Winter Wrens in western United States (Oregon) and the east (New York and Maine). Borror (1975) stressed several basic organizational differences characterizing populations of Rufous-sided Towhees from eastern United States versus populations from two other regions, the Plains States and the United States West Coast. Becker *et al.* (1980a) found that Blue Tit songs from Morocco and Tenerife were similar in organization, but these differed in some fundamental respects from the orderings found in Blue Tit songs from Central Europe and Spain. The Carolina Chickadee has a simple song pattern with limited syllable reordering possibilities. The standard song pattern is an alternating sequence of four whistled syllables, two high pitched (H) and two low pitched (L), in the following order: HLHL (Ward, 1966). But Ward's regional survey uncovered other orderings that were often typical of a regional area, e.g., LHL, HLLH, and HILL (where I is a whistle of intermediate pitch). Finally,

Mundinger (1975, p. 413) reported basic syntactical differences in House Finch songs from two neighboring populations. The songs of these populations are now known to belong to two different song institutions. One institution is characterized by songs with a very stereotyped syllable ordering and only two or three themes; the other song institution has many different themes (at least six) composed of various permutations and combinations of phraselike blocks of syllables.

Fifth, correlations between certain environmental factors, such as dense vegetation, and physical features in songs, such as highest frequencies, may eventually prove to be an important generalization. Hunter and Krebs (1979) review the evidence. In addition to the Great Tit, the Small Tree Finch (*Camarhynchus parvulus*) of the Galapogos (Bowman, 1979), Australian zebra finches (*Poephila* spp.; Zann, 1975), and the Rufous-collared Sparrow (Nottebohm, 1975) have had variations in various physical features in their songs or calls correlated with environmental effects.

The effect of insular distribution on song variation is another kind of environmental effect that was initially pointed out years ago by Marler (1959; Marler and Boatman, 1951), and has been reviewed and theoretically strengthened by Thielcke (1969). For example, the Goldcrest, Firecrest, Short-toed Tree Creeper, and Great Tit all show more song diversity where their population densities are low and habitat provides insular distributions (Thielcke and Linsenmair, 1963; Thielcke, 1969; Becker, 1977). At the local level a similar effect was found by Ward (1966) in two Florida populations of the Carolina Chickadee.

IV. DISCUSSION

This overview was done with several purposes in mind. One was to compile a literature review on geographic variation. Tables I and II represent this literature review. The attempt to be comprehensive may have come close, but the literature is now so vast that this first purpose was not entirely fulfilled. In particular, several of the earlier works reviewed by Thielcke (1969) were not included, as they had already been treated. Also, the East European and Russian literature was omitted, and there must be important gaps in the remaining literature.

A second purpose was to spotlight tradition when interpreting song variation, and to show that traditional patterns of vocal behavior are not really as comparable to morphological traits as we once thought. This is partly due to the effects of cultural evolution, which is an important factor whenever traditional inheritance is involved but is apparently not a factor in morphological variation.

Another purpose was to reach some generalizations through comparison. Several have emerged: (1) A song or call learning capacity is taxonomically widespread; a vocal learning ability may be a primitive passerine trait. (2) Syllable sharing is widespread in nature; the absence of syllable sharing is present but uncommon. (3) Most, but not all, species exhibiting syllable sharing also exhibit song dialects; dialects are taxonomically widespread and seem to be the rule rather than the exception. (4) Microgeographic studies of variation tend to focus on the results and processes of cultural evolution; macrogeographic studies tend to emphasize the results and processes of biological evolution. (5) The microgeographic perspective primarily examines variation in phonetic variation; macrogeographic variation tends to involve regional differences or similarities in syllable repertoires (i.e., lexical variation). (6) Within passerines, the spatial distribution of regional variation can take several forms, the most common is one in which different regions are characterized by qualitatively different syllable vocabularies. (7) Very distinctive intraspecific variations in syntax are found in a number of species. (8) Certain environmental factors can affect physical features of songs (for an original source, see Hunter and Krebs, 1979). (9) Insular distribution is correlated with increased variation in song patterns; where birds are continuously distributed and breeding densities are high, song diversity is less (for a general review, see Thielcke, 1969).

A fourth purpose was to determine if the linguistic methods of dialect geography could be applied to avian vocal behavior generally. At the local level a general application was successful, by and large. Given the proper data in the form of maps of bird positions and a thoroughly representative set of sonagrams for each individual bird, isoglosses can be drawn which may reveal dialect boundaries. The method was successfully applied to those 11 species for which published literature provided adequate data sets (Table I, column 4).

The method of dialect geography is valuable for a number of reasons. One of the most important is that it provides an objective, operational definition of a song or call dialect (see Section II,B). If such a definition meets with general acceptance and is consistently applied, then future studies of microgeographic variation should provide more valuable, and interpretable, comparative material.

Turning to regional analyses, the isogloss method that characterizes dialect geography also holds promise for use in macrogeographic studies, although it has not met with general application as yet. Isogloss analysis has been applied to only two or three species. It was used to define House Finch song institutions (Mundinger, 1980). And in the course of this survey I found only two other data sets adequate for application of the method: The thoroughly studied White-crowned Sparrow (Baker, 1975; Baptista, 1975, 1977; Orejuela and Morton, 1975; Baptista and King, 1980); and a regional study of Cardinal song patterns (Lemon, 1966). The Cardinal example poses a problem which is discussed below. The White-crowned Sparrow data strongly support the hypothesis that this species has song institutions (Fig. 10) (Mundinger, unpublished data). Several other species have distinct syllable lexicons in different geographic regions. Of these, the Redwing (Bjerke, 1974; Espmark, 1981), Chiffchaff (Thielcke and Linsenmair, 1963), Willow Tit (Thönen, 1962), Carolina Chickadee (Ward,

1966), and Bobolink (Avery and Oring, 1977) have been studied in such a way as to suggest that their regional lexicons do not change gradually over distance. The Solitary Vireo (Borror, 1973), Winter Wren (Kreutzer, 1974; Kroodsma, 1980) and Marsh Warbler (Lemaire, 1975) also have distinctively different regional lexicons but the sampling was such that gradual change in the lexicon with distance, which may be incompatible with the concept of a song institution, cannot be ruled out. Additional studies and better sampling techniques are needed to determine if song institutions are a general phenomenon.

This survey also revealed a number of species for which geographic variation in song yielded a spatial pattern that seems to be incompatible with the song institution concept. In particular, the concept may not apply to species such as the Indigo Bunting or the Cirl Bunting which lack regionally unique syllable lexicons and where species-typical limitations, presumably due to genetic and not cultural factors, are important in determining a limited variety of syllable forms. In Kreutzer's (1979) interpretation of macrogeographic variation in the Cirl Bunting, chance factors, by which I assume he means the chance occurrence of the independent development of similar syllable types in different regions, are identified as important causal mechanisms. This is certainly not compatible with the concept of cultural institutions. For other species, such as the Cardinal, the question is more equivocal. Cardinal song patterns change via cultural evolutionary processes (Lemon, 1966, 1975), but at the same time species-typical constraints and independent development also seem to be causal factors of macrogeographic variation (Lemon, 1966). The question of whether or not the concept of song institution can be fruitfully applied to species such as the Cardinal is probably best left open until we have better evidence and a better understanding of how cultural evolution affects its song patterns regionally.

In sum, the concept of the song institution appears to be valuable and potentially generalizable, but not universally applicable. The song dialect concept seems not to be universally applicable either, i.e., there are nondialect species. Yet these concepts can be very useful tools in the arsenal of ethologists interested in variation and evolution. For example, the operationally definable song institution, song dialect, and subdialect provide a three-step hierarchy for describing the spatial distribution of certain kinds of vocal variants: institutions are regional populations of lexical variants; dialects are local populations of phonetic variants, bounded by major isophone bundles; and subdialects are neighborhood populations of phonetic variants marked by minor isophone bundles. For the student of evolutionary change these concepts are exceptionally well suited for the analysis of cultural evolution, and for examining the complex interactions between cultural and biological evolution. One example of how these concepts can be used in evolutionary interpretation is the final topic of this discussion. This is a reappraisal of dialect function, a question of considerable interest to evolutionists, but one that has not yet been adequately answered.

A Reappraisal of Dialect Function

Early in the history of thinking about dialects, Andrew (1962) proposed that song dialects were just incidental by-products of the song-learning process, a view that sees dialects as functionless epiphenomena. However, a variety of possible functions has subsequently been postulated by a number of investigators. Thielcke (1969) suggested that dialects may function to reduce vocal variability and thereby enhance the effectiveness of the sound signal, but he neither specified the process nor discussed possible mechanisms. Marler and Tamura (1962) suggested a relationship between song dialects and the genetic constitution of populations, mentioning potential mechanisms such as philopatry and female choice. Both of these mechanisms have since been examined by others. For example, Baker and Mewaldt (1978) collected data supporting a relationship between dialects and philopatry. They found that White-crowned Sparrows (*Z. l. nuttalli*), primarily banded as juveniles but a few as nestlings, rarely dispersed across their natal dialect boundary and into a neighboring dialect area, but seemed to move parallel to that boundary or away from it into the interior of their natal dialect area. This issue is not settled, however. Recently, Petrinovich *et al.* (1981) described a different pattern of juvenile dispersal in another local population of the White-crowned Sparrow (*Z. l. nuttali*). This alternative pattern does not support the philopatry hypothesis. Shifting the focus to the mechanism of female choice, Nottebohm (1969) proposed that *Zonotrichia* dialects function in assortative mating, with consequent enhancement of inbreeding and local adaptation. But the empirical evidence bearing on this hypothesis is somewhat equivocal [see Milligan and Verner (1971) and Baker *et al.* (1981a) on female choice; see Petrinovich *et al.* (1981) for a contrasting view; see also Handford and Nottebohm (1976) on dialects, inbreeding, and local adaptation]. Jenkins (1977) interpreted his data quite differently, suggesting that the adaptive function of Saddleback dialects is to enhance outbreeding. Payne and Payne (1977) concluded that Senegal Indigo-birds, a lek species, have dialects that function in sexual selection. More recently, Payne (1981) linked Senegal Indigo-bird dialects to social adaptation and, apparently, to various social functions. Following another line of reasoning, Treisman (1978) associated dialects with a lessening of destructive conflict between kin, but Trainer (1980) disputes the argument. Following a line of reasoning similar to that of Treisman, Espmark (1981) links dialects to kin groups and to a reduction of effort in territorial behavior. Finally, Baker *et al.* (1981b) present data supporting the hypothesis that dialect boundaries in White-crowned Sparrows are maintained by male–male aggressive interactions, with the inferred function of dialects being that of a population marker that may thereby affect the genetic constitution of local populations.

Some of the hypotheses summarized above were not formulated strictly in terms of dialect function. In several instances relationships rather than functions were discussed. This distinction may be important because assigning a function to dialects may be illogical and incorrect. There are at least two reasons for this assertion.

First, for many biologists the concept of biological function is synonymous with adaptive (or ecological) function. It seems logical that if dialects have adaptive functions then they should also be considered biological adaptations. A biological adaptation is the product of natural selection acting on those alternative alleles that code for different forms of the trait in question. So, as long as the question is restricted to the function of the *capacity* for dialects it is logical to relate this capacity to adaptive functions since capacities for song learning are presumably gene coded and therefore subject to natural selection. But if the discussion is about specific dialects, such as the Berkeley dialect of the White-crowned Sparrow, then questions about adaptive function are not appropriate. Specific song dialects are animal cultures (Marler and Tamura, 1964; Baptista, 1975; Dawkins, 1976; Jenkins, 1977; Slater and Ince, 1979; Bonner, 1980; Mundinger, 1980) and the various song patterns identifying these dialects are products of cultural evolution, not biological evolution (Lemon, 1975; Slater and Ince, 1979; Mundinger, 1980). Specific song dialects are traditions passed on by vocal imitation, they are not specific behavioral patterns passed on by alleles. Therefore, natural selection cannot favor one dialect pattern as against an alternative. From this line of reasoning I am led to conclude that specific song dialects cannot be adaptations, as biologists use that term, and hence cannot have adaptive functions.

The conclusion that specific song dialects cannot be adaptations does not entirely eliminate functional possibilities. Specific, acquired, behavior patterns could also have socially useful functions, even though they are not adaptations in the narrow, biological, sense of that term. Consider this familiar linguistic example. Specific phonemes, such as /r/ and /l/, are acquired. They are not products of natural selection. Therefore the social/communicative functions of /r/ and /l/ sounds, e.g., a semantic function of providing the distinction between the words "*ra*ke" and "*la*ke," are not adaptive functions (in the narrow sense) although they are socially useful functions. Such social functions are not automatically a part of human speech behavior. Some people who have acquired other phonemic systems in which the /r/ and /l/ distinction does not appear, e.g., the Japanese language, experience difficulty in making the /r/–/l/ distinction when they subsequently acquire English as a second language late in life. Their failure to perceive "ra" and "la" sounds as separate categories reduces their ability to communicate fully in their second language (Strange and Jenkins, 1978).

This linguistic example was introduced because it is a relatively familiar one. It is also useful in that it suggests how analogous mechanisms may operate in some bird species. For example, Central European Chiffchaff males respond to playback of Spanish Chiffchaff song as though it were the song of an alien species (Thielcke *et al.*, 1978). It is not known if this failure to recognize the Spanish Chiffchaff song pattern is the result of genetic differences between Spanish and Central European Chiffchaff populations or whether it is due to differences in early song learning experiences which then affect perceptual processing, or if it is some combination of genetic and social processes. To the degree that it is due to early experience alone, we would then have an example of specific, acquired, song patterns that have social/communicative functions (in the appropriate song community) but presumably have no adaptive functions. This hypothetical example was introduced to illustrate one possible mechanism (i.e., acquisition of specific perceptual categories when specific song patterns are learned) for Thielcke's early suggestion that song dialects function to enhance the sound signal. Such a function would be a socially useful function, not necessarily an adaptive function. Payne (1981) lists several other behaviors (e.g., matched countersinging, copying one's neighbor) as mechanisms of social adaptation and dialect formation. Linking social behavior to dialects in this way is an important consideration. But it may then tempt some to ascribe various social functions to dialects, and there is at least one other reason for rejecting such assertions.

This second reason is that song dialects are group phenomena. A song dialect is a pool of vocal variants expressed by a number of individual birds. The total pool is not possessed by any one of those individuals. Stated another way, a dialect is a holistic concept, like a deme or a species. Holistic concepts have communal properties (see Bonner, 1980) and differ from individually held traits such as feathers, wings, or bills, which do have functions. Do group entities such as species, demes, and dialects truly have functions? If they do not then are they epiphenomena? Perhaps the answer lies somewhere in between. Dialects, demes, and species are useful operational concepts. They can be identified in nature. But they do not have to have functions. Rather, it is better that functional questions be asked of those individual traits, capacities, behaviors, etc., that yield the boundaries delimiting species, demes, and dialects. For the species, factors such as the hybrid sterility of individuals, or geographic barriers to individual dispersal, etc., function in reproductive isolation. In relation to demes, factors such as assortative mating or philopatry (which are dependent on individual behavior) may lead to the formation of local gene pools and local adaptations. Although poorly understood as yet, individual behavioral capacities, such as song-learning mechanisms, or the social bonding capacities that lead to father–son bonds (Nicolai, 1959; Immelman, 1969; Kroodsma, 1974) or neighbor–neighbor bonds (Jenkins, 1977), appear to operate in the establishment and/or maintenance of song dialect boundaries. I have also suggested that individually acquired syllable patterns may have socially useful functions for specific individuals operating in specific communication systems. Therefore it would seem to be these individually expressed capacities and behaviors that have specific adaptive or social functions and not the dialect per se.

The conclusion that dialects do not have functions does not preclude inquiries into how dialects are related to important biological or social/communication phenomena. For example, it may be entirely appropriate to inquire into how dialects are related to the genetic structuring of local populations. But such questions are not easy to assess, as evidenced by the different interpretations surrounding that classic and well-studied dialect species, the White-crowned Sparrow. I believe such questions are difficult to outline and answer because they are essential about interrelations between cultural and biological evolution and culture–gene interactions. And such relationships are notoriously difficult to understand and state precisely.

Perhaps by focusing on those individually held behavioral factors responsible for dialect boundaries some progress can be made in this area in the future. For example, the sensitive period for Saddleback song learning is apparently open-ended. This would permit individual Saddlebacks to change dialect membership in later years, with the attendant enhancement of outbreeding as Jenkins (1977) has suggested. In contrast, individual White-crowned Sparrows (*Z. l. nuttalli*) seem to have temporally constrained sensitive periods. Such a different, individually held, learning mechanism may well be expected to have quite different effects on the genetic structuring of White-crowned Sparrow populations. But although the precise nature of the relationships may differ, we can still state generally that the dialects of Saddlebacks, White-crowned Sparrows, and perhaps other (but not necessarily all?) species are related to the genetic structure of their local populations. This is equivalent to stating that there is a relationship between the product of cultural evolution (dialects) and the product of biological evolution (genetic structure of populations of birds).

ACKNOWLEDGMENTS

I thank Peter Marler, Donald Kroodsma, and Edward Miller for their comments and suggestions on early drafts of this paper, and Edgar Gregersen for his comments on the short glossary. My son and daughter, Tom and Ann, helped me with figures and translations, respectively. Special thanks are given to individuals, publishing houses, societies, and journals for copyright releases of figures: The Cooper Ornithological Society and *The Condor* for permission to reproduce Fig. 1 from Baptista and King, 1980, Figs. 5 and 6 from Mundinger, 1975, and for permission to adapt Figs. 1 and 2 from Wiley, 1971; The Indiana University Press for permission to reproduce Fig. 32 from Hans Kurath's book, "Studies in Area Linguistics," 1972; Elsevier North Holland and *Ethology and Sociobiology* for permission to reproduce Figs. 1, 2, and 5 from Mundinger, 1980; and Verlag Paul Parey and *Zeitschrift für Tierpsychologie* for permission to adapt Fig. 1 from Kroodsma, 1974. Time for researching and writing this paper was provided by support from a grant from the Harry Frank Guggenheim Foundation, from NSF Grant #DEB-7906861, and from a City University of New York sabbatical year fellowship.

SHORT GLOSSARY OF LINGUISTIC TERMS

Dialect—A local speech variant. A variant of a given spoken language characterizing a regionally restricted speech community. Dialect boundaries are delineated through the analysis of lexical, morphological, and phonological isoglosses (e.g., see Kurath, 1972, pp. 24 ff.).

Ideolect—An individually distinctive pattern of speech.

Isogloss (also called *heterogloss*)—A line on a map that separates the geographical distribution of one type of speech variant from others.

Isolex—A map line (isogloss) delimiting the distribution of an item in the lexicon.

Isophone—A phonological isogloss. The question of handling and interpreting human phonological variants is complex (Kurath, 1972, pp. 30 ff.). Basically, it appears that isophones are map lines that mark the distribution of pronunciation variants.

Language—The lexicon and the rules for combining lexical items into the functional communication system used by a particular people. A criterion often used to differentiate different human languages is that they be mutually unintelligible.

Lexicon—The vocabulary of a language.

Phone—A speech sound. An objectively recognizable component of spoken words, e.g., the "t" sound in the word "*t*in."

Phonology—The study of speech sounds, especially their articulation.

Phoneme—A family of speech items that signals a difference in meaning. The symbol / / is used to designate a phoneme. For example, substitution of the phoneme /r/ for the phoneme /l/ changes the meaning of the word "lake" to "rake." A given phoneme may be represented by several speech sounds (allophones). Thus the "r" sound representing the phoneme /r/ may differ depending on its position in a word, e.g., compare articulations of *r*at and ta*r*. In some dialects of English the final "r" sound may not be pronounced at all, as in the Bostonian pronunciation of tar ("tah").

We do not know if, or how, birds conceive of the meanings of their sound signals, therefore we cannot yet (and possibly can never) discuss avian phonemes. But we can discuss avian phones, since these objective sounds are accessible to ethological description and analysis.

Syntax—Orderly arrangement. The part of grammar that treats the expression of word relations according to established usage in the language under study.

REFERENCES

Adkisson, C. S., and Conner, R. N. (1978). Interspecific vocal imitation in White-eyed Vireos. *Auk* **95,** 602–606.

Andrew, R. J. (1962). Evolution of intelligence and vocal mimicking. *Science* **137,** 585–589.

Avery, M., and Oring, L. W. (1977). Song dialects in the Bobolink (*Dolichonyx oryzivorus*). *Condor* **79,** 113–118.

Bagg, A. M. (1958). A variant form of the Chickadee's "fee-bee" call. *Mass. Audubon* **43,** 9.

Baker, M. C. (1974). Genetic structure of two populations of White-crowned Sparrows with different song dialects. *Condor* **76,** 351–356.

Baker, M. C. (1975). Song dialects and genetic differences in White-crowned Sparrows (*Zonotrichia leucophrys*). *Evolution* **29,** 226–241.

Baker, M. C., and Mewaldt, L. R. (1978). Song dialects as barriers to dispersal in White-crowned Sparrows, *Zonotrichia leucophrys nuttalli*. *Evolution* **32,** 712–722.

Baker, M. C., Spitler-Nabors, K. J., and Bradley, D. C. (1981a). Early experience determines song dialect responsiveness of female sparrows. *Science* **214,** 819–821.

Baker, M. C., Thompson, D. B., Sherman, G. L., and Cunningham, M. A. (1981b). The role of male vs. male interactions in maintaining population dialect structure. *Behav. Ecol. Sociobiol.* **8,** 65–69.
Baptista, L. F. (1974). The effects of songs of wintering White-crowned Sparrows on song development in sedentary populations of the species. *Z. Tierpsychol.* **34,** 147–171.
Baptista, L. F. (1975). Song dialects and demes in sedentary populations of the White-crowned Sparrow (*Zonotrichia leucophrys nuttalli*). *Univ. Calif., Berkeley, Publ. Zool.* **105,** 1–52.
Baptista, L. F. (1977). Geographic variation in song and dialects of the Puget Sound White-crowned Sparrow. *Condor* **79,** 356–370.
Baptista, L. F., and King, J. R. (1980). Geographical variation in song and song dialects of montane White-crowned Sparrows. *Condor* **82,** 267–284.
Baptista, L. F., and Wells, H. (1975). Additional evidence of song-misimprinting in the White-crowned Sparrow. *Bird Banding* **46,** 269–272.
Becker, P. H. (1974). Der Gesang von Winter-und-Sommergoldhähnchen (*Regulus regulus, R. ignicapillus*) am westlichen Bodensee. *Vogelwarte* **27,** 233–243.
Becker, P. H. (1977). Geographische Variation des Gesanges von Winter-und-Sommergoldhähnchen (*Regulus regulus, R. ignicapillus*). *Vogelwarte* **29,** 1–37.
Becker, P. H. (1978). Der Einfluss des Lernens auf einfache und komplexe Gesangsstophen der Sumpfmeise (*Parus palustris*). *J. Ornithol.* **119,** 388–411.
Becker, P. H., Thielcke, G., and Wüstenberg, K. (1980a). Der Tonhöenverlauf ist entscheidend für das Gesangserkenned beim mitteleuropäichen Zilpzlap (*Phylloscopus collybita*). *J. Ornithol.* **121,** 229–244.
Becker, P. H., Thielcke, G., and Wüstenberg, K. (1980b). Versuche zum angenommenen Kontrastverlust im Gesang der Blaumeise (*Parus caeruleus*) auf Teneriffa. *J. Ornithol.* **121,** 81–95.
Bensen, J. R. (1975). A study of the vocal behavior of the American Robin. Master's Thesis, Oregon State Univ., Corvallis.
Bergmann, H. H. (1976a). Konstitutionsbedingte Merkmale in Gesängen und Rufen europäischer Grasmücken (Gattung:*Sylvia*). *Z. Tierpsychol.* **43,** 315–329.
Bergmann, H. H. (1976b). Inseldialekte in den Alarmrufen von Weissbart und Samptkoff grasmucke (*Sylvia cantillans* und *S. melanocephala*). *Vogelwarte* **28,** 245–257.
Bertram, B. (1970). The vocal behavior of the Indian Hill Mynah, *Gracula religiosa. Anim. Behav. Monogr.* **3,** 81–192.
Bertram, B. C. R. (1977). Variation in the wing-song of the Flappet Lark. *Anim. Behav.* **25,** 165–170.
Bitterbaum, E., and Baptista, L. F. (1979). Geographical variation in songs of California House Finches (*Carpodacus mexicanus*). *Auk* **96,** 462–474.
Bjerke, T. (1974). Geografisk sang variajon hos Roødvingetrost, *Turdus iliacus. Sterna* **13,** 65–76.
Bloomfield, L. (1933). "Language." Holt, New York.
Bonner, J. T. (1980). "The Evolution of Culture in Animals." Princeton Univ. Press, Princeton, New Jersey.
Borror, D. J. (1956). Variation in Carolina Wren songs. *Auk* **73,** 211–229.
Borror, D. J. (1959). Songs of the Chipping Sparrow. *Ohio J. Sci.* **59,** 347–356.
Borror, D. J. (1967). Songs of the Yellowthroat. *Living Bird* **6,** 141–161.
Borror, D. J. (1973). Yellow-green Vireo in Arizona, with notes on vireo songs. *Condor* **74,** 80–86.
Borror, D. J. (1975). Songs of the Rufous-sided Towhee. *Condor* **77,** 183–195.
Borror, D. J., and Gunn, W. W. H. (1965). Variation in White-throated Sparrow songs. *Auk* **82,** 26–47.
Bowman, R. I. (1979). Adaptive morphology of song dialects in Darwin's Finches. *J. Ornithol.* **120,** 353–389.

Bradley, R. A. (1977). Geographic variation in the song of Belding's Savannah Sparrow (*Passerculus sandwichensis beldingii*). *Bull. Fla. State Mus. Biol. Sci.* **22,** 57–100.
Bradley, R. A. (1981). Song variation within a population of White-eyed Vireos. *Auk* **98,** 80–87.
Bremond, J. C. (1976). Specific recognition in the song of Bonelli's Warbler (*Phylloscopus bonelli*). *Behaviour* **58,** 99–116.
Burr, T. A. (1974). Vocalization in a population of Green-tailed Towhees, *Chlorura chlorura.* Master's Thesis, Utah State Univ., Logan.
Conrads, K. (1966). Der Egge-Dialekt des Buchfinken (*Fringilla coelebs*)—Ein Beitrag zur geographischen Gesangsvariation. *Vogelwelt* **87,** 176–182.
Conrads, K. (1976). Studien an Fremddialect-Sängern und Dialect-Michsängern des Ortolans (*Emberiza hortulana*). *J. Ornithol.* **117,** 438–450.
Conrads, K., and Conrads, W. (1971). Regionaldialeckte des Ortolans (*Emberiza hortulana*) in Deutschland. *Vogelwelt* **92,** 81–100.
Dawkins, R. (1976). "The Selfish Gene." Oxford Univ. Press, London and New York.
DeWolfe, B. B., and DeWolfe, R. H. (1962). Mountain White-crowned Sparrows in California. *Condor* **64,** 378–389.
DeWolfe, B. B., Kaska, D. D., and Peyton, L. J. (1974). Prominent variations in the songs of Gambel's White-crowned Sparrows. *Bird Banding* **45,** 224–252.
Dixon, K. L. (1969). Patterns of singing in a population of the Plain Titmouse. *Condor* **71,** 94–101.
Dixon, K. L., and Stefanski, R. A. (1970). An appraisal of the song of the Black-capped Chickadee. *Wilson Bull.* **82,** 53–62.
Dowsett-Lemaire, F. (1979). The imitative range of the song of the Marsh Warbler *Acrocephalus palustris,* with special reference to imitations of African birds. *Ibis* **121,** 453–568.
Eberhardt, C., and Baptista, L. F. (1977). Intraspecific and interspecific song mimesis in California Song Sparrows. *Bird Banding* **48,** 193–205.
Egli, W. (1971). Investigaciones sobre el canto de *Zonotrichia capensis chilensis. Bol. Mus. Nac. Hist. Nat. Chile* **32,** 173–190.
Emlen, S. T. (1971). Geographic variation in Indigo Bunting song (*Passerina cyanea*). *Anim. Behav.* **19,** 407–408.
Emlen, S. T., Rising, J. D., and Thompson, W. T. (1975). A behavioral and morphological study of sympatry in the Indigo and Lazuli buntings of the Great Plains. *Wilson Bull.* **87,** 145–177.
Espmark, Y. (1981). Dialecter i rödvingetrastens *Turdus iliacus* sang. *Var Fagelvarld* **40,** 81–90.
Ewert, D. N. (1978). Song of the Rufous-sided Towhee. Ph.D. Thesis, City Univ. of New York, New York.
Ewert, D. N. (1979). Development of song of a Rufous-sided Towhee raised in acoustic isolation. *Condor* **81,** 313–316.
Ewing, G. (1977). The song of the Reed Bunting (*Emberiza schoeniclus*). Ph.D. Thesis, Univ. of London.
Falls, J. B. (1969). Functions of territorial song in the White-throated Sparrow. *In* "Bird Vocalizations" (R. A. Hinde, ed.), pp. 207–232. Cambridge Univ. Press, London and New York.
Feekes, F. (1977). Colony-specific song in *Cacicus cela* (Icteridae, Aves): the password hypothesis. *Ardea* **65,** 197–202.
Ficken, M. S., Ficken, R. W., and Witkin, S. R. (1978). Vocal repertoire of the Black-capped Chickadee. *Auk* **95,** 34–48.
Forsythe, D. M. (1974). Song characteristics of sympatric and allopatric Indigo and Painted bunting populations in the southeastern United States. Ph.D. Thesis, Clemson Univ., Clemson, South Carolina.
Gill, F. B., and Murray, B. G., Jr. (1972). Song variation in sympatric Blue-winged and Golden-winged warblers. *Auk* **89,** 625–643.
Goldman, P. (1973). Song recognition by Field Sparrows. *Auk* **90,** 106–113.

Goldstein, R. B. (1978). Geographic variation in the "Hoy" call of the Bobwhite. *Auk* **95,** 85–94.

Gompertz, T. (1961). The vocabulary of the Great Tit. *Br. Birds.* **54,** 369–394, 409–418.

Gompertz, T. (1968). Results of bringing individuals of two geographically isolated forms of *Parus major* into contact. *Vogelwelt* **1,** 63–92.

Grabowski, G. L. (1979). Vocalizations of the Rufous-backed Thrush (*Turdus rufopalliatus*) in Guerrero, Mexico, *Condor* **81,** 409–416.

Grant, P. R. (1966). The coexistence of two wren species of the genus *Thryothorus. Wilson Bull.* **78,** 266–277.

Grimes, L. G. (1974). Dialects and geographical variation in the song of the Spendid Sunbird *Nectarinia coccinigaster. Ibis* **116,** 314–329.

Güttinger, H. R. (1974). Gesang des Grünlings (*Chloris chloris*). Lokale Unterschiede und Entwicklung bei Schallisolation. *J. Ornithol.* **115,** 321–337.

Güttinger, H. R. (1976). Variable and constant structures in Greenfinch songs (*Chloris chloris*) in different locations. *Behaviour* **60,** 304–318.

Güttinger, H. R., and Achermann, J. (1972). Die Gesangsentwicklung des Kleinelsterchens (*Spermestes cucullata*). *J. Ornithol.* **113,** 37–48.

Handford, P., and Nottebohm, F. (1976). Allozymic and morphological variation in population samples of Rufous-collared Sparrow, *Zonotrichia capensis,* in relation to vocal dialects. *Evolution* **30,** 802–817.

Hansen, P. (1978). Variations in Yellowhammer songs. *Biophon* **6**(1), 7–8.

Harcus, J. L. (1977). The functions of vocal duetting in some African birds. *Z. Tierpsychol.* **43,** 23–45.

Harris, M. A., and Lemon, R. E. (1972). Songs of Song Sparrows (*Melospiza melodia*): individual variation and dialects. *Can. J. Zool.* **50,** 301–309.

Hatch, J. J. (1967). Diversity of the song of Mockingbirds (*Mimus polyglottos*) reared in different environments. Ph.D. Thesis, Duke Univ., Durham, North Carolina.

Heckenlively, D. B. (1970). Songs in a population of Black-throated Sparrows. *Condor* **72,** 24–36.

Heckenlively, D. B. (1976). Variation in cadence of Field Sparrow Songs. *Wilson Bull.* **88,** 588–602.

Heinemann, D. (1981). Song dialects, migration, and population structure of Puget Sound White-crowned Sparrows. *Auk* **98,** 512–521.

Heinz, G. H., and Gysel, L. W. (1970). Vocalization behavior of the Ring-necked Pheasant. *Auk* **87,** 279–295.

Hooker, T., and Hooker, B. I. (1969). Duetting. *In* "Bird Vocalizations" (R. A. Hinde, ed.), pp. 185–205. Cambridge Univ. Press, London and New York.

Howard, R. D. (1974). The influence of sexual selection and interspecific competition on Mockingbird song (*Mimus polyglottos*). *Evolution* **28,** 428–438.

Hunter, M. L., and Krebs, J. R. (1979). Geographical variation in the song of the Great Tit (*Parus major*) in relation to ecological factors. *J. Anim. Ecol.* **48,** 759–785.

Immelmann, K. (1969). Song development in the Zebra Finch and other estrildid finches. *In* "Bird Vocalizations" (R. A. Hinde, ed.), pp. 61–74. Cambridge Univ. Press, London and New York.

Ince, S. A., Slater, P. J. B., and Weismann, C. (1980). Changes with time in the songs of a populations Chaffinches. *Condor* **82,** 285–290.

Isaac, D., and Marler, P. (1963). Ordering of sequences of singing behavior of Mistle Thrushes in relationship to timing. *Anim. Behav.* **30,** 344–374.

James, R. D. (1976). Unusual songs with comments on song learning among vireos. *Can J. Zool.* **54,** 1223–1226.

Jenkins, P. F. (1977). Cultural transmission of song patterns and dialect development in a free-living bird population. *Anim. Behav.* **25,** 50–78.

Kaiser, W. (1965). Der Gesang der Goldammer und die Verbreitung ihrer Dialekte. *Falke* **12,** 40–42, 92–93, 131–135, 169–170, 188–191.

King, A. P., West, M. J., and Eastzer, D. H. (1980). Song structure and song development as potential contributors to reproductive isolation in cowbirds. *J. Comp. Physiol. Psychol.* **94,** 1028–1030.

King, J. R. (1972). Variation in the song of the Rufous-collared Sparrow, *Zonotrichia capensis,* in northwestern Argentina. *Z. Tierpsychol.* **30,** 344–373.

Konishi, M. (1964). Song variation in a population of Oregon Juncos. *Condor* **66,** 423–436.

Konishi, M. (1965). Effects of deafening on song development in American Robins and Black-headed Grosbeaks. *Z. Tierpsychol.* **22,** 584–599.

Kramer, H. G., and Thompson, N. S. (1979). Geographic variation in the bell calls of the Blue Jay (*Cyanocitta cristata*). *Auk* **96,** 423–425.

Kreutzer, M. (1972). Les variations dans les chants, territoriaux de *Troglodytes troglodytes* et leurs consequences comportementales. *C. R. Hebd. Seances Acad. Sci., Ser. D* **275,** 2423–2425.

Kreutzer, M. (1974). Stereotypie et variations dans les chants de proclamation territoriale chez le Troglodyte (*Troglodytes troglodytes*). *Rev. Comp. Anim.* **8,** 270–286.

Kreutzer, M. (1979). Etude du chant chez le Bruant Zizi (*Emberiza cirlus*). *Behaviour* **71,** 291–321.

Kroodsma, D. E. (1971). Song variation and singing behavior in the Rufous-sided Towhee, *Pipilo erythropthalmus oregonus*. *Condor* **73,** 303–308.

Kroodsma, D. E. (1972). Variations in songs of Vesper Sparrows in Oregon. *Wilson Bull.* **84,** 173–178.

Kroodsma, D. E. (1973). Coexistence of Bewick's Wrens and House Wrens in Oregon. *Auk* **90,** 341–358.

Kroodsma, D. E. (1974). Song learning, dialects, and dispersal in the Bewick's Wren. *Z. Tierpsychol.* **35,** 352–380.

Kroodsma, D. E. (1975). Song patterning in the Rock Wren. *Condor* **77,** 294–303.

Kroodsma, D. E. (1980). Winter Wren singing behavior: A pinnacle of song complexity. *Condor* **82,** 180–188.

Kroodsma, D. E., and Verner, J. (1978). Complex singing behaviors among *Cistothorus* wrens. *Auk* **95,** 703–716.

Kurath, H. (1972). "Studies in Areal Linguistics." Indiana Univ. Press, Bloomington.

Lanyon, W. E. (1957). The comparative biology of the meadowlarks (*Sturnella*) in Wisconsin. *Publ. Nuttall Ornithol. Club* No. 1.

Lanyon, W. E. (1958). Geographical variation in the vocalizations of the Western Meadowlark. *Condor* **60,** 339–341.

Lanyon, W. E. (1960). The Middle American populations of the Crested Flycatcher *Myiarchus tyrannulus*. *Condor* **62,** 341–350.

Lanyon, W. E. (1966). Hybridization in meadowlarks. *Bull. Am. Mus. Nat. Hist.* **134,** 1–25.

Lanyon, W. E. (1978). Revision of the *Myiarchus* flycatchers of South America. *Bull. Am. Mus. Nat. Hist.* **161,** 429–621.

Lemaire, F. (1974). Le chant de la Rousserole verderolle (*Acrocephalus palustris*): etende du repertoire imitatif, construction rhythmique et musicalité. *Gerfaut* **64,** 3–28.

Lemaire, F. (1975). Dialectal variations in the imitative song of the Marsh Warbler (*Acrocephalus palustris*) in western and eastern Belgium. *Gerfaut* **65,** 95–106.

Lein, M. R. (1978). Song variation in a population of Chestnut-sided Warblers (*Dendroica pensylvanica*): its nature and suggested significance. *Can. J. Zool.* **56,** 1266–1283.

Lein, M. R. (1979). Song pattern of the Cypress Hills population of White-crowned Sparrows. *Can. Field-Nat.* **93,** 272–275.

Lemon, R. E. (1965). The song repertoire of Cardinals (*Richmondena cardinalis*) at London, Ontario. *Can. J. Zool.* **43,** 559–569.

Lemon, R. E. (1966). Geographic variation in the song of Cardinals. *Can. J. Zool.* **44,** 413–42.
Lemon, R. E. (1967). The response of Cardinals to songs of different dialects. *Anim. Behav.* **15,** 538–545.
Lemon, R. E. (1968). Coordinated singing in Black-crested Titmice. *Can. J. Zool.* **46,** 1163–1167.
Lemon, R. E. (1971a). Differentiation of song dialects in Cardinals. *Ibis* **113,** 373–377.
Lemon, R. E. (1971b). Analysis of songs of Red-eyed Vireos. *Can. J. Zool.* **49,** 847–854.
Lemon, R. E. (1975). How birds develop song dialects. *Condor* **77,** 385–406.
Lemon, R. E., and Chatfield, C. (1973). Organization of song of Rose-breasted Grosbeaks. *Anim. Behav.* **21,** 28–44.
Lemon, R. E., and Harris, M. A. (1974). The question of dialects in the songs of White-throated Sparrows. *Can. J. Zool.* **52,** 83–92.
Lemon, R. E., and Herzog, A. (1969). The vocal behavior of Cardinals and Pyrrhuloxias in Texas. *Condor* **71,** 1–15.
Lemon, R. E., and Scott, D. M. (1966). On the development of song in young Cardinals. *Can. J. Zool.* **44,** 191–199.
Lindauer, M. (1971). "Communication among Social Bees." Harvard Univ. Press, Cambridge, Massachusetts.
McGregor, P. K. (1980). Song dialects in the Corn Bunting. *Z. Tierpsychol.* **54,** 285–297.
Marler, P. (1952). Variations in the song of the Chaffinch (*Fringilla coelebs*). *Ibis* **94,** 458–472.
Marler, P. (1959). Developments in the study of animal communication. *In* "Darwin's Biological Work" (P. R. Bell, ed.), pp. 150–206. Cambridge Univ. Press, London and New York.
Marler, P., and Boatman, D. J. (1951). Observations on the birds of Pico, Azores. *Ibis* **93,** 90–99.
Marler, P., and Isaac, D. (1960). Song variation in a population of Brown Towhees. *Condor* **62,** 272–283.
Marler, P., and Isaac, D. (1961). Song variation in a population of Mexican Juncos. *Wilson Bull.* **73,** 193–206.
Marler, P., and Mundinger, P. C. (1971). Vocal learning in birds. *In* "The Ontogeny of Vertebrate Behavior" (H. Moltz, ed.), pp. 389–450. Academic Press, New York.
Marler, P., and Mundinger, P. C. (1975). Vocalizations, social organization, and breeding biology of the Twite, *Acanthis flavirostris*. *Ibis* **117,** 1–17.
Marler, P., and Tamura, M. (1962). Song "dialects" in three populations of White-crowned Sparros. *Condor* **64,** 368–377.
Marler, P., and Tamura, M. (1964). Culturally transmitted patterns of vocal behavior in sparrows. *Science* **146,** 1483–1486.
Marler, P., Kreith, M., and Tamura, M. (1962). Song development in hand-raised Oregon Juncos. *Auk* **79,** 12–30.
Martens, J. (1979). Gesang und Verwandtschaft des Steinortolan (*Emberiza buchanani*). *Nat. Mus.* **109,** 337–343.
Martens, J. (1980). Lautausserungen, verwandtschaftliche Beziehungen und Verbreitungsgeschichte asiatishcer Laubsänger (*Phylloscopus*). *Adv. Ethol.* **22,** 1–77.
Martin, D. J. (1977). Songs of the Fox Sparrow. I. Structure of song and its comparison with song in other Emberizidae. *Condor* **79,** 209–221.
Martin, D. J. (1979). Songs of the Fox Sparrow. II. Intra- and interpopulation variation. *Condor* **81,** 173–184.
Martindale, S. (1980a). A numerical approach to the analysis of Solitary Vireo songs. *Condor* **82,** 199–211.
Martindale, S. (1980b). On the multivariate analysis of avian vocalization. *J. Theor. Biol.* **83,** 107–110.
Messmer, E., and Messmer, I. (1956). Die Entwicklung der Lautäusserungen und einiger Verhaltensweisen der Amsel *Turdus merula*. *Z. Tierpsychol.* **13,** 341–441.

Metzmacher, M., and Mairy, F. (1972). Variations geographiques de la figure finale du chant du Pinson des Arbres, *Fringilla coelebs coelebs* L. *Gerfaut* **62,** 215–243.

Milligan, M., and Verner, J. (1971). Inter-population song dialect discrimination in the White-crowned Sparrow. *Condor* **73,** 208–213.

Mirsky, E. N. (1976). Song divergence in hummingbird and junco populations on Guadalupe Island. *Condor* **78,** 230–235.

Møller, A. P. (1977). Variationer i Gulspurvens sang. *Feltornithologen* **3,** 166–167.

Mulligan, J. A. (1963). A description of Song Sparrow song based on instrumental analysis. *Proc. 13th Int. Ornithol. Congr.* **1,** 273–284.

Mulligan, J. A. (1966). Singing behavior and its development in the Song Sparrow, *Melospiza melodia. Univ. Calif., Berkeley, Publ. Zool.* **81,** 1–76.

Mulligan, J. A., ed. commentator (1975). *AIBS Symp. Bird Song Dialects, Washington Univ., St. Louis, Mo.* 6 tape cassettes.

Mundinger, P. C. (1970). Vocal imitation and individual recognition of finch calls. *Science* **168,** 480–482.

Mundinger, P. C. (1975). Song dialects and colonization in the House Finch, *Carpodacus mexicanus,* on the east coast. *Condor* **77,** 407–422.

Mundinger, P. C. (1979). Call learning in the Carduelinae: Ethological and systematic considerations. *Syst. Zool.* **28,** 270–283.

Mundinger, P. C. (1980). Animal cultures and a general theory of cultural evolution. *Ethol. Sociobiol.* **1,** 183–223.

Nicolai, J. (1959). Familientradition in der Gesangsentwicklung des Gimpels (*Pyrrhula pyrrhula* L.). *J. Ornithol.* **100,** 39–46.

Nicolai, J. (1973). Das Learnprogramm in der Gesangsausbildung der Strohwitwe *Tetaenura fischeri* Reichenow. *Z. Tierpsychol.* **32,** 113–138.

Nottebohm, F. (1969). The song of the Chingolo, *Zonotrichia capensis,* in Argentina: Description and evaluation of a system of dialects. *Condor* **71,** 299–315.

Nottebohm, F. (1975). Continental patterns of song variability in *Zonotrichia capensis:* Some possible ecological correlates. *Am. Nat.* **109,** 605–634.

Nottebohm, F. (1976). Phonation in the Orange-winged Amazon parrot, *Amazona amazonica. J. Comp. Physiol.* **108,** 157–170.

Orejuela, J. E., and Morton, M. L. (1975). Song dialects in several populations of mountain White-crowned Sparrows (*Zonotrichia leucophrys oriantha*) in the Sierra Nevada. *Condor* **77,** 145–153.

Payne, R. B. (1973a). Wingflap dialects in the Flappet Lark *Mirafra rufocinnamomea. Ibis* **115,** 270–274.

Payne, R. B. (1973b). Parasitic Indigobirds of Africa. *Orinthol. Monogr.* **11.**

Payne, R. B. (1978). Microgeographic variation in songs of Splendid Sunbirds *Nectarinia coccinigaster:* Population phenetics, habitats, and song dialects. *Behaviour* **65,** 282–308.

Payne, R. B. (1981). Population structure and social behavior: models for testing ecological significance of song dialects in birds. *In* "Natural Selection and Social Behavior: Recent Research and New Theory" (R. D. Alexander and D. W. Tinkle, eds.), pp. 108–119. Chiron, New York.

Payne, R. B., and Budde, P. (1979). Song differences and map distances in a population of Acadian Flycatchers, *Epidonax virescens. Wilson Bull.* **91,** 29–41.

Payne, R. B., and Payne, K. (1977). Social organization and mating success in local song populations of Village Indigobirds, *Vidua chalybeata. Z. Tierpsychol.* **45,** 113–173.

Petrinovich, L., Patterson, T., and Baptista, L. F. (1981). Song dialects as barriers to dispersal: A reappraisal. *Evolution* **35,** 180–188.

Pickstock, J. C., Krebs, J. R., and Bradbury, S. (1980). Quantitative comparison of sonagrams using an automatic image analyser: Application to song dialects of Chaffinches *Fringilla coelebs*. *Ibis* **122,** 103–109.

Poulsen, H. (1951). Inheritance and learning in the song of the Chaffinch (*Fingilla coelebs*). *Behaviour* **3,** 216–228.

Rice, J. O., and Thompson, W. L. (1968). Song development in the Indigo Bunting. *Anim. Behav.* **16,** 462–469.

Rich, T. (1981). Microgeographic variation in the song of the Sage Sparrow. *Condor* **83,** 113–119.

Robinson, F. N. (1974). The function of vocal mimicry in some avian displays. *Emu* **74,** 9–10.

Robinson, F. N. (1975). Vocal mimicry and the evolution of bird song. *Emu* **75,** 23–27.

Romanowski, E. (1978). Der Gesang von Sumf- und Weidenmeise (*Parus palustris* und *Parus montanus*)—Variation und Funktion. *Vogelwarte* **29,** 235–253.

Rothstein, S. I., and Fugle, G. N. (1980). Vocal variation in a social parasite, the Brown-headed Cowbird. Report under co-operative aid agreement. *U.S. For. Serv., PSW* No. 60.

Russel, K. (1976). Migrant Golden-winged Warbler with a bivalent repertoire. *Auk* **93,** 178–179.

Samson, F. B. (1978). Vocalizations of Cassin's Finch in northern Utah. *Condor* **80,** 203–210.

Samuel, D. E. (1971). Vocal repertoires of sympatric Barn and Cliff swallows. *Auk* **88,** 839–855.

Seibt, U. (1975). Instrumentaldialekte der Klapperlercke *Mirafra rufocinnamomea* (Salvadori). *J. Ornithol.* **116,** 103–107.

Shiovitz, K. A., and Thompson, W. L. (1970). Geographic variation in song composition of the Indigo Bunting, *Passerina cyanea*. *Anim. Behav.* **18,** 151–158.

Sick, H. (1939). Über die Dialektbildung beim Regenruf des Buchfinken. *J. Ornithol.* **87,** 568–592.

Slater, P. J. B., and Ince, S. A. (1979). Cultural evolution in Chaffich song. *Behaviour* **71,** 146–166.

Smith, G. T., and Robinson, F. N. (1976). The Noisy Scrub-bird: An interim report. *Emu* **76,** 37–42.

Smith, S. T. (1972). Communication and other social behavior in *Parus carolinensis*. *Publ. Nuttall Ornithol. Club* No. 11.

Smith, W. J., Pawlukiewicz, J., and Smith, S. T. (1978). Kinds of activities correlated with singing patterns of the Yellow-throated Vireo. *Anim. Behav.* **26,** 862–884.

Snow, B. K. (1977). Territorial behavior and courtship of the male Three-wattled Bellbird. *Auk* **94,** 623–645.

Snow, D. W. (1968). The singing assemblies of Little Hermits. *Living Bird* **7,** 47–55.

Sparling, D. W. (1979). Evidence for vocal learning in prairie grouse. *Wilson Bull.* **91,** 618–621.

Sparling, D. W., and Williams, J. D. (1978). Multivariate analysis of avian vocalizations. *J. Theor. Biol.* **74,** 83–107.

Strange, W., and Jenkins, J. J. (1978). The role of linguistic experience in the perception of speech. *In* "Perception and Experience" (R. D. Walk and H. L. Pick, eds.), pp. 125–169. Plenum, New York.

Tasker, R. R. (1955). Chipping Sparrow with song of Clay-colored Sparrow at Toronto. *Auk* **72,** 303.

Thielcke, G. (1965). Gesangsgeographische Variation des Gartenbaumläufers (*Certhia brachydactyla*) im Hinblick auf das Artbildungsproblem. *Z. Tierpsychol.* **22,** 542–566.

Thielcke, G. (1969). Geographic variation in bird vocalizations. *In* "Bird Vocalizations" (R. A. Hinde, ed.), pp. 311–340. Cambridge Univ. Press, London and New York.

Thielcke, G. (1972). Waldbaumläufer (*Certhia familiaris*) ohmen artfremdes Signal nach und reagieren darauf. *J. Ornithol.* **113,** 287–295.

Thielcke, G. (1973). Uniformierung des Gesanges der Tannenmeiser (*Parus ater*) durch Lernen. *J. Ornithol.* **114,** 443–454.

Thielcke, G., and Linsenmair, K. E. (1963). Zur Geographischen Variation des Gesanges des Zilpzalps, *Phylloscopus collybita,* in Mittel-und Südwesteuropa mit einem Vergleich des

Gesanges der Fitis, *Phylloscopus trochilus*. *J. Ornithol.* **104,** 372–402.

Thielcke, G., and Thielcke, H. (1960). Akustisches Lernen verschieden alter schallisolierter Amseln (*Turdus merula* L.) und die Entwicklung erlenter Motive ohne and mit künstlichem Einfluss von Testosteron. *Z. Tierpsychol.* **17,** 211–244.

Thielcke, G., Wüstenberg, K., and Becker, P. H. (1978). Reaktionen von Zilpzalp und Fitis (*Phylloscopus collybita, P. trochilus*) auf verschiedne Gesangsformen des Zilpzalps. *J. Ornithol.* **119,** 313–326.

Thönen, W. (1962). Stimmgeographische, Ökologische und verbreitungsgeschichtliche Studien über die Mönchsmeise (*Parus montanus*). *Ornithol. Beob.* **59,** 101–172.

Thomas, B. T. (1979). Behavior and breeding of the White-bearded Flycatcher (*Conopias inornata*). *Auk* **96,** 767–775.

Thompson, W. L. (1968). The songs of five species of *Passerina*. *Behaviour* **31,** 261–287.

Thompson, W. L. (1970). Song variation in a population of Indigo Buntings. *Auk* **87,** 58–71.

Thompson, W. L., and Jane, P. L. (1969). An analysis of catbird song. *Jack-Pine Warbler* **47,** 115–125.

Thorpe, W. H. (1958). The learning of song patterns by birds, with especial reference to the song of the Chaffinch, *Fringilla coelebs*. *Ibis* **100,** 535–570.

Thorpe, W. H. (1972). Duetting and antiphonal song in birds. *Behaviour, Suppl.* No. 18.

Todt, D. (1970). Die antiphonen Paargesänge des ostafrikanischen Grassänger *Cisticola hunteri prinioides* Neumann. *J. Ornithol.* **111,** 332–356.

Trainer, J. M. (1980). Comments on a kin association model of bird song dialects. *Anim. Behav.* **28,** 310–311.

Treisman, M. (1978). Bird song dialects, repertoire size, and kin association. *Anim. Behav.* **26,** 814–817.

Tretzel, E. (1965). Imitation und Variation von Schäferpfiffen durch Haubenlerchen (*Galerida c. cristata* L.). Ein Beispiel für spezielle Spottmotiv-Prädisposition. *Z. Tierpsychol.* **22,** 784–809.

Tretzel, E. (1967). Imitation und Transposition menschlicher Pfiffe durch Amseln (*Turdus m. merula* L.). Ein weiterer Nachweiss relativen Lernens und akustischer Abstraktion bei Vögeln. *Z. Tierpsychol.* **24,** 137–161.

Verner, J. (1975). Complex song repertoire of male Long-billed Marsh Wrens in eastern Washington. *Living Bird* **14,** 263–300.

Ward, R. (1966). Regional variation in the song of the Carolina Chickadee. *Living Bird* **5,** 127–150.

Ward, W. V. (1964). The songs of the Apapane. *Living Bird* **3,** 97–117.

Wickler, W. (1972). Aufbau und Paarspezifität des Gesansduettes von *Laniarius funebris*. *Z. Tierpsychol.* **30,** 464–476.

Wildenthal, J. L. (1965). Structure in primary song of the Mockingbird (*Mimus polyglottos*). *Auk* **82,** 161–189.

Wiley, R. H. (1971). Song groups in a singing assembly of Little Hermits. *Condor* **73,** 28–35.

Wiley, R. H. (1976). Communication and spacial relationships in a colony of Common Grackles. *Anim. Behav.* **24,** 570–584.

Wilkinson, R., and Howse, P. E. (1975). Variation in the temporal characteristics of the vocalizations of Bullfinches, *Pyrrhula pyrrhula*. *Z. Tierpsychol.* **38,** 200–211.

Williams, L., and MacRoberts, M. H. (1977). Individual variation in songs of Dark-eyed Juncos. *Condor* **79,** 106–112.

Williams, L., and MacRoberts, M. H. (1978). Song variation in Dark-eyed Juncos in Nova Scotia. *Condor* **80,** 237–304.

Zann, R. (1975). Inter- and intraspecific variation in the calls of three species of grassfinches, *Poephila* (Gould) (Estrildidae). *Z. Tierpsychol.* **39,** 85–125.

7

Genetic Population Structure and Vocal Dialects in *Zonotrichia* (Emberizidae)

MYRON CHARLES BAKER

I. INTRODUCTION

A. Evolutionary Rate

Ecological literature is replete with studies that describe the distribution of plant and animal species in space and time; often these species are important food resources for other organisms. Over approximately the same historical period, ethologists obtained detailed knowledge of the structure of many animal so-

ACOUSTIC COMMUNICATION IN BIRDS
VOLUME 2

ISBN 0-12-426802-1

cieties. Then, more recently, ecological geneticists have provided numerous descriptions of the spatial patterns of gene frequencies in natural populations. Emerging in contemporary work is the realization that these three research areas complement one another, suggesting a theoretical construct: resource patterns effect social structure and the societies so produced effect structure in the gene pool.

Communicative interactions are the essence of all animal societies, so it is appropriate to examine communication as a way to gain insight into social structure and thereby generate hypotheses about genetic structure. Random mating is rare in vertebrates (a likely exception is the American eel, *Anguilla rostrata;* Jones, 1968), so our task is to focus on communicative signals that may cause or maintain departures from panmixia, particularly in relation to population subdivision.

It is worth pointing out a suggestive pattern in the class Aves, with respect to the approach to genetic population structure by study of vocal communication (Nottebohm, 1972). Birds can be divided into passerines and non-passerines. These groupings differ in an interesting way. Passerines include many species and average about 100 species per family, whereas non-passerines average about 40 species per family (Mayr and Amadon, 1951). Further subdivision of passerines into sub-oscines (non-songbirds) and oscines (songbirds) reveals that oscines average 111 species per family (as high as 150 depending upon families recognized) and the sub-oscines average 77 species per family (Mayr and Amadon, 1951). These differences in taxonomic diversity may be related to vocal learning, which is much more prominent in passerines than non-passerines, and in oscines than sub-oscines. Dialects in the songs of oscine species, suggestive evidence of vocal learning, have been widely documented by many workers.

A simple and early hypothesis is that song dialects act as partial barriers to gene migration (Nottebohm, 1969), with each dialect population acting as a deme. This view was also favored in the subsequent work of Payne (1973) and Baptista (1975). A consequence of a deme–dialect correspondence is that each dialect population could evolve somewhat independently. Dialect groups could evolve to protect local environmental adaptations from being swamped, or could follow a founder effect with subsequent emergence of coadapted gene complexes. In any case, the overall genetic variance of the *meta*-population comprising the several dialect subpopulations would be increased and highly responsive to evolutionary pressures (Fisher, 1930/1958).

The purpose of this chapter is to explore the relationship between song dialects and genetic population structure in sparrows of the genus *Zonotrichia.* The most intensive work on this question is on some populations of a nonmigratory White crowned Sparrow (*Zonotrichia leucophrys nuttalli;* Baker, 1974, 1975; Baker and Mewaldt, 1978; Baptista, 1975). Some relevant comparative data exist for several other *Zonotrichia* species, however, and these will be incorporated where useful.

Nine species are considered, each to an extent commensurate with existing data, and represent an attractive series of species upon which to build a more comprehensive theory of the causes and consequences of song variation (Table I, Fig. 1). Species in this group that are highly fragmented racially tend to have high levels of song variation on a geographic basis. Rufous-collared (see Table I for Latin binomens), White-crowned, and Song sparrows are species in which we find song variation over large geographic scale distances as well as local population differences, the latter commonly called dialects. Less complete information is available for Fox Sparrows, but existing data suggest it to have broad-scale variation in song and probable dialects (Martin 1977, 1979).

TABLE I

Some Important Characteristics of *Zonotrichia* Sparrows[a]

Sparrow species	Indication of morphological variation	Migratory status	Song variation (songs/bird)
White-throated (*Z. albicollis*)	1 race, plumage polymorphic	Virtually all migrate	Usually one
Golden-crowned (*Z. atricapilla*)	1 race, plumage polymorphic	Virtually all migrate	Usually one
Harris' (*Z. querula*)	1 race, plumage polymorphic	All migrate	Usually one
Lincoln's (*Z. lincolnii*)	3 races	Virtually all migrate	Often two
Swamp (*Z. georgiana*)	3 races	Virtually all migrate	Two to five
White-crowned (*Z. leucophrys*)	5 races	Most migrate	Usually one
Fox (*Z. iliaca*)	18 races	Virtually all migrate	Two to three
Rufous-collared (*Z. capensis*)	22 races	Few migrate, some altitudinally	Usually one
Song (*Z. melodia*)	34 races	Many but not all migrate	Five to ten

[a] Sources: American Ornithologists' Union (1957); Chapman (1940); Bent (1968); Miller (1956); Lowther (1961); Morony *et al.* (1975); D. E. Kroodsma (unpublished data).

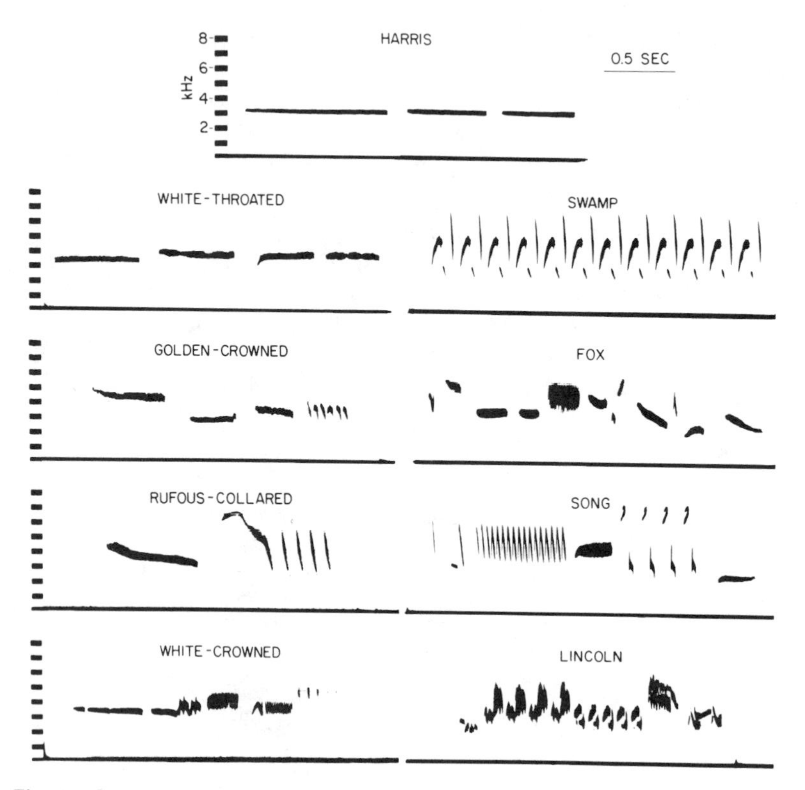

Fig. 1. Sonagrams of nine species of *Zonotrichia* (analyzing filter bandwidth, 300 Hz; prepared from tapes provided by the Laboratory of Ornithology, Cornell University, courtesy of James L. Gulledge, and from recordings made by M. C. Baker and G. L. Sherman).

By contrast, Harris', Golden-crowned, Lincoln's, and Swamp sparrows are less differentiated racially. Lacking descriptive sonagraphic studies of geographic variation in song, one can only assert that these species very likely have such variability. Only for White-throated Sparrows do we have sufficient data to say that song variation appears to be less than in multirace species. Within the distances over which one might anticipate dialects to occur, they seem to be lacking (Borror and Gunn, 1965; Lemon and Harris, 1974).

On the whole, then, there may be a trend toward increasing morphological variability with increasing song variability. Neither the variability within populations nor between populations has been adequately analyzed for either songs or morphology in these species. For certain other organisms, however, it appears that those characters that are highly variable within populations are usually highly divergent between populations (Farris, 1970; Kluge and Kerfoot, 1973; Sokal, 1976; Johnson and Mickevich, 1977; Pierce and Mitton, 1979). Such a correlation between evolutionary rate and within-population variability has been shown for morphological traits as well as allozymes. The most incisive analysis of the manner in which patterns of population differentiation could have originated and how these differing modes of origin produce variation within and among populations has been performed recently by Sokal (1978).

B. Interpretations of Singing Behavior

The definition of dialect is somewhat arbitrary and it can be argued that there is often no compelling reason to assert that some species have dialects but others do not. What seems to be meant by such an assertion is that where there is no fine scale or microgeographic level of population difference in song, there are no dialects. But perhaps the size of a region comprising comparative homogeneity of song should not be the decisive criterion. Instead, one could adopt an operational approach and, for example, record songs of individuals living differing distances from a local population and then ask: At what distance from the local group does a song stimulus elicit a response differing from that caused by playback of a local song? Such a question could be addressed to females as well as to males.

Since this chapter considers the possible evolutionary consequences of communication signals, it is appropriate at the outset to consider anew the question of how a signal can be used in multifunctional contexts. All the species considered here have a song that most workers suspect (or assume) is used for both territorial defense and mate attraction. Other functions have been postulated but are not explored here. A key observation is that when a male becomes mated his rate of singing decreases, often dramatically (Armstrong, 1963). An important experiment was performed to clarify the mate attraction function of song in White-throated Sparrows (Wasserman, 1977). First, it was noted that color-banded males decreased singing rates once mated. When the female mate was removed, the song rate of her mate rose immediately to that rate characteristic of an unmated male, but it decreased again when he attracted a new mate. There are at least two simple interpretations. Either the newly mated male has new time demands and therefore less time available for singing, or the difference between the earlier high rate of singing and the lower present rate can be assigned to the mate-attraction function. Clearly, an increase was not needed for territorial defense since that was accomplished during the mated period with less song. Although the time demand hypothesis is not rejected, I find it may be less compelling. Males can and do sing while engaged in other activities such as foraging; they simply stop for a few seconds, sing, then continue. It is possible that a male may signal his unmated status by a high rate of song production, over and above the level of singing that is actually necessary for maintaining a territory.

A problem then arises in the interpretation of the observed increased rate of singing by a mated male who is the subject of a playback experiment. Such an increase typically occurs when an experimenter plays a song of a nonneighboring male to a target individual (Brooks and Falls, 1975). The usual interpretation of the male's higher rate of song during playback is that he is exhibiting increased aggression toward an interloper; it is viewed as heightened defense of territory. A different view, with the results of Wasserman's experiment in mind, is that the experimental male increases his singing as an extra effort to retain his mate in the face of competition. This interpretation also provides an alternative explanation for the observation that the experimental male sings at a high rate to both neighbor and nonneighbor when the playback speaker resides in the middle of the territory rather than on the border (e.g., Wunderle, 1978; Falls and Brooks, 1975). The common interpretation is that the neighbor's song is not recognized when it comes from an unusual location. It could be that a neighbor in its normal place is not a competitor for the resident's female but is a serious threat when in a location other than normal.

II. POPULATION GENETIC CONSEQUENCES OF NONRANDOM MATING

Patterns of nonrandom mating influence the genetic architecture of a population in several ways. Theory exists to guide our thinking on these influences, and here I treat briefly several major concepts which concern the impact of mating systems on genetic population structure. In a later section I apply these methods to some data.

Most population genetics theory considers processes that occur in simple randomly mating populations of infinite size. Random mating and infinite size are tempting assumptions, for they greatly simplify mathematical considerations. Practical needs of animal breeders, however, have stimulated investigations into the influences of nonrandom mating patterns in finite populations on the gene pool of a population (Wright, 1921). Wright's work systematically lays out the consequences of different systems of mating, but the application of these results is only slightly helpful to the researcher of natural populations. Useful to the field biologist is the concept of effective population number (N_e), also attributable to Wright (1931).

In a population of N_m males and N_f females, the effective number, N_e, is:

$$N_e = \frac{4N_mN_f}{N_m + N_f} \tag{1}$$

If, for example, a population consists of 10 males and 100 females, N_e is approximately 36. This tells us that this nonmonogamous breeding system, e.g., polygyny or promiscuity, is equivalent genetically to 18 randomly mated pairs. Both stochastic and deterministic influences on gene frequencies in a population are strongly affected by population size, but it is the effective size N_e that is important and not the census number of individuals in the population.

Effective size may also be reduced below the census size by population fluctuations, according to the relationship:

$$N_e = \frac{n}{\sum_{i=1}^{n}} \quad N_i \tag{2}$$

in which n is the number of generations and N_i the number of individuals counted each generation. As a harmonic mean the value of N_e is mainly determined by the small values of population size (Crow and Kimura, 1970).

In a population of constant size, the number of progeny per parent may also influence the genetically effective size due to the variance in numbers of gametes contributed to the next generation. When the variance in progeny number exceeds the mean, the effective number is found by:

$$N_e = \frac{4N - 2}{V_k + 2} \tag{3}$$

in which V_k is the variance of K, the number of gametes provided by N parents (Wright, 1938). It is likely that the variance exceeds the mean in most natural populations, thus reducing the effective size.

The effective size of a population is also affected strongly by gene dispersal. The distribution of distances from the sites of hatching to the sites of breeding by a cohort of individuals forms the zygote-to-zygote dispersal distribution which is an indirect measure of gene flow. The shape of the dispersal distribution predicts the genetic neighborhood population from which there is a high probability of gametes uniting to form the next generation (Wright, 1943).

If a dispersal distribution is skewed right with a short mean dispersal distance and a small variance, for example, this implies a very restricted pool of birds from which an individual may find a mate. Wright's theory of neighborhood size lets us make statements about inbreeding effects in a population, and consequently we can calculate a first approximation to N_e in a population knowing only the zygote-to-zygote dispersal distribution.

In effect, what is done in the calculation of neighborhood size is to take a given location in a population with a known dispersal distribution and imagine rotating the distribution around the point. This generates a probability density surface which we can imagine superimposed onto the breeding habitat. Knowing the density of breeders in this circle gives the neighborhood population. The physical structure of the environment plays a modifying role since the area inscribed is seldom uniform or uniformly populated. It is perhaps common, therefore, that the area-continuity assumption of the neighborhood analysis is not met, and dealing with truncated dispersal distributions and habitat discontinuities will always necessitate careful application of the theory.

Few good published data on dispersal distributions in birds exist. In the genus *Zonotrichia,* a reasonably accurate quantitative account of dispersal exists for Song Sparrows (Nice, 1937/1964; Johnston, 1956; Halliburton and Mewaldt, 1976) and for a population of White-crowned Sparrow (Z. *leucophrys nuttallii*

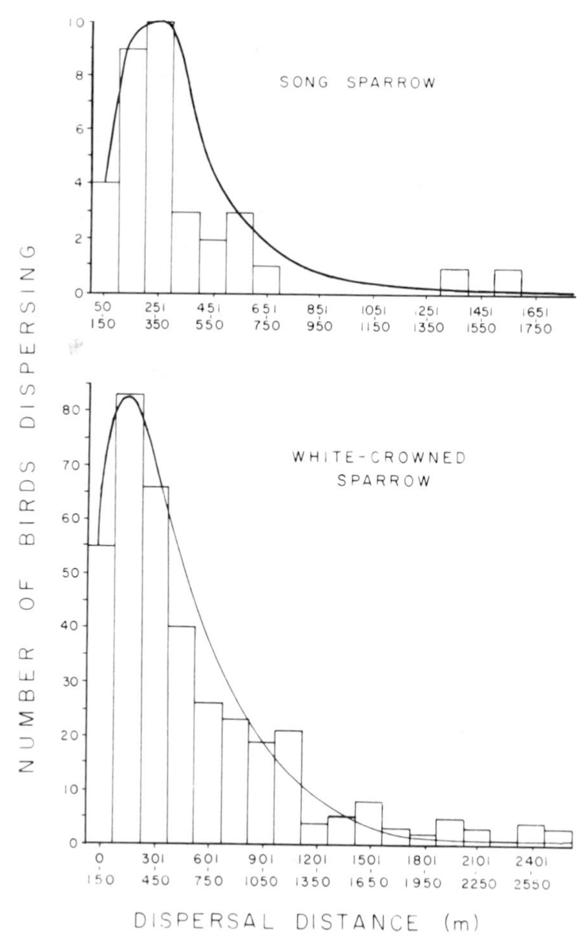

Fig. 2. Upper: Distance from point of hatching at which young Song Sparrows settle for breeding (curve drawn by eye). Lower: Distance from point of first capture as fledglings, at which young White-crowned Sparrows settle for breeding. Curve fitted to gamma distribution. (From Baker and Mewaldt, 1978; Miller, 1947.)

Baker and Mewaldt, 1978) (Fig. 2). In these two species, using the method just described, there is a neighborhood population of roughly 100 adults (Miller, 1947; Baker, 1981).

Neighborhood size is only one population parameter that will influence genetic structure. For songbirds in particular, we need to ask: What role does the existence of vocal dialects play in the analysis of genetic structure? A possibility exists that vocal dialects, indeed any kind of discontinuity in communication signals, may imply discontinuity in the pattern of gamete combination.

Most evolutionists are familiar with the conventional thinking about genetic integrity at the species level and the classical processes of isolating mechanisms, hybridization, and introgression. There is good reason to suppose that similar, probably less absolute, processes occur at other levels within populations. Among *Zonotrichia,* for example, species have different songs (Fig. 1), and morphological races within a species have different songs (Fig. 3). Song mapping of some of the morphological races of *Zonotrichia* reveals dialects, and within dialects one encounters a mosaic of heterogeneity (Baptista, 1975). Indeed, other researchers have found what may be called subdialects (King, 1972), and we are now searching for familial clusters which one might predict from the song-learning process and dispersal patterns (Treisman, 1978).

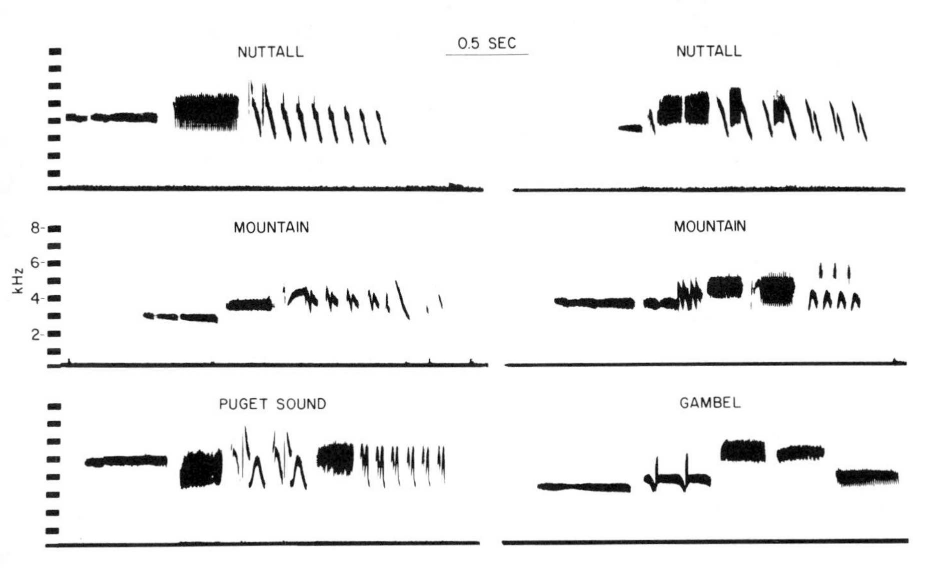

Fig. 3. Sonagrams of *Zonotrichia leucophrys nuttalli* (Nuttall), *Z. l. oriantha* (Mountain), *Z. l. pugetensis* (Puget Sound), and *Z. l. gambelii* (Gambel). The two dialects of *Z. l. nuttalli* and *Z. l. oriantha* were recorded 55 and 240 km apart, respectively. (Analyzing filter bandwidth, 300 Hz.)

As a working hypothesis, I suggest that these hierarchical groupings of vocal patterns imply that there may be hierarchical patterns of genetic structure. Just as in species-level studies, we can expect to find differences in the degree of isolation among song groups (involving perhaps recent contact and hybridization-like processes) and encounter difficulties in assessing the impact on the gene pool. There can be, for example, little doubt that some individuals disperse prior to the completion of song learning. When such an event occurs into an unoccupied habitat the result may be the formation of a new dialect population. A speculative argument along these lines has been made previously to provide a historical model of dialect origin (Baker, 1975).

If the dispersal takes place within an existing population but out of the natal dialect, the individiual may be at a major disadvantage (go unmated, for instance), or there may be gene flow the effects of which may be difficult to assess. We need much more information on the length of the critical period for song learning in a variety of species. It would be tempting, but premature, to suppose that a protracted critical period implies greater rates of gene flow through populations.

On the other hand, there can also be little doubt that many *Zonotrichia* breed near the site of hatching, probably an adaptation which tells us that genomes on the average are adapted to the local environment. What must be emphasized is that behavioral structures, such as dialect groupings, may sometimes reflect important dispersal discontinuities, or may sometimes reflect recency of contact, perhaps in disturbed environments, where dispersal may be more continuous. Recency of contact between species may often be characterized by considerable hybridization (Remington, 1968). It is abundantly clear, however, that song recording alone is not going to take us to a deeper level of understanding.

III. *F* STATISTICS AND POPULATION MODELS

Wright's F statistics are useful in approaching the question of subdivided populations in terms of gene frequency differentiation (Wright, 1965). If a total population (t) is composed of several subpopulations (s), then the inbreeding coefficient of any individual (i) in relation to t is:

$$F_{it} = F_{st} + (1 - F_{st})\, F_{is} \tag{4}$$

in which F_{st} is the inbreeding coefficient of the subpopulation relative to the total population and F_{is} that of the individual relative to the subpopulation. The inbreeding coefficient is the probability of homologous genes being identical by descent (Wright, 1922).

The interpretation of these coefficients is that F_{is} is the effect of inbreeding within the local group, such as by consanguinity, whereas F_{st} is the effect of genetic drift, which causes heterogeneity of gene frequencies among subpopulations. *F* statistics can be calculated from several types of data, including direct observations of the frequency of consanguineous matings and from the gene frequency data obtainable from electrophoretic studies.

The concepts of effective size (N_e) and *F* statistics can be combined to allow predictions about genetic structure in a population from knowledge of dispersal and other variables.The first approximation to N_e, as mentioned earlier, is from the dispersal distribution which allows calculation of the neighborhood area. Information about consanguinity patterns within a neighborhood population, mating systems, mean and variance in number of progeny, fluctuations in population size, and degree of overlap in generations substantially improves the accuracy of the estimate of N_e.

Once N_e is estimated from a set of data, its expected influence on the gene pool can be examined. A simple model of structure is Wright's isolation-by-distance model (Wright, 1943), which is a quantitative statement of the fact that individuals living near each other tend to be more genetically alike than individuals living farther apart. Under such circumstances, and with alleles of small selective value, random differentiation can occur in a population. The extent of differentiation depends upon the value of N_e and the geographic distribution of the population. With small values of N_e and a linear habitat, differentiation can be substantial. With large N_e and continuous two dimensional habitat, differentiation is much less (Crow and Kimura, 1970).

For most bird species, perhaps more relevant is the stepping-stone model of population structure in which the population is distributed in habitat patches and most exchanges occur between adjacent patches (Kimura and Weiss, 1964). When such patches are arranged in a one-dimensional array, genetic differentiation among them can be large, depending upon the rate of gene migration. Dialect populations are often a combination of stepping-stone and isolation-by-distance models.

The size and spatial arrangement of dialect areas vary within and between species. Information is available for *Zonotrichia* sparrows and is supportive of this point. A small sample of dialect regions in the White-crowned Sparrow (*Z. l. nuttalli)* indicates that dialects are about 5–15 km long by 0.5 to 4 km wide (M. C. Baker and D. B. Thompson, unpublished data). Along the outer coastal scrub environment in central California, these dialect populations are strung along in a linear array, much like linked sausages. Physical breaks may occur at the entrances of major lagoons and bays, but as often as not dialects merely abut one another with usually narrow overlap zones. The short-distance migrant race, *Z. l. pugetensis,* has larger dialect regions; these birds inhabit a coastal strip not greatly different in width from *Z. l. nuttalli* but a dialect may be unchanged from 75–150 km distance (Baptista, 1977). In the long-distance migrant, *Z. l. gambelii,* dialect variation is more obscure (DeWolfe *et al.*, 1974) and may not be present at all. *Zonotrichia l. gambelii* and the Golden-crowned Sparrow, both migratory far-northern breeders, would repay more intensive study.

The mountain race of the White-crowned Sparrow, *Z. l. oriantha*, provides an interesting contrast. Sierra Nevada populations exhibit dialect patches of a relatively localized nature, often changing from one meadow to the next (Orejuela and Morton, 1975; Banks, 1964). In this case, population sizes are frequently small and prone to extinction; a breeding group of less than five to ten pairs is not unusual (DeWolfe and DeWolfe, 1962). In the Colorado Rockies, on the other hand, dialect areas and population sizes are often much larger, and sometimes correspond to entire drainage basins (Baker, 1975).

Variations in the structure of dialect groups among the North American *Zonotrichia* species are paralleled by those found in populations of the Rufous-collared Sparrow in Argentina. Dialect regions are enormous in some areas, whereas in others dialects change abruptly over small distances (Nottebohm, 1969). Homogeneity of *Z. capensis* song over hundreds of square kilometers and dialect changes over distances of a few kilometers were described only on the basis of intervals between successive notes in the terminal trill of songs (Nottebohm, 1969). Both King (1972) and Nottebohm (1969) noted that within dialect areas one can identify clusters of neighboring males that sing similar themes. Such themes, described from similarities in the introductory portions of songs, are spatially distributed as patches in a mosaic.

There is clearly enormous variation in size of dialect areas in *Zonotrichia* species, as dialects are usually defined, and it is doubtful that comparisons with other species will turn up much increase in this range. Such variation in size of dialect populations may be expected if dialects are mapping on a heterogeneous environment.

Most studies of spatial variation in song suffer from a lack of objective quantification. Workers either resort to a visual inspection of a large array of sonagrams or perform a few single-variable statistics to map out dialect areas. Such techniques may be quite accurate in the hands of experienced naturalists but comparative studies by several different workers could produce misleading interpretations. We know so little about the communicative significance of the song variations that it seems important to use multivariate statistical procedures and distance measures applied routinely in other areas of quantitative biology (e.g., Payne, 1978). Such a quantitative approach must be complemented by field experiments that ask whether the birds discriminate among dialects or subdialects or themes.

IV. HYPOTHESIS TESTING IN SONG DIALECTS RESEARCH

Progress toward understanding the biological significance of song dialects in birds has been modest. By contrast, human speech dialects and language differences have been more thoroughly explored by physical anthropologists to interpret cultural and genetic population structure (Howells, 1966; Spuhler, 1972; Friedlaender, 1975). In a most provocative analysis of Yanomama Indian populations, a highly significant correspondence was found between genetic and linguistic data (Spielman *et al.*, 1974).

For avian vocal dialects, the first major stimulus for heightened curiosity was the work of Marler and Tamura (1962). They performed the first mapping of song dialect patterns of *Z. l. nuttalli* in the San Francisco Bay area of California. Nottebohm (1969) followed on this start by studying song of Argentinian *Z. capensis* at different locations, and his detailed analysis and focused reasoning about the possible meanings of dialect patterns served as a major incentive to many subsequent studies. Most of the work undertaken by others after these pioneering studies was duplicative rather than breaking new ground. It was clear from the Marler and Nottebohm efforts, for example, that studies of dispersal, female choice, male–male interactions, and genetic population structure would be necessary to test the pertinent hypotheses.

A. Dispersal

Studies of dispersal in the transition area between two dialects can answer the question of whether there is a reduction of migration between dialects. The important dispersal function is the distribution of distances from birth place to breeding site for a sample of individuals. This approach has been taken for only a single dialect transition at the present time (Baker and Mewaldt, 1978). The results indicated that there was less migration between dialects than expected from a free-exchange model (Fig. 4).

A problem for dispersal studies is capturing juveniles at the earliest possible age. In some situations, it is necessary to trap fledglings at a young age rather than band nestlings. In the White-crowned Sparrows referred to above, about 50 to 80% of the nests that are located and visited by observers for the purpose of banding nestlings are subsequently preyed upon, perhaps as a result of the observer's visit (Bart, 1977; Lenington, 1979). Such a disturbance could create a dispersal sink for surrounding populations and befuddle any conclusions about gene flow.

An alternative approach of banding juveniles was used by Baker and Mewaldt (1978) and must be considered in the light of what is known of the general perinatal events in White-crowned Sparrows. Following a 10- to 12-day nestling life, juveniles fledge and spend the next 20 days or so being provisioned by the parents, gradually completing growth of the plumage, and becoming independent in obtaining food. About 20 days later, postjuvenal molt begins. This is a general body molt replacing the speckled, streaked breast so characteristic of young. Our records of the initial captures, at which time the young were banded, indicate that 50% of the birds were not yet molting and an additional 30% were

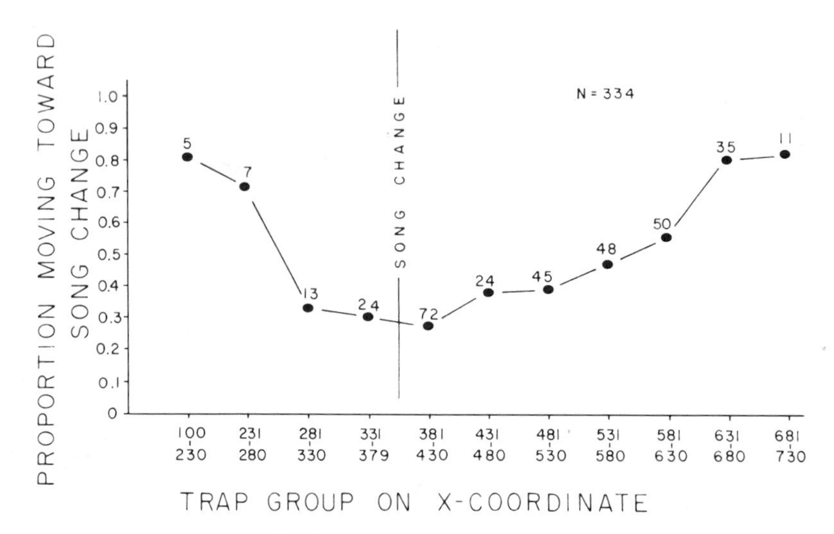

Fig. 4. The relationship of a dialect transition area to dispersal in White-crowned Sparrows. Points represent proportion of birds breeding at a site located toward the song change, relative to their sites of fledging. The abscissa is in meters. (From Baker and Mewaldt, 1978.)

scored as very light body molt. These data, along with our small sample of banded nestlings recaptured several times in early fledgling life, imply that little significant movement had occurred prior to banding the fledglings.

Moreover, about 75% of our sample was banded prior to mid-July. Experience with a series of mist nets in permanent locations also suggests that a major burst of dispersal activity occurs in the latter half of July. The nets capture few juveniles throughout May and June with a gradual increase in early July followed by a large increase in numbers in mid- to late July and early August, indicating significant population-wide movements of juveniles. In collaboration with Professor L. R. Mewaldt and the Point Reyes Bird Observatory, studies on early movements of fledged young and their time in local residence are continuing with intensive observations on banded nestlings as well as saturation trapping and recapture.

B. Female Mate Choice

Behavioral mechanisms that are potentially involved in maintaining a system of dialects could be based on either female mate choice, male–male aggressive interactions, or both. The possibility of female mate choice as a mechanism was suggested by the work of Konishi (1965), who induced female White-crowned Sparrows to sing by use of testosterone implants; they sang the song characteristic of their natal dialect. Thus, the females in normal circumstances seemed to obtain a "template" during an auditory song-learning phase of development. That they might use this stored information in selecting a mate seemed an obvious but speculative and untested possibility. Experiments are currently in progress in my laboratory that address this possibility. In a laboratory setting, we control the song learning process in females and at a later time test their preferences for different dialects. Results of experiments performed on females from a Colorado population of *Z. l. oriantha* indicate that these females respond strongly to their natal dialect and virtually not at all to an alien dialect (M. C. Baker *et al.*, 1982a).

Pertinent here are recent field results of Baptista and Morton (1980) and of Petrinovich *et al.* (1981). In *Z. l. nuttalli* in San Francisco and in *Z. l. oriantha* in the Sierra Nevada range, it was found that when some of the mated females in one population were injected with testosterone they developed a song resembling another dialect. It is not yet known what proportion of a population follows this pattern or how general the phenomenon is. Experiments using a technique similar to that of L. F. Baptista (unpublished data) and Petrinovich *et al.* (1981) but applied to females obtained in the Point Reyes area, Marin County, California, gave different results. Thus far in experiments completed, females of two different dialects sing the song type characteristic of the local resident males and not an alien dialect (D. F. Tomback and M. C. Baker, unpublished data). To analyze the population genetic consequences we need two types of information. First, we need a statistical approach that describes the frequency of the assortative mating, its year-to-year fluctuation, and whether it occurs throughout the population or mostly at the borders. Second, we need information on the fertility of matings to determine any possible differential between positive and negative assortative matings. For the time being, it is important to note that the results obtained by Baptista and Morton (1980) and Petrinovich *et al.* (1981) do seem to demonstrate that the female migrants learned their song before dispersal from their natal dialect (see also Baker and Mewaldt, 1981).

Dispersal studies indicate that males are more philopatric than females (Baker and Mewaldt, 1978). Taken together with the results of female song induction noted above, we obtain a natural analog to the theoretical model of Rohlf and Schnell (1971). This model simulates genetic differentiation among subpopulations of various effective sizes and arrayed in differing geographic patterns. A further variation is considered which allows for different types of mating patterns. Pertinent to the present discussion is the model in which one sex is fixed in space and the mate is selected at random from the surrounding population. Geographic variation in allelic frequencies can be substantial under such conditions.

Depending upon the selection coefficients involved, the recency of contact between dialect populations, and the local reproductive success, we should expect that differing degrees of assortative mating will be encountered. It is important to keep in mind the work of O'Donald (1974, 1976; O'Donald and Davis, 1977) showing that to maintain the distinct color phase polymorphism in Arctic Skuas, whose selection coefficients have been estimated, less than 50% of the females in the population need to exhibit mating preferences for dark and intermediate phase males; the rest can mate at random.

This particular line of thought emphasizes the statistical nature of gene flow among and within dialect populations. We are not dealing with species' borders but rather the varying degrees to which populations are made more or less viscous, with respect to gene flow, by dialects. The issue is a broader one than can be examined here in detail. It is only comparatively recently that attempts are being made to obtain an integrated view of how dispersal, population subdivision, and mating systems affect genetic population structure (Baker and Marler, 1980).

C. Male–Male Interactions

We next review several hypotheses and some data that suggest an important role for male–male interactions in maintaining a system of dialects. For example, if males of one dialect population exclude males of another dialect by aggressive reaction, then a song playback experiment should reveal increased territorial aggression by a resident male when he receives a playback of a foreign song dialect. An early playback study on *Z. l. nuttalli* (Milligan and Verner, 1971) asked the question: Do males of different dialects discriminate their home dialect from other dialects? The historical timing of that study (the research was done in 1964–1965) was just following the spectrographic quantification of dialects by Marler and Tamura (1962), and of importance then was to find out if the birds themselves paid any attention to the observed song differences among populations. The study suggested that males of a given dialect area reacted more aggressively to a song from their own dialect region than to songs from distant dialects. Viewing this result from a post hoc basis argues against the effect of male–male interaction as being responsible for any important reduction in cross-dialect migration. Indeed, it could imply the opposite effect; males of a foreign dialect should find it easy to invade and establish a territory in another dialect population.

More recently (Baker *et al.*, 1981b) (Fig. 5) it has been discovered that a song from a contiguous dialect area (Buzzy) can elicit a greater singing response in a target male compared to a song from a nonneighboring individual from his own (Clear) dialect. This result is extremely interesting and reopens the possibility that male–male interactions are an important influence on the exchange of individuals between contiguous dialect areas. In the same set of experiments, song from a dialect area 55 km away (Bodega) elicited a weaker response from

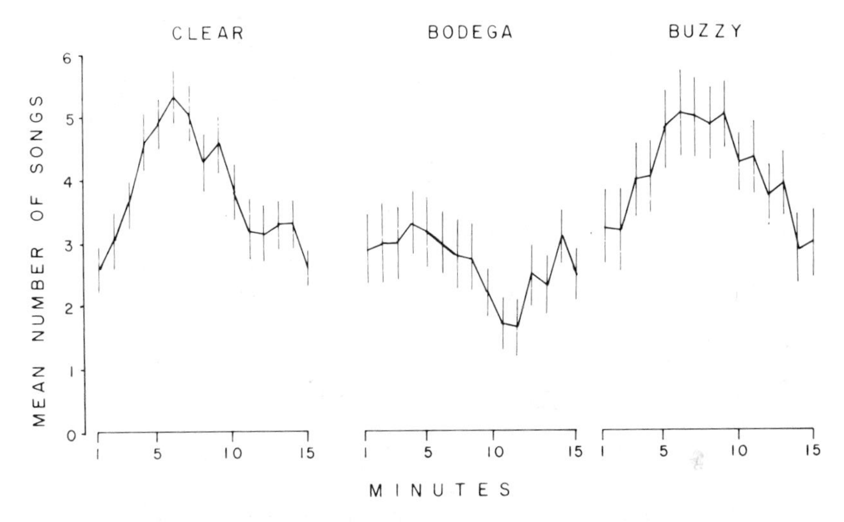

Fig. 5. Response of experimental "Clear" males to playback of three different dialects. Clear is the home dialect, Bodega the distant dialect, and Buzzy the dialect contiguous with Clear. Vertical lines are ± 1 SE. Twenty songs were played during the first 5 min, while the remaining 10 min were silent. (From Baker *et al.*, 1981b.)

territorial males compared to their home dialect. This latter finding confirms the results of Milligan and Verner (1971) who tested dialect responses only among populations far distant from one another.

An additional study has now been completed on the adjacent Clear and Buzzy dialects. In the new study, experiments were conducted on both Clear and Buzzy subjects with the same stimulus songs. One group of males in each dialect received playback of the local song type and another group in each dialect received playback of the song type from the neighboring dialect. Results similar to those of the earlier (Baker *et al.*, 1981b) study were obtained; males sang more frequently to stimuli representing the neighboring dialect than to those representing the home dialect (M. C. Baker *et al.*, unpublished data). In contrast to these results, similar experiments in another dialect in the same dialect system along the coast of the Point Reyes National Seashore have obtained different results. A greater singing rate was elicited by the home dialect in relation to two neighboring ones (D. F. Tomback *et al.*, unpublished data). Other playback experiments are in progress to search further beyond these interesting results.

A somewhat different view of the male role in maintenance or production of dialect systems is the "vocal convergence" or "social adaptation" hypothesis which assumes an advantage to singing a song similar to that of an established male (Payne, 1978). At least for one case (Payne, 1978) the song-copying process leading to convergence among males was thought to occur after post-juvenile dispersal, although no data on dispersal were available. Clearly, though, the same advantages one could imagine to accrue to postdispersal males would also be gained by males who do not disperse. The distinctive element of the social adaptation hypothesis is its focus on male–male interactions as the causal evolutionary pressure for vocal similarity among males.

Relatively greater homogeneity among songs of males within a population, whether the population be classified as a subdialect, a dialect, or some larger unit, as compared to the divergence among populations, raises the possibility that something akin to normalizing selection may be causal. The concept of normalizing selection is most commonly applied to morphological traits, but this is not a necessary restriction of the idea. Selection is the differential reproduction by phenotypes in a population; song is part of the phenotype. Under normalizing selection, a bird which learned and then sang a "different" song, an extreme in one tail of the distribution, would be less successful than birds with songs nearer the average. Selection may favor conformity. Success may be measured as the capacity to obtain a territory and attract a mate in the case of males, or in the case of a female, to recognize a male from the same natal population possessing a similar coadapted genome.

A cultural method of song transmission from generation to generation need not imply the absence of genetic influences. Learning experiments clearly indicate that a model song must be within the species' typical range. For example, it now appears that the actual morphology of syllables is an essential feature of song that is learned selectively by maturing young Swamp Sparrows (Marler and Peters, 1977). Whether there is any similar predisposition to learn selectively the population-characteristic syllables (dialect, subdialect, or racial) is not known.

Selective learning is important to song dialect research and to the understanding of population discrimination effects. There is evidence that perception at the level of syllables is important (Harris and Lemon, 1974). The results indicate that when some of the syllables in the song used as a playback stimulus are rare or absent from the local target population, the subjects gave a lower response than to a stimulus song possessing mostly common local syllables. The result parallels that on White-crowned Sparrows (Milligan and Verner, 1971).

D. Genetic Relationships among Dialects

For certain evolutionary considerations it is more important to ascertain the genetic consequences or correlates of dialect systems than to evaluate behavioral mechanisms. The most general hypothesis about the ecological genetics of dialect systems concerns the possible genetic differences between dialect populations. If members of different dialect populations are not randomly interbreeding, then either by natural selection differences or by genetic drift, or both, the two populations may diverge to a measurable extent. Nottebohm and Selander

(1972) carried out allozyme studies on *Z. capensis* samples from Argentina that were preliminary to the more intensive work of Handford and Notebohm (1976). The latter study did not find a substantial and consistent correlation between trill rate (the authors' dialect marker in the songs) and allelic frequency changes along an environmental gradient in a mountainous region of Argentina. Tissue samples were collected at five sites, three within one dialect and two within an adjacent dialect (Handford and Nottebohm, 1976). Pooling the samples within dialects and performing a simple comparison of the frequency of the most common allele at each locus indicate that there is a significant difference between dialects for two of the four loci for which complete data are reported (Handford and Nottebohm, 1976, Table 3). In spite of my crude reanalysis of their data, the results from *Z. capensis* do not give clear cause to reject particular hypotheses about the genetic impact of dialect patterns. An extremely important point, however, is found in their report (Fig. 6) that shows a major dialect change just

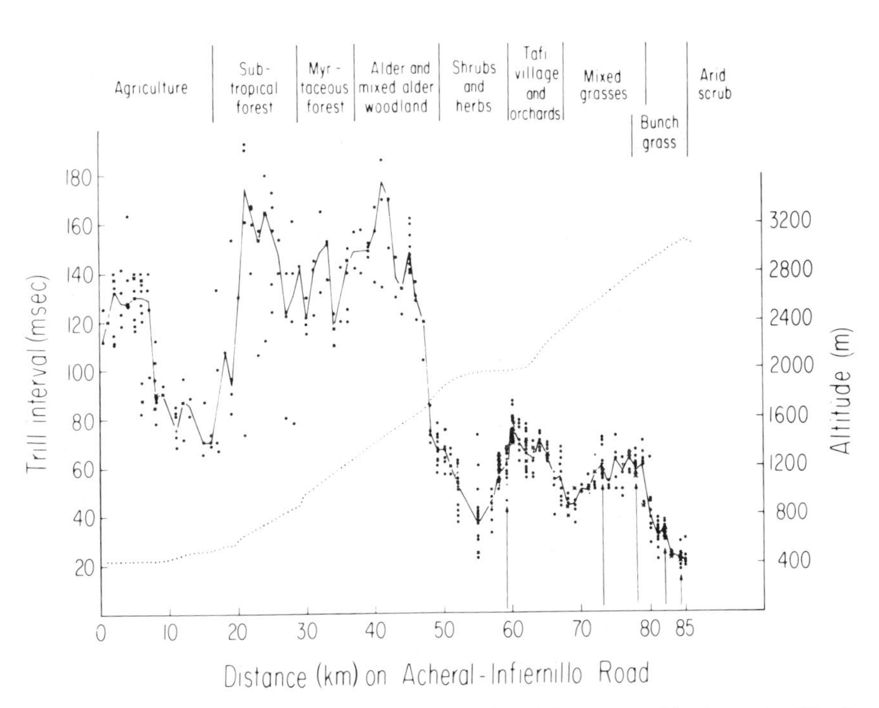

Fig. 6. Changes in trill interval in Rufous-collared Sparrows with change in altitude (dotted line) and vegetation zone. Arrows indicate five sites of samples used in electrophoretic analyses. (From Handford and Nottebohm, 1976.)

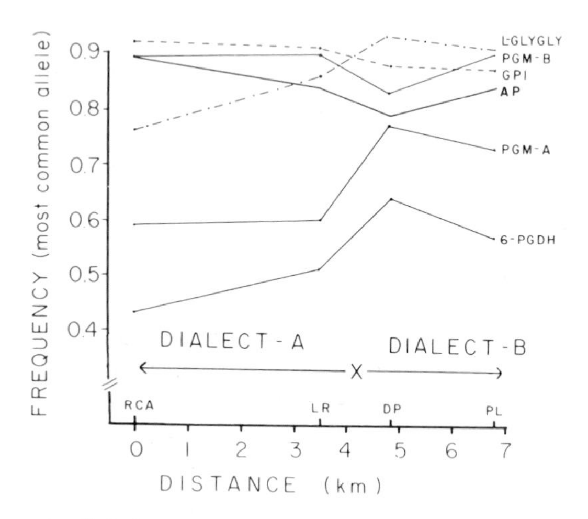

Fig. 7. Changes in allelic frequencies at six loci in White-crowned Sparrows sampled at four locations. Loci are indicated at the right end of each line and study sites along the ordinate. Song dialect change takes place between sites LR and DP. (From Baker, 1975.)

downhill from their tissue collection sites, a dialect change vastly more distinct than that which lay within their sampling locations. The researchers were unable to work on the major dialect transition for logistical reasons, but given the wealth of information on these populations there is an ideal opportunity to perform a further electrophoretic study at the major song change.

The only other studies designed specifically to examine correlated patterns of behavioral structure (dialects) and genetic structure are those of Baker (1974, 1975). In the most relevant of the two studies (Baker, 1975), there were significant changes in allelic frequencies at three of six loci across a song dialect transition in *Z. l. nuttalli* populations resident in coastal California (Fig. 7). Recent work has extended sampling to nine sites across four dialects in the same system (Baker *et al.*, 1982b). The results indicate that dialects more distant from one another are genetically more different and that this pattern is caused by the isolating influence of dialects. The electrophoretic studies of the four dialects indicate that the gene frequencies do not consistently follow a clinal pattern along the coast but may either increase or decrease across successive dialect transitions. Statistical tests indicate that all four dialect populations show significant genetic differentiation from one another.

Heterogeneity among samples within a single dialect population of *Z. l. oriantha* was also described previously (Baker, 1975). The sample sites lay along an altitudinal gradient in Colorado and also did not yield gene frequencies of an obvious clinal pattern. Unfortunately no other dialect was sampled so the pertinent comparions of within- and between-dialect genetic variation could not be made.

One can ask how the genetic heterogeneity in a system of dialects compares to other structured populations. As a measure of heterogeneity, F_{st} values (Wright, 1965) are a convenient summary. When calculated for four samples across two dialects of *Z. l. nuttalli* (Baker, 1981), the mean F_{st} was 0.01 and is similar in magnitude to F_{st} values computed among villages of Yanomama Indians (0.06), among districts on a reservation of Papago Indians (0.02), and among barns on the same farm populated by *Mus musculus* (0.02; Selander and Kaufman, 1975, Table 9).

V. THE SEARCH FOR STRUCTURE WITHIN DIALECT POPULATIONS

It may be appropriate to think of song dialect populations as lying on a continuum of hierarchically structured populations. We know that at the first level, the species, one finds minor gene exchange with other species. Races, or subspecies, of *Zonotrichia* species are morphologically distinguishable, but there is undoubtedly gene flow between these races where they are in breeding contact. It is probably safe to say that the named races represent a second level in the hierarchy of population structure. The third level is that of the dialect population. Here there is even less evidence that these groups are isolated sufficiently to evolve at all independently. Last, the most unresolved level of population structure is that of subpopulations within dialects. One often finds clusters of birds singing similar "themes" within dialect populations. These may be analogous to the "dialects" described by Payne (1980) in indigo-birds and it is conceivable that these clusters will turn out to contribute to genetic heterogeneity as well. What I wish to pursue here, however, is one approach to thinking about genetic structure within dialects that can be elaborated from existing data on *Z.l. nuttalli*.

As outlined earlier, there are certain theoretical expectations about genetic structure in a population that can be predicted from characteristics such as the shape of the dispersal function, the variance in progeny numbers, and the degree of overlap in generations. These characteristics have been estimated for a population of White-crowned Sparrows (Baker and Mewaldt, 1978; Baker *et al.*, 1981a).

Considering the dispersal function (Baker and Mewaldt, 1978), one can derive an expression for the size of a neighborhood population defined as that group of individuals whose gametes may, with high probability, combine to form the next generation. If N_e is the effective population size, B the standard deviation of the dispersal function, N the number of potential parents in a strip of habitat of length

$2B$, and D the density of breeding birds per unit distance, it can be shown that (Baker, 1981):

$$\frac{1}{N_e} = \frac{2B}{N} \int_{-\infty}^{+\infty} \frac{1}{B^2} e^{-2|x|/B} \, dx \tag{5}$$

which reduces to:

$$\frac{1}{N_e} = \frac{2}{N} \tag{6}$$

Since $D = N/2B$, $N_e = DB$.

Data available to estimate the parameters (Baker and Mewaldt, 1978; Baker *et al.*, 1981a) provide a calculated N_e of 100 birds. This first approximation of N_e is the neighborhood population and would be equivalent to the effective size if mating were at random, progeny number distributed as a Poisson variate, generations discrete, and population size constant. Information on fluctuations in population size is lacking. Other demographic data are available, however, and provide assessment of two other influences on N_e (Baker *et al.*, 1981a). Results from a study of reproduction provide a variance estimate for number of offspring and this value can be used to adjust the effective population size estimate from Eq. (3). By this adjustment, $N_e = 74$ birds.

Finally, the effect of overlapping generations on N_e can be evaluated from the formulation provided by Nei and Imaizumi (1966):

$$N_e = N_m \lambda \tag{7}$$

in which N_m is the number born per year that are able to reach the mean reproductive age, and λ is the mean reproductive age. These parameters are estimated from life table statistics (Baker *et al.*, 1981a) and result in N_e of 36 birds. Therefore, the genetically effective population size of *Z. l. nuttalli* is small. On theoretical grounds, such a small effective size implies that the genetic structure of the population can be strongly influenced by sampling error leading to genetic drift (Baker, 1981). Small effective size also has an important impact on the probability of fixation, and rate of spread, of new mutations in the population.

VI. DIALECTS AND AREA EFFECTS

Dialect systems in *Zonotrichia* in particular, and perhaps for oscine birds in general, may profitably be compared to the "area effects" that have been described in several mollusc and plant species. The most completely analyzed area effects are found in the snail *Cepaea nemoralis* in England and in another land snail, *Partula,* found in the Society Islands of the South Pacific.

In these taxa, an area effect refers to geographic areas in which the genetic makeup of the population is different from that of adjacent surrounding populations. In a region of about 50 km^2 in southern England, for example, are four areas populated by *C. nemoralis* which differ in the shell phenotype in marked ways. Allozyme studies of these same areas reveal that the biochemical loci vary concordantly with the shell characters (Johnson, 1976).

The most convincing explanation of area effects at the present time is that of Clarke (1966, 1968). He theorizes that an area represents a complex of interacting and coadapted genes selectively adjusted to the average habitat of the area. Clarke's genetic model of this pattern suggests that the result could be achieved by a single gene locus in a population evolving until a sharp geographic discontinuity is produced. Clarke's model visualizes alleles being differentially favored in different parts of the geographic range with morph-ratio clines produced by subsequent migration. Coadaptation takes place by selection in favor of modifier genes compatible with locally frequent alleles whose frequency is increased by the spreading of the modifiers. The end result would be differentiated populations with abrupt discontinuity at borders. The discontinuity would have no necessary correspondence to a sharp environmental change. In a general outline, this corresponds roughly to the early "thought" model of Nottebohm (1969) to explain the origin of vocal dialects. The area-effect phenomenon seems less compatible with the founder principle (Goodhart, 1963), but data to resolve this point conclusively are lacking.

A picture similar to that found in *Cepaea*, but on a much reduced scale, has been described for *Partula* (Clarke, 1968). In this instance, however, areas of significantly different genotypic proportions may be as few as 20 meters apart. The races of *Zonotrichia* species are, in some respects, analogous to area effects and it may be worth considering, for heuristic purposes, that song dialect groups represent area effects also, but at a lesser level of genetic divergence and with a more fluid and dynamic behavioral system as the isolating mechanism.

In conclusion, my theme has been that a central interest in vocal dialects can be enriched by employing a conceptual framework synthesized from theoretical and ecological genetics and from behavior and bioacoustics. In this way, we can see common ground with interesting problems in other disciplines and, perhaps, explore more effectively the behavioral and evolutionary processes involved in avian song dialects.

ACKNOWLEDGMENT

This work was supported in part by National Science Foundation Grant DEB-78-22657 to the author.

REFERENCES

American Ornithologists' Union (1957). "Check-list of North American Birds." Am. Ornithol. Union, New York.

Armstrong, E. A. (1963). "A Study of Bird Song." Oxford Univ. Press, London and New York.

Baker, M. C. (1974). Genetic structure of two populations of White-crowned Sparrows with different song dialects. *Condor* **76,** 351–356.

Baker, M. C. (1975). Song dialects and genetic differences in White-crowned Sparrows (*Zonotrichia leucophrys). Evolution* **29,** 226–241.

Baker, M. C. (1981). Effective population size in a sparrow: Some possible implications. *Heredity* **46,** 209–218.

Baker, M. C., and Marler, P. (1980). Behavioral adaptations that constrain the gene pool in vertebrates. *In* "Evolution of Social Behavior: Hypotheses and Empirical Tests" (H. Markl, ed.), Dahlem Konferenzen, pp. 59–80. Verlag Chemie, Weinheim.

Baker, M. C., and Mewaldt, L. R. (1978). Song dialects as barriers to dispersal in White-crowned Sparrows, *Zonotrichia leucophrys nuttalli. Evolution* **32,** 712–722.

Baker, M. C., and Mewaldt, L. R. (1981). Reply to "Song dialects and dispersal in White-crowned Sparrows: A re-evaluation." *Evolution* **35,** 189–190.

Baker, M. C., Mewaldt, L. R., and Stewart, R. M. (1981a). Demography of White-crowned Sparrows (*Zonotrichia leucophrys nuttalli). Ecology* **62,** 636–644.

Baker, M. C., Thompson, D. B., Sherman, G. L., and Cunningham, M. A. (1981b). The role of male vs. male interactions in maintaining population dialect structure. *Behav. Ecol. Sociobiol.* **8,** 65–69.

Baker, M. C., Spitler-Nabors, K. J., and Bradley, D. C. (1982a). The response of female Mountain White-crowned Sparrows to songs from their natal dialect and an alien dialect. *Behav. Ecol. Sociobiol.* **10,** 175–179.

Baker, M. C., Thompson, D. B., Sherman, G. L., Cunningham, M. A., and Tomback, D. F. (1982b). Allozyme frequencies in a linear series of song dialect populations. *Evolution* **36** (in press).

Banks, R. C. (1964). Geographic variation in the White-crowned Sparrow, *Zonotrichia leucophrys. Univ. Calif., Berkeley, Publ. Zool.* **70.**

Baptista, L. F. (1975). Song dialects and demes in sedentary populations of the White-crowned Sparrow *(Zonotrichia leucophrys nuttalli). Univ. Calif., Berkeley, Publ. Zool.* **105.**

Baptista, L. F. (1977). Geographic variation in song and dialects of the Puget Sound White-crowned Sparrow. *Condor* **79,** 356–370.

Baptista, L. F., and Morton, M. L. (1980). Song dialects and mate selection in montane White-crowned Sparrows. *Am. Ornithol. Union, Annu. Meet., Fort Collins, Colo.* Abstr. No. 4.

Bart, J. (1977). Impact of human visitations on avian nesting success. *Living Bird* **16,** 187–192.

Bent, A. C. (1968). "Life Histories of North American Cardinals, Grosbeaks, Buntings, Towhees, Finches, Sparrows, and Allies." Dover, New York.

Borror, D. J., and Gunn, W. H. J. (1965). Variation in White-throated Sparrow songs. *Auk* **82,** 26–47.

Brooks, R. J., and Falls, J. B. (1975). Individual recognition by song in White-throated Sparrows. I. Discrimination of songs of neighbors and strangers. *Can. J. Zool.* **53,** 879–888.

Chapman, F. M. (1940). Post-glacial history of *Zonotrichia capensis. Bull. Am. Mus. Nat. Hist.* **77,** 381–438.

Clarke, B. (1966). The evolution of morph ratio clines. *Am. Nat.* **100,** 389–402.

Clarke, B. (1968). Balanced polymorphism and regional differentiation in land snails. *In* "Evolution and Environment" (E. T. Drake, ed.), pp. 351–368. Yale Univ. Press, New Haven, Connecticut.

Crow, J. F., and Kimura, M. (1970). "An Introduction to Population Genetics Theory." Harper, New York.

DeWolfe, B. B., and DeWolfe, R. H. (1962). Mountain White-crowned Sparrows in California. *Condor* **64,** 378–389.

DeWolfe, B. B., Kaska, D. D., and Peyton, L. J. (1974). Prominent variation in the songs of Gambel's White-crowned Sparrows. *Bird-Banding* **45,** 224–252.

Falls, J. B., and Brooks, R. J. (1975). Individual recognition by song in White-throated Sparrows. II. Effects of location. *Can. J. Zool.* **53,** 1412–1420.

Farris, J. S. (1970). On the relationship between variation and conservatism. *Evolution* **24,** 825–827.

Fisher, R. A. (1958). "The Genetical Theory of Natural Selection." Dover, New York. (Orig. publ., 1930.)

Friedlaender, J.S. (1975), "Patterns of Human Variation. The Demography, Genetics, and Phenetics of Bougainville Islanders." Harvard Univ. Press, Cambridge, Massachusetts.

Goodhart, C. B. (1963). "Area effects" and non-adaptive variation between populations of *Cepaea* (Mollusca). *Heredity* **18,** 459–465.

Halliburton, R., and Mewaldt, L. R. (1976). Survival and mobility in a population of Pacific coast Song Sparrows *(Melospiza melodia gouldii)*. *Condor* **78,** 499–504.

Handford, P., and Nottebohm, F. (1976). Allozymic and morphological variation in population samples of Rufous-collared Sparrow, *Zonotrichia capensis,* in relation to vocal dialects. *Evolution* **30,** 802–817.

Harris, M. A., and Lemon, R. E. (1974). Songs of Song Sparrows: Reactions of males to songs of different localities. *Condor* **76,** 33–44.

Howells, W. W. (1966). Population distances: Biological, linguistic, geographical, and environmental. *Curr. Anthropol.* **7,** 531–540.

Johnson, M. S. (1976). Allozymes and area effects in *Cepea nemoralis* on the western Berkshire Downs. *Heredity* **36,** 105–121.

Johnson, M. S., and Mickevich, M. F. (1977). Variability and evolutionary rates of characters. *Evolution* **31,** 642–648.

Johnston, R. F. (1956). Population structure in salt marsh Song Sparrows. Part I: Environment and annual cycle. *Condor* **58,** 24–44.

Jones, F. R. H. (1968), "Fish Migration." Arnold, London.

Kimura, M., and Weiss, G. H. (1964). The stepping stone model of population structure and the decrease of genetic correlation with distance. *Genetics* **49,** 561–576.

King, J. R. (1972). Variation in the song of the Rufous-collared Sparrow, *Zonotrichia capensis,* in northwestern Argentina. *Z. Tierpsychol.* **30,** 344–373.

Kluge, A. G., and Kerfoot, W. C.(1973). The predictability and regularity of character divergence. *Am. Nat.* **107,** 426–443.

Konishi, M. (1965). The role of auditory feedback in the control of vocalization in the White-crowned Sparrow. *Z. Tierpsychol.* **22,** 770–783.

Lemon, R. E., and Harris, M. (1974). The question of dialects in the songs of White-throated Sparrows. *Can. J. Zool.* **52,** 83–98.

Lenington, S. (1979). Predators and blackbirds: The "uncertainty principle" in field biology. *Auk* **96,** 190–192.

Lowther, J. K. (1961). Polymorphism in the White-throated Sparrow, *Zonotrichia albicollis* (Gmelin). *Can. J. Zool.* **39,** 281–292.

Marler, P., and Peters, S. (1977). Selective vocal learning in a Sparrow. *Science* **198,** 519–521.

Marler, P., and Tamura, M. (1962). Song "dialects" in three populations of White-crowned Sparrows. *Condor* **64,** 368–377.

Martin, D. J. (1977). Songs of the Fox Sparrow. I. Structure of song and its comparison with songs in other Emberizidae. *Condor* **79,** 209–221.

Martin, D. J. (1979). Songs of the Fox Sparrow. II. Intra- and interpopulation variation. *Condor* **81,** 173–184.

Mayr, E., and Amadon, D. (1951). A classification of recent birds. *Am. Mus. Novit.* No. 1496.

Miller, A. H. (1947). Panmixia and population size with reference to birds. *Evolution* **1,** 186–190.

Miller, A. H. (1956). Ecologic factors that accelerate formation of races and species of terrestrial vertebrates. *Evolution* **10,** 262–277.

Milligan, M., and Verner, J. (1971). Inter-population song dialect discrimination in the White-crowned Sparrow. *Condor* **73,** 208–213.

Morony, J. J., Jr., Bock, W. J., and Farrand, J., Jr. (1975). "Reference List of the Birds of the World." Am. Mus. Nat. His., New York.

Nei, M., and Imaizumi, Y. (1966). Genetic structure of human populations. II. Differentiation of blood group gene frequencies among isolated populations. *Heredity* **21,** 183–190, 344.

Nice, M. M. (1964). "Studies in the Life History of the Song Sparrow. I: A Population Study of the Song Sparrow and Other Passerines." Dover, New York. (Orig. publ., 1937).

Nottebohm, F. (1969). The song of the Chingolo, *Zonotrichia capensis,* in Argentina: Description and evaluation of a system of dialects. *Condor* **71,** 299–315.

Nottebohm, F. (1972). The origins of vocal learning. *Am. Nat.* **106,** 116–140.

Nottebohm, F., and Selander, R. K. (1972). Vocal dialects and gene frequencies in the Chingolo Sparrow *(Zonotrichia capensis). Condor* **74,** 137–143.

O'Donald, P. (1974). Polymorphism maintained by sexual selection in monogamous species of birds. *Heredity* **32,** 1–10.

O'Donald, P. (1976). Mating preferences and sexual selection in the Arctic Skua. II. Behavioural mechanisms of the mating preferences. *Heredity* **39,** 111–119.

O'Donald, P., and Davis, J. W. F. (1977). Mating preferences and sexual selection in the Arctic Skua. III. Estimation of parameters and tests of heterogeneity. *Heredity* **39,** 121–132.

Orejuela, J. E., and Morton, M. L. (1975). Song dialects in several populations of mountain White-crowned Sparrows *(Zonotrichia leucophrys oriantha)* in the Sierra Nevada. *Condor* **77,** 145–153.

Payne, R. B. (1973). Behavior, mimetic songs and song dialects, and relationships of the parasitic Indigobirds (*Vidua*) of Africa. *Ornithol. Monogr.* **11.**

Payne, R. B. (1978). Microgeographic variation in songs of Splendid Sunbirds, *Nectarinia coccinigaster:* Population phenetics, habitats, and song dialects. *Behaviour* **65,** 282–308.

Payne, R. B. (1980). Population structure and social behavior: Models for testing the ecological significance of song dialects in birds. *In* "Natural Selection and Social Behavior: Recent Research and New Theory" (R. D. Alexander and D. W. Tinkle, eds.), pp. 108–120. Chiron, New York.

Petrinovich, L., Patterson, T., and Baptista, L. F. (1981). Song dialects as barriers to dispersal: A re-evaluation. *Evolution* **35,** 180–188.

Pierce, B. A., and Mitton, J. B. (1979). The relationship of genetic variation within and among populations: An extension of the Kluge–Kerfoot phenomenon. *Syst. Zool.* **28,** 63–70.

Remington, C. L. (1968). Suture-zones of hybrid interaction between recently joined biotas. *Evol. Biol.* **2,** 321–428.

Rohlf, F. J., and Schnell, G. D. (1971). An investigation of the isolation-by-distance model. *Am. Nat.* **105,** 295–324.

Selander, R. K., and Kaufman, D. W. (1975). Genetic structure of populations of the brown snail (*Helix aspersa*). I. Microgeographic variation. *Evolution* **29,** 385–401.

Sokal, R. R. (1976). The Kluge–Kerfoot phenomenon reexamined. *Am. Nat.* **110,** 1077–1091.

Sokal, R. R. (1978). Population differentiation: Something new or more of the same? *In* "Ecological Genetics: The Interface" (P. F. Brussard, ed.), pp. 215–239. Springer-Verlag, Berlin and New York.

Spielman, R. S., Migliazza, E. C., and Neel, J. V. (1974). Regional linguistics and genetic differences among Yanomama Indians. *Science* **184,** 637–644.

Spuhler, J. N. (1972). Genetic, linguistic and geographical distances in native North America. *In* "The Assessment of Population Affinities in Man" (J. S. Weiner and J. Huizinga, eds.), pp. 72–95. Oxford Univ. Press (Clarendon), London and New York.

Treisman, M. (1978). Bird song dialects, repertoire size, and kin association. *Anim. Behav.* **26,** 814–817.

Wasserman, F. E. (1977). Mate attraction function of song in the White-throated Sparrow. *Condor* **79,** 125–127.

Wright, S. (1921). Systems of mating. II. The effects of inbreeding on the genetic composition of a population. *Genetics* **6,** 124–143.

Wright, S. (1922). Coefficients of inbreeding and relationship. *Am. Nat.* **56,** 330–338.

Wright, S. (1931). Evolution in Mendelian populations. *Genetics* **16,** 97–159.

Wright, S. (1938). Size of population and breeding structure in relation to evolution. *Science* **87,** 430–431.

Wright, S. (1943). Isolation by distance. *Genetics* **28,** 114–138.

Wright, S. (1965). The interpretation of population structure by F-statistics with special regard to systems of mating. *Evolution* **19,** 395–420.

Wunderle, J. M., Jr. (1978). Differential response of territorial Yellowthroats to the songs of neighbors and non-neighbors. *Auk* **95,** 389–395.

8

Individual Recognition by Sounds in Birds

J. BRUCE FALLS

I. INTRODUCTION

Recognition of individuals is important in many of the varied and subtle social interactions found among vertebrates (Wilson, 1975). In view of the elaborate development of acoustic signaling in birds, individual recognition by sound

ACOUSTIC COMMUNICATION IN BIRDS
VOLUME 2

ISBN 0-12-426802-1

should be widespread. Considerable evidence to this effect has accumulated, much of it since Beer's (1970) excellent review. However, the broad outlines have not changed substantially.

Research in this area can be subdivided by the nature of the sender–receiver pair under investigation. While mate recognition and parent–young recognition have sometimes been investigated together, studies of neighbor recognition have been conducted separately. To varying degrees, methods of study, behavioral mechanisms, and biological functions of recognition differ among these categories. In this chapter, I describe and discuss evidence of individual recognition between mates, between parents and young, and between territorial neighbors. Finally, I discuss more general issues. First, however, I want to consider approaches and methods.

Close observers of bird behavior have long realized that individuals as well as species can be recognized by their calls and songs (Nice, 1943; Robinson, 1949; see also references in Beer, 1970). The use of sophisticated recording and analyzing equipment has confirmed and extended the finding that bird sounds are often individually distinctive (Borror, 1960). For present purposes I use the term *individual variation* where it has been demonstrated that significant variation exists among the sounds of different individuals beyond that found in repeated sounds of the same individual.

If we recognize individual birds by their sounds, it seems likely that other birds, adapted to receive these signals, do so as well. Studies of hearing and learning suggest that birds can detect the details of their own vocalizations (see Chapter 4, Volume 1, and Chapter 1, this volume) Thus, it seems likely that a function of these sounds is to convey individual identity (Craig, 1908; Marler, 1956). Obvious as this may seem, the question has to be put to the birds.

Observation has suggested, either by the context in which a sound is used (Wiley, 1976) or by the way others respond to it, that it identifies the sender. For example, female Prairie Warblers (*Dendroica discolor*) respond only to songs of their own mates, even when they cannot see them (Nolan, 1978). Experimentation with playback has amply demonstrated that in some circumstances birds respond differently to the voices of different individuals. I shall use the term *individual recognition* where it has been shown that recipients discriminate among similar sounds of different individuals in the absence of other identifying cues.

Individual variation is a prerequisite for individual recognition. The process of recognition requires adequate identifying information, whether or not signals are selected to serve this function, as well as the ability of receivers to detect and learn individual differences. An ideal signal for individual recognition would be highly stereotyped within each individual but vary noticeably among individuals. By contrast, signals used for species recognition are likely to be stereotyped both within and among individuals, while those used to express differences in motivation may be highly variable. Since individuals differ in heredity and experience, we should expect some tendency toward individual variation in most signals.

While the occurrence of individual variation is suggestive, it does not show that recognition also occurs. There may be insufficient variation to guarantee that most individuals are separable by their sounds. Furthermore, birds may not detect or may not pay attention to variations that we find in sound analysis. Thus, while individual variation is virtually universal, individual recognition may not be. In this review, I emphasize studies where individual recognition by sound has been clearly demonstrated.

II. METHODS

A. Individual Variation

To determine the extent of individual variation in sound signals, various measurements can be made on sonagrams, oscillograms, power spectra, or other displays, or displays can be compared visually. Once measurement is completed, analysis of variance, or ANOVA (or a nonparametric equivalent), is commonly used to compare variation in a particular sound feature within and across individuals (Miller, 1978). Although some individual variation is to be expected (Section I), ANOVA quantifies the number of features with significant interindividual variation and the extent of variation in each, as shown by the magnitude of the F ratio. The Coefficient of Variation (CV = standard deviation/mean × 100%) is useful for comparing the extent of variability across individuals in different features where measurements differ in magnitude (Hutchison *et al.*, 1968). A feature showing a large CV across individuals is most likely to be useful for individual recognition. Jouventin *et al.* (1979) use the ratio CV across individuals/CV within individuals to compare individuality in calls of different species. To combine individual variation in several features in a single value, Beecher (1982a) employs the information measure H, which is based on variance measures and can be derived approximately from the square root of the F ratio. Values of H for each separate variable are added and information shared by intercorrelated variables is subtracted to obtain an overall estimate of the information "capacity" of a "signature" system. This measure should be useful for comparing the individuality of different signals.

Several methods are available to measure the similarity of individual signals. For example, sounds of the same or different individuals can be compared using measures of correlation. Thus, White *et al.* (1970) measured a series of ordinates along the amplitude–time profile of Gannet (*Morus bassanus*) calls. By aligning pairs of calls, they showed that sounds of the same individual were more highly correlated than those of different individuals. Complex sounds which are difficult to measure can be compared using overlaid grids to determine the degree of overlap in sonagrams (Bertram, 1970; E. H. Miller, 1979). Digitization of such data can provide measures of similarity or difference but determining the best alignment of traces may present problems.

The most direct way to assess the potential for individual recognition is to determine the extent to which individuals in a population can be identified accurately by their sounds. Using particular features, all possible pairs of birds for which adequate samples exist can be compared by means of univariate tests. Those that have significantly different means may be considered identifiable. However, the possibility of identifying an individual by a single call will depend not just on means but also on overlap resulting from intraindividual variation. Since separation among individuals will improve as more variables are included in the analysis, sounds of different birds may form distinct clusters in a multivariate plot (Bailey, 1978). Graphical methods or multivariate analysis can be used to assess the percentage of sounds that can be classified as belonging to the correct individual. Many authors simply report that sounds of different individuals may be distinguished in sonagrams. This represents a subjective assessment of the signal as a whole. Such judgments can be quantified by having "naive" observers classify sonagrams. Again, the percentage correctly identified may be taken as a measure of individuality (Brooke, 1978). For a further comparison of methods the reader is referred to Bailey (1978).

B. Individual Recognition

While field observations of behavior may yield circumstantial evidence, the most direct way to demonstrate individual recognition by sound is to compare responses of birds to recordings of different individuals. Once discrimination has been found, it may be possible to determine which features of the sound are involved by selectively altering the recordings used for playback until the response changes. These experimental approaches make it possible to exclude visual cues to recognition and obtain an adequate sample of comparable data. While laboratory experiments would allow maximum control of extraneous variables, typical responses may not occur under artificial conditions, leading to negative results. Recording and playback make it possible to experiment in the field where birds may be expected to perform normally in a rich natural environment.

Presentations should be as natural as possible with respect to location, time of day and season, loudness, and temporal features of the program. Since identifying characteristics may be subtle, equipment and recordings should be of good quality. Most loudspeakers are more directional than birds, but this may help to reduce complications resulting from responses of individuals other than the experimental subject. In most cases it is important to identify the responding bird(s) and individual marking should be considered. Except where an experiment is specifically designed to study habituation, short sequences should be used to minimize lasting changes in response. Initial responses are more likely to reflect natural behavior than prolonged encounters between a bird and an unresponsive sound source. Sequential presentation of the sounds to be compared (controlled for order) is useful where location is critical, while simultaneous (or near simultaneous) presentation from different directions may allow more subtle distinctions to be made.

If different experimental sounds are presented on different occasions, it may be useful to include a period of observation before the playback begins, to control for the effects of time-related variables. In this case some measures of response may be expressed as differences between the values obtained before and during or after the playback. Measures of response commonly include: latency from the beginning of playback to the first occurrence of some behavior, the occurrence or number of different vocalizations, movements, or visual displays, and the closeness of approach or time spent within some criterion distance of the loudspeaker. Detailed knowledge based on other data may enable the experimenter to score the overall response on some scale of behavioral intensity (Emlen, 1971; Falls and McNicholl, 1979). Otherwise, it is probably more informative (and less subject to bias) to treat each variable separately and, if desired, to examine the combined response using multivariate methods (Falls and Brooks, 1975). For further discussion of playback methods for studying individual recognition see Falls (1969), Beer (1970, 1979), Wiley and Wiley (1977), and examples presented in the following sections.

III. RECOGNITION BETWEEN MATES

A. Introduction

In order to reproduce successfully, members of a breeding pair must coordinate their efforts, a task which should be easier if they can recognize one another. Thus, mate recognition should be widespread, especially where both sexes contribute substantial parental care. Not only is the need great in such cases but the opportunity to learn one another's individual characteristics is afforded by a protracted relationship. In long-lived species, individual recognition may allow the same pairs to reform from year to year. In breeding colonies, where nest sites are crammed together, the chances of confusion seem particularly great and the ability to recognize mates should be at a premium, if only as a means of locating the proper nest. The combination of monomorphic adult plumage and vocal greeting displays, which occurs in many colonial species, suggests that sound is likely to play a major role in mate recognition. Since colonial seabirds exemplify all these characteristics, they have been the subjects of much of the research on vocal recognition of mates.

B. Evidence of Mate Recognition

1. Colonial Seabirds

In penguins (which are monomorphic), mates are often separated for weeks during the breeding season, as each in turn goes to sea to gather food. When a bird returns to the colony, it engages in a noisy mutual display with its mate. Penney (1968) obtained suggestive evidence that sound may be involved in mate recognition in the Adélie Penguin (*Pygoscelis adeliae*), which has fixed nest sites. First, he exchanged a few neighboring birds and found that when a mate returned, only the correct partner, now on the wrong site, responded. This result does not rule out visual recognition. However, playback of a previous mate's calls the following year elicited an obvious response from one female which returned to the previous territory. Later studies (Jouventin and Roux, 1979) showed that adult calls of this species have individually distinctive temporal patterns.

The Emperor Penguin (*Aptenodytes forsteri*) is peculiar in moving its egg and chick about rather than occupying a fixed site. Thus, it cannot use topographic cues to locate a partner and may be especially dependent on sound for recognition. The call ("chant de coeur") of this species consists of wave trains (pulses) separated by silent intervals. Jouventin (1972) showed that each individual had a consistent temporal pattern different from its neighbor, and these differences could be detected by ear. Emperor calls showed much greater individuality than those of the more sedentary Adélie, Gentoo (*Pygoscelis papua*), and King (*Aptenodytes patagonica*) penguins, although these latter species also have similar individualistic calls. A brooding adult Emperor Penguin will respond selectively to recorded calls of its mate by leaving the colony and approaching a loudspeaker (Jouventin *et al.*, 1979). Experiments with altered calls confirmed that temporal pattern is important for recognition. Thus, in penguins we begin to see associations between individuality of calls, individual recognition, and variations in the social systems of different species.

The Manx Shearwater (*Puffinus puffinus*) provides one of the most convincing examples of mate recognition by sound. This species enters its nest burrow at night, precluding the use of visual cues. The male, which arrives first in the spring, normally returns to the same burrow but sometimes occupies a different one. When the female arrives she calls from the air, eliciting calls from males in burrows. Brooke (1978) showed experimentally that the male's call may permit burrow location and reestablishment of the pair bond from year to year. Using playback in burrows at night when one adult was incubating, he found that females replied selectively to their mates' calls while males did not preferentially answer their own females. Human subjects could match sonagrams of calls by the same male.

The Gannet was the subject of the first investigation to provide both clear experimental evidence and analysis of the sounds involved. A compact group of 17 pairs was photographed while individual calls were played from a loudspeaker located outside the group (White, 1971). Responses were scored using a scale based on activity and orientation. Calls of both males and females were individually recognized by their respective mates but there was no evidence of recognition of neighbors or other individuals. Directional cues were ruled out because of the unnatural location from which the sounds were broadcast. Analysis of the calls indicated that the pattern of overall amplitude with time was the most individually distinctive feature, with the most characteristic parts being the first few peaks and troughs (White and White, 1970; White *et al.*, 1970) (Section II,A). It was suggested that Gannets might use the temporal patterning of the call rather than amplitude *per se,* a conclusion also reached by Jouventin (1972) who found that amplitude was more variable than the temporal pattern of pulses in calls of individual Gannets.

Larid gulls provide examples of mate recognition in colonial nesters with semiprecocial young which both parents guard and feed over an extended period. Tinbergen (1953) reported that a sleeping Herring Gull (*Larus argentatus*) is awakened by its mate's "long" call but ignores calls of other adults. Similar observations have been made on Ring-billed Gulls (*L. delawarensis*) (P. Fetterolf, personal communication) and Laughing Gulls (*L. atricilla*) and experimental evidence is also available for the latter species (Beer, 1970). Using a loudspeaker located equidistant from two nests, Beer showed that adults responded only to long calls and "ke-hahs" of their mates, not distinguishing between calls of neighbors and strangers. Sonagrams showed individuality in both these calls.

Black-legged Kittiwakes (*Rissa tridactyla*) nest on narrow ledges which they defend against individuals other than their mates. Pairs usually remain mated for years even on different nest sites and those that do are usually more successful than newly formed pairs (Coulson, 1966). A pair may perform as many as 100,000 calls together each season. Playback experiments similar to Beer's on Laughing Gulls showed that kittiwakes recognize their mates' calls but not those of neighbors (Wooller, 1978). Call recognition may assure the mate of a landing space free from attack. Since early arriving individuals respond strongly to playback of their previous mates' calls, recognition may also assist pairs to reform each year. Analysis of the calls indicated that overall length, and length and tonal quality of the final part, might facilitate recognition. Wooller demonstrated constancy of calls in an individual recorded over a 6-year period.

Clear experimental evidence of mate recognition has also been obtained for the Least Tern (*Sterna albifrons*). Birds responded selectively to playback of the "purrit-tit-tit" calls of their own mates (Mosely, 1979). This call, which is normally given when mates approach one another, exhibits individual variation in both temporal and spectral features.

2. *Other Species*

When quail are visually separated, their calls may be recognized by their mates or, during the nonbreeding season, by other covey members. Playback to two pairs of California Quail (*Lophortyx californica*) showed that each bird responded only to calls of its own mate (Williams, 1969). Individual variation was found in the frequency of the first syllable of the call. Bobwhites (*Colinus virginianus*) also respond selectively to their mates' calls (Stokes, 1967). Bailey (1978) was able to separate the calls of six Bobwhites on a three-dimensional plot based on measurements of the fundamental frequency trace.

Mate recognition of flight calls in cardueline finches was investigated by Mundinger (1970). Incubating and brooding female American Goldfinches (*Carduelis tristis*) are fed by their mates and respond to male flight calls by soliciting displays. Playback of the mate's call followed by that of another male showed that the female responded only to the former. The calls were individually distinctive and those of mates were more alike than calls of other adults. This was also found in the Twite (*Acanthis flavirostris*) (Marler and Mundinger, 1975). Convergence was demonstrated by comparing the flight calls of Siskins (*C. spinus*) and Pine Siskins (*C. pinus*) before and after mating (Mundinger, 1970).

Remarkable examples of motor and sensory learning between mates are provided by duetting song, in which both members of a pair sing together (see Chapter 4, this volume). The point of interest here is that individual pairs have more or less distinctive duets. Isolated individuals have been shown to respond to playback or imitations of the partner's portion of the song and, in some cases, to render the entire performance including the part usually sung by the mate. Playback of songs of mates, neighboring and strange males of the Eastern Whipbird (*Psophodes olivaceus*), showed that females respond antiphonally only to their mates' songs (Watson, 1969). Duetting seems to coordinate the activities of pairs, and sometimes larger groups, as well as serving the functions of territorial song.

Male Zebra Finches (*Poephila guttata*) have individually distinctive, stereotyped songs (Immelmann, 1969). In the wild they mate for life; the male sings during courtship and near the nest but is not territorial. Captive females, separated from their mates for 2 to 3 days, respond preferentially to recordings of their mates' songs over songs of familiar neighbors (D. B. Miller, 1979a). Six females, each of which had reared broods with more than one male, still showed some preference for their previous mates' songs.

To date, the only species in which both mate recognition and neighbor–stranger discrimination have been demonstrated is the Red-headed Woodpecker (*Melanerpes erythrocephalus*) (D. Crusoe, unpublished data). Females flew toward a loudspeaker that broadcast their mates' breeding calls in preference to a second speaker (in a different direction) broadcasting calls of a strange male.

C. Discussion

Judging from the variety of species and social contexts in which it has been demonstrated, mate recognition by sound is widespread among birds that maintain long pair bonds. There does not appear to be any evidence from strongly polygynous or lekking species. Particularly in the latter case mating is very transitory. If mate recognition does occur here, it should be more likely in females, which make a greater reproductive investment than males.

IV. RECOGNITION BETWEEN PARENTS AND YOUNG

A. Introduction

Recognition between parent and offspring has been studied by ethologists interested in the development of behavior as well as by those who view it as an example of kin recognition. Like mate recognition, it has often been studied in colonial species. Questions investigated have included the occurrence and means of recognition, the age of the young at which it appears, and the question of whether the young or the parent recognizes the other. Enough critical work has accumulated to examine the results in the light of differences in the nest sites and life histories of different species. The most comprehensive studies to date have been on gulls and terns (Beer, Evans, and others) and on swallows (Beecher).

B. Evidence of Parent–Young Recognition

1. *Penguins, Shearwaters, and Alcids*

After a period during which they remain with a parent, young penguins associate in groups while both parents gather food. However, chicks continue to be fed by their own parents, which requires some form of parent–young recognition. Adult Adélie Penguins returning from the sea with food generally call from the natal territory; the young come running and engage in a mutual vocal display with the parents before being fed (Thompson and Emlen, 1968). Strange chicks that approach adults may be driven off. While parents could be responding to visual cues or the behavior of their own young, chicks show little response to silent adults. In a creche they responded selectively to playback of their own parents' calls by moving quickly to the natal territory (Penney, 1968). When young Emperor Penguins were removed from adults, the parents readily retrieved their calling chicks but were only successful in five of ten cases when the chicks were muted by taping their bills (Jouventin and Roux, 1979). A parent that adopted a mute chick left it when its own young called. In both species adult calls are individually distinctive (Section III,B). Jouventin and Roux (1979) showed that the calls of young Adélie Penguins vary greatly with age, becoming more complex and individually distinctive before the young leave the nest. This contrasts with the Emperor Penguin, which does not have a fixed nest site. In this species the calls of the young are stereotyped from the time of hatching until the young go to sea. Taken together these results suggest mutual recognition of calls by both parents and young.

An instructive exception to parent–young recognition occurs in the Manx Shearwater (Brooke, 1978). Chicks were unable to recognize playbacks of their parents' calls. In this species the chick is immobile in an isolated burrow and does not even solicit from its parents, who can locate it simply by returning to the correct burrow.

Two alcids, the Common Murre (*Uria aalge*) and the Razorbill (*Alca torda*), which often nest on the same cliffs, were the subjects of thorough studies by Tschantz (1968) and Ingold (1973), respectively. While murres nest in crowded aggregations on ledges, Razorbill nests are more widely separated. At an early age murre chicks respond only to the "luring" call of parents and respond selectively to playbacks of parental calls in field trials. Laboratory experiments showed that if they were exposed to a particular luring call in the egg they approached that call selectively after hatching. Tschantz investigated the properties of the luring call which elicited chick response using playback of altered and artificial calls in a two-choice experiment. Changes in pulse duration and cadence caused the greatest decrease in response. Correspondingly, individual differences in these features were most stable in normal adult calls. Among vocalizations of the young, "weeping" and "water" calls proved to be different among individuals and became increasingly distinctive with age. The behavior of parents suggested that they could recognize these calls of their own chicks by 10 days after hatching.

Young Razorbills are first separated from their parents at 16 to 23 days of age, when they plunge from the breeding ledges and join other young at sea. Ingold exposed adults to playback of the "leap" calls of their own and strange chicks. Parents did not distinguish among calls of 4-day-old chicks but responded selectively to calls of chicks at least 10 days old, giving "acceptance" calls only in answer to their own young. Thus, by the time the young leave, parents are prepared to recognize and feed them in the sea.

2. *Gulls and Terns*

Parent–young recognition has been investigated in several species by exchanging broods at different ages and observing adult behavior. If at some stage the adults no longer treat strange chicks as their own, this is taken as evidence that they have begun to recognize their own young. However, the behavior of the young toward the adults may play a part in any such change. Although vocalizations may be involved, this type of result does not rule out visual identification. However, if the adults continue to accept strange young, it does suggest that they have no means of identifying their own young. In general, the age at which recognition appears varies widely among species and correlates with the stage when young are capable of leaving the nest area (Davies and Carrick, 1962). For example, a dramatic contrast was found by Lashley (1913) between the ground-nesting Sooty Tern (*Sterna fuscata*), which rejected strange chicks after about 4 days of age, and the tree-nesting Brown Noddy (*Anous stolidus*), which did not discriminate until chicks were about 14 days old, the age at which they left the nest. Black-legged Kittiwakes, which nest on narrow ledges so that young cannot mix until they fledge, show no discrimination between their own and strange chicks (Cullen, 1957). By contrast, ground-nesting Herring Gulls, whose young can mix much earlier, will not accept strange chicks after 5 days of age (Tinbergen, 1953). However, Herring Gulls occupy a variety of habitats and von Rautenfeld (1978) showed that a cliff-nesting population would still accept transfers of young at 1 to 2 weeks of age. These results raise questions concerning the extent to which differences in parent–offspring recognition are species-specific or under direct environmental influence. Franklin's Gull (*L. pipixcan*), which nests on widely separated floating platforms so that young are unable to wander, showed no discrimination on the part of parents or young until between 7 and 14 days after hatching (Burger, 1974). By 14 days strange young were no longer accepted and by 16 days the young themselves attacked intruding adults. Chicks showed no discrimination when exposed to playback of long calls of parents and other adults, but details of experiments are lacking.

I now turn to cases where call recognition has been clearly demonstrated. The most thorough studies to date are those of Beer on the Laughing Gull. In this species, chicks remain close to the nest until 3 or 4 days after hatching. Over the next day or two the adults lure them with food on short excursions away from and back to the nest. During this "early" period, feeding is accompanied by the parents' "crooning" calls, which are difficult to locate and rather variable within and among individuals (Beer, 1970). After the sixth day ("middle" period), adults land with food several yards from the nest and attract the young with crooning and ke-hah calls. Until the eighth day chicks are seldom left unattended. Gradually they are left alone more often, until they are visited only briefly for feeding. From 12 days of age until fledging (after 28 days) chicks wander more and, if they go beyond the family territory, are chased by strange adults. They appear to recognize and approach their parents, which use long calls and ke-hahs more frequently in this "late" stage. These two calls, which are easily located, can be used to distinguish different individuals by ear. They show individuality with respect to temporal pattern and tonal quality. Long calls of mates are often similar.

Tapes including all three call types of particular adults were played to young of different ages. Captive chicks in the middle period (6 to 12 days of age) responded to playback of parental calls by calling as well as orienting toward and approaching the loudspeaker but turned away and withdrew when calls of strange adults were played (Beer, 1969, 1970). This result could represent no more than discrimination between familiar and unfamiliar calls. However, when the experiment was repeated using parental calls and calls of familiar neighbors, chicks again recognized calls of their parents (Beer, 1970). This experiment was then repeated using chicks in the early and late age groups. Early chicks (1–3 days) responded very strongly to parental calls but also showed a weak positive response to neighbors' calls. Older groups responded less strongly, and showed no response to neighbors. All three groups showed significant discrimination between calls of parents and other adults.

Young chicks showed positive responses to crooning but avoidance of long calls and ke-hahs. This trend was reversed in the middle and later groups, when young are more mobile and parents direct ke-hahs and long calls to them. Older chicks continued to respond negatively to these calls given by other adults which would normally have attacked them. Thus, as the chicks mature and the behavior of their parents changes, their responses to calls also change, while they continue to recognize the voices of their parents.

The role of early experience was investigated by raising incubator-hatched chicks and exposing them to recorded calls at 6 to 8 days of age. These chicks fled from parental calls. Even after being restored to their parents for another 6 days, their response to parental calls was neutral, while their siblings responded positively.

Learning of parental calls by chicks is a countinuing process (Beer, 1975). Older chicks did not respond to parental long calls recorded during the nestling period but did respond when played similar calls recorded near the time of playback. Also, playback of chick-directed long calls elicited stronger positive responses than did adult-directed calls. In addition to suggesting the subtle interplay of the learning process, these experiments show how easily misleading negative results can be obtained if the context of an experiment is inappropriate.

Adult Laughing Gulls responded to playback of chick calls of the three age groups mentioned above but did not discriminate between calls of their own and strange chicks (Beer, 1979). The result was the same with both successive and simultaneous presentations and whether or not the adult was accompanied by one of its own chicks.To test for vocal interactions between parents and young, Beer placed two unrelated chicks in covered boxes equidistant from a parent of one of them. When young nestlings were used, both chicks called and the adult called but stayed at the nest, showing no discrimination. However, with young of the middle group only the parent's own chick responded to its calls (mainly ke-hahs), and the parent showed a clear response, orienting toward and approaching its own chick, even regurgitating food on the box in some cases. With older chicks there was less calling on the part of both adult and chick but once an adult called, its own chick answered, setting up a continuing interchange with the parent. Parents seldom approached closely at this stage but one adult that did approach a strange chick attacked the box containing it. Considering the negative results of the playback experiments, it seems likely that the discrimination shown by adults in this case, which occurred with chicks as young as 2 days of age, was the result of differential vocal responses by the chicks rather than recognition of any intrinsic differences in their calls. Of course, recognition of young by adults cannot be ruled out by negative results. Moreover, visual recognition may be possible under natural conditions.

In summary, Beer's experiments show clearly that very young Laughing Gull chicks can recognize the calls of their parents. Early exposure to the calls is necessary but learning continues as the young develop and the parents' behavior changes. For their part, adults identify their own young not by individual characteristics of their calls but by the differential response of the young to their parents.

Playback experiments on the Black-billed Gull (*L. bulleri*) of New Zealand are of special interest because it nests in closely packed colonies subject to flooding (Beer, 1966; Evans, 1970a). As a result, young from different nests intermingle as early as 1 to 2 days after hatching and leave the nest in family groups at 3 to 4 days of age. When Evans exposed two unrelated chicks as young as 3 days of age to "mew" calls of their respective parents in sequence, each responded selectively to its own parents' call by approaching the speaker and peeping. The mew call is thought to be homologous with the croon of the Laughing Gull but differs from it in showing individually identifying characteristics (Beer, 1970). Thus, the young Black-billed Gull, which is likely to encounter strange adults at an early age, is able to recognize its parents' feeding calls. At a comparable age, the Laughing Gull chick is less selective in its response. Living in a colony where nests are widely spaced, it is unlikely to meet other adults until it is older. It will probably not respond to the croons of neighbors since they will be fainter than those of its parents (Beer, 1970).

A further contrast is provided by the Ring-billed Gull, in which young do not leave the nesting area until at least 5 days of age. Evans' (1970b) studies of this species also illustrate the importance of experimental methodology. With sequential presentation of parental and other mew calls, chicks from 3 to 5 days of age showed no selective response, although older chicks (10 to 12 days old) did discriminate. Using two loudspeakers, however, with almost simultaneous presentation of calls, chicks as young as 4 to 5 days oriented toward calls of their parents. Evans argued that the difference in response was adaptive, since a young chick with a parent can compare calls (as in the simultaneous presentation), while a chick that has been abandoned should not be so discriminating. The other point of interest here is that recognition of parental calls appears at an age coinciding with an increase in chick mobility. Both recognition and mobility occur later in this species than in the Black-billed Gull.

Adult Ring-billed Gulls recognize their own chicks (Miller and Emlen, 1975). However, calls may not be involved, because muted 10- to 12-day-old chicks were still recognized, while others whose head patterns had been altered were treated differently by their parents.

There is some evidence of recognition by sound in terns. Two broods of Common Terns (*Sterna hirundo*) discriminated between playback of "mew" calls of parents and strangers (Stevenson *et al.*, 1970). When placed in a treadmill between territories, young Arctic Terns (*S. paradisaea*) showed a preference for their parents' voices (they could not see them) from the second day after hatching (Busse and Busse, 1977). Hutchison *et al.* (1968) reported that the "fish" call of the Sandwich Tern (*Thalasseus sandvicensis*), which is given when the parent returns with food, shows individual variation in several features.However, no experiments were carried out. The only evidence suggesting parental recognition of offspring is for the Royal Tern *(Thalasseus maximus)*, a species in which older chicks associate in creches (Buckley and Buckley, 1972). When young were exchanged between neighboring nests, one parent recognized its silent chick while the other three did so only after their chicks vocalized.

3. *Swallows and One Species of Jay*

A detailed comparison of parent–offspring recognition in the highly colonial Bank Swallow (*Riparia riparia*) and its more solitary relative the Rough-winged Swallow (*Stelgidopteryx ruficollis*) has been carried out by Beecher and his colleagues.Bank Swallow chicks remain in the nest cavity for 14 days, are then fed at the burrow entrance, and begin to fly when 18 to 19 days old. Parents feed their young simply by returning to the correct burrow until the young begin to fly; then they feed them on the wing or retrieve them when they visit the wrong burrow. Flying young give a two-note "lost" call to which only their own parents respond. Later, parents feed their own young in creche-like assemblages on power lines. Young visiting strange burrows are occasionally fed but more often ignored or rejected. The cost to the residents of feeding strange young is very small, most "mistakes" being made by males. Thus, adult Bank Swallows recognize their own young in several situations where young from different nests intermingle (Beecher *et al.*, 1981a). The age at which recognition appears was investigated by exchanging partial broods between burrows. Exchanged chicks up to 15 days of age were accepted; older chicks either left themselves or were rejected. Three young that had been adopted at an early age later accidentally visited their original burrows and were rejected by their parents.

Further studies showed that recognition is based on calls of the young (Beecher *et al.*, 1981b). Begging calls of dependent young were replaced by two-note "signature" calls between 15 and 17 days of age. Naive human observers could distinguish sonagrams of these two call types and match signature calls (but not immature calls) of the same individual. Measurements revealed individual variation in several features of signature calls. Calls of young within the same brood proved to be similar. Playback from burrows on either side of the home burrow, at an age when young are beginning to visit strange burrows, showed that parents could distinguish signature calls of their own and other young.

Taken together these results clearly show that adult Bank Swallows can identify their own young by calls that develop just before the young become mobile and mingle with strangers. Beecher (1982b) suggests that recognition of parents by young develops later.

Although Rough-winged Swallows occupy a similar habitat and may even nest in association with Bank Swallows, they do not face the same problems of parent–offspring recognition. Their young fledge as a family group and remain together, neither visiting other burrows nor associating in creches. However, young Bank Swallows may visit nearby rough-wing nests. Interspecific exchanges showed that adult Rough-winged Swallows are not discriminating and will feed young Bank Swallows, while Bank Swallows reject young rough-wings at an early age (Beecher, 1982b). Intraspecific exchanges in Rough-winged Swallows were still accepted when young were ready to fly (Hoogland and Sherman, 1976). Calls of young rough-wings proved to be simpler and lacking in individuality compared with the "signature" calls of Bank Swallows and observers had difficulty classifying them. Analysis of eight variables of the two species' calls showed fewer and weaker individual differences in the Rough-winged Swallow (Beecher, 1982b).

Thus, Rough-winged Swallows, whose young do not mingle with those of other broods, seem unable to recognize their own young. Beecher (1982b) argues that differences between the two species are not facultative but result from specializations (distinctive calls, recognition, rejection) found only in Bank Swallows.

Barn Swallows (*Hirundo rustica*) and Tree Swallows *(Iridoprocne bicolor)* nest in loose aggregations. Young exchanged intraspecifically at five points in the nesting cycle were accepted by adults (Burtt, 1977). However, when nestling Barn Swallows were tethered outside the nest just before fledging, adults interacted differently with their own and strange young, directing more aggression toward strangers. After fledging, young of both species join mixed flocks and are fed by their parents. Since young show no discrimination and vocalize even to silent adults, it may be that parents recognize young by their calls. Burtt suggests that adults learn the calls of nestlings but only show discrimination after the young fledge.

Parents and young of the Piñon Jay (*Gymnorhinus cyanocephala*) recognize each other's calls by the time young fledge at 21 days of age and join a creche (Balda and Balda, 1978). Exchanged nestlings are accepted before they develop individually distinctive begging calls at 14 days of age (McArthur, 1982 and personal communication). Between 15 and 20 days after hatching, both young and parents show increasing discrimination of each other's recorded calls. Nestlings and fledglings beg loudly at the approach of their parents. In the nestling period this may enable adults to learn the calls of their young. These authors suggest that the ability of young to recognize their parents' calls shortens the time spent in noisy begging and feeding, reducing the probability of attracting predators.

C. Discussion

1. Timing of Recognition

The most important point to emerge from this review of parent–young recognition is that recognition develops when and where it is needed to restrict parental care to a bird's own offspring and, conversely, that it is retarded or lacking in the absence of such need. The crucial point in time is when young of different broods begin to mingle.

In species with altricial young this time depends on the nesting situation. For murres on crowded ledges it comes shortly after hatching, and young learn parental calls in the egg. For most species with discrete nest sites the important point is when the young fledge. Thus, for swallows parental recognition of young begins just before or after young leave the nest. For penguins the comparable stage is when young leave to join a creche. There are exceptions that prove the rule. The Rough-winged Swallow, a solitary nester, does not develop recognition of its young. The Emperor Penguin, lacking a fixed nest site, is likely to prove an exception showing recognition at an early age.

In species such as gulls, with semiprecocial young that require extended care, often in colonial situations, the problem is more complex. Young can mingle before fledging, although a variety of constraints reduces the probability that a parent will care for the wrong chick. Conover *et al.* (1980) report that, even before adult Ring-billed Gulls recognize their own chicks, they are more likely to protect chicks which are similar in size to their own and found where their own chicks are likely to be. Although mingling is a more gradual process here than in altricial birds, we still see relationships between the age when recognition occurs and the life history of the species. Evans' studies of Black-billed and Ring-billed gulls provide an example. Physical barriers to mingling are an important factor in gulls and other colonial species. We see this in Franklin's Gulls, with isolated floating nests, and in kittiwakes, whose young are confined to isolated ledges. Compared with other gulls, the former shows delayed development of parent–young recognition while the latter shows no recognition before fledging.

2. *Why Recognition May Be Lacking*

Where recognition is lacking, is the absence of immediate need an adequate explanation? In the Herring Gull it appears that parent–young recognition may be facultative, depending on the environment. However, in other cases, necessary adaptations may have failed to develop in the absence of selection. Consider the different degrees of individuality in calls of young Adélie as compared with Emperor penguins or in calls of young Rough-winged as compared with Bank swallows. In each case, there is reason to suppose that the species with the less distinctive call is subject to less selection for individual recognition. It is more difficult to determine whether in the absence of need mechanisms of perception and learning involved in the recognition process are lacking or simply not activated.

There is a tendency to regard the variation that underlies recognition and the process of recognition as "neutral" characters, but both may entail costs. Identification usually depends on variation in signals that have other functions (Smith, 1977) and this variation may result in some loss of effectiveness of the primary function. Discrimination requires time for learning and comparison and this may necessitate greater use of a signal than might otherwise be required, exposing young to predation (McArthur, 1982). In the same vein, time involved in discrimination by parents might better be invested in making more feeding trips (Beecher *et al.*, 1981a). There may even be disadvantages in recognition itself. A parent could misidentify and reject its own offspring (Beecher, 1982b). To the extent that a young bird can receive parental care from, or avoid attacks by, other adults, it may benefit from anonymity, especially if it loses its parents. While some of these costs may be trivial, taken together they help to explain why recognition between parents and offspring develops only when there is a genuine need.

3. *Which Recognizes the Other, Parent or Young?*

Of those species that have been studied in detail, penguins, alcids, swallows, and the Piñon Jay provide examples where parents recognize their young or recognition is mutual. In each case the young mingle with others, often in creches, and are fed by their own parents. In gulls and terns exchanges of nestlings suggest that parents recognize their young, but the only carefully controlled experiment to test this, performed by Beer with Laughing Gulls, yielded negative results. Moreover, the experimental evidence concerning call recognition in gulls and terns is nearly all on the side of young recognizing parents. The only evidence of parental recognition of chick calls is for the Royal Tern, in which young join in a creche.

Before we try to account for these apparent differences between gulls and other groups, it must be recognized that the evidence is incomplete. Only a few species have been studied in detail and often only one side of the relationship has been investigated. More balanced investigation may reveal additional cases of mutual recognition. Most of the evidence relates to recognition of calls, but vision may also play a part, so that negative evidence must be accepted with reservations. Interactions between parents and young may be subtle, as Beer's work shows, and more careful studies are needed to resolve these problems.

Although in general parents and young have a strong mutual interest, there are possible asymmetries in the costs and benefits to each. Thus, young may benefit if they can elicit care from adults other than their parents, but adults do not benefit by feeding the young of others (here we leave aside the special case of helpers at the nest). Hence, we should expect selection to favor parental recognition of offspring whether or not young recognize parents. This argument would apply with particular force where young mingle in creches. Where parents recognize their own young there may be strong selection for young to recognize parents, since young that approach the wrong adults may be neglected, driven off, or even killed. Thus, it is not difficult to explain cases where parents recognize young or where mutual recognition occurs.

How can we explain the situation which apparently occurs in gulls where young recognize their parents without reciprocation? Here adults are territorial while young are semiprecocial and able to wander at an early age. Even if the adult is unable to recognize its young as such, other clues, including their behavior, may enable it to effectively channel its parental care. Thus, recognition by the young may actually reduce selection for recognition by the parent. Still, such a system would leave the way open for chicks to benefit by begging from any adult. Beer occasionally observed Laughing Gull chicks "beating the system" but found that they normally directed their behavior toward their parents. Thus, it appears that young gulls are under direct selection to recognize their parents, even in the absence of a precise reciprocal mechanism. Beer suggests that the intimate interaction with parents, which his work demonstrates, may be necessary to normal social development of the young, including the acquisition of competence in communication. Even in terms of satisfaction of immediate needs there may be costs to the young in failing to recognize its parents. Indeed, the penalties for error are much greater for a young bird than for a parent which has other young. A chick which crosses a territory boundary runs a strong risk of being attacked. In the unlikely event that it is accepted by a strange adult it has given up membership in its own family group to join a group the size of which it has increased (Beecher, 1982b). Considering the identifying signals available for recognition, the calls of the adults are stereotyped and selected for recognition by mates while those of the young are undergoing development. Unless there is strong selection for adults to recognize young, it may be easier for young to learn the calls of parents.

4. Development of Recognition

The development of parent–offspring recognition undoubtedly involves an intimate mix of inherited tendencies and various types of learning. While a detailed review is beyond the scope of this chapter, a few points can be made. Relatively little work has been done on the acquisition of the sound signals implicated in experiments on parent–offspring recognition. Similarity of signals between mates in Laughing Gulls suggests convergence by learning similar to that described by Mundinger (1970) for finches. Similarity among the calls within a brood of Bank Swallows may arise by common inheritance or mimicry but experimental evidence that might resolve this question is still inconclusive (Beecher, 1982b). In these cases similarity of calls could make it easier for chicks and adults, respectively, to learn to recognize them.

Where the development of individual recognition has been investigated in either young or adults it appears to involve learning during one or more critical periods. Thus, Beer's work demonstrates the importance of early experience in young Laughing Gulls. Learning occurs in stages as the young are exposed to different adult calls. In a series of studies with Ring-billed Gulls, Evans shows that stimulation received in the egg is followed after hatching by exposure to adult calls. In a series of studies with Ring-billed Gulls, Evans shows that stimulation received in the egg is followed after hatching by exposure to adult calls which are reinforced by visual stimuli and feeding by the parents (see Evans, 1980, for further references). Although the term imprinting has been applied to this learning process, we find that early studies of imprinting were interpreted in terms of the learning of species characteristics (Beer, 1970). Considering the literature reviewed here, it seems more likely that imprinting of young normally results in responses to the individual characteristics of their parents. As an example, female Zebra Finches respond selectively to the songs of their father following early exposure and 2 months of isolation (D. B. Miller, 1979b). Oddly enough, few studies of individual recognition deal with Anatidae or gallinaceous birds, two groups that feature strongly in the literature on imprinting. It appears that the subjects of individual recognition and imprinting have developed separately (but see Bateson, 1979).

The observation that adult Bank Swallows and Piñon Jays recognize calls of their young within a day or two of the appearance of the appropriate vocalizations points to learning on the part of the parents. Perhaps more convincing is

Beecher's observation that Bank Swallow chicks adopted by other adults at an early age are rejected by their own parents if they visit the home burrow after recognition has developed. Preliminary evidence suggests that there are species-specific constraints on what an adult Bank Swallow can learn (young rough-wings are rejected at an early age), including the possibility that it can recognize only a limited number of different chicks (studied by augmentation of broods) (Beecher, 1982).

Since parents and young share half their genes it is conceivable that recognition could be based on inherited similarities. That is, recognition might occur without previous experience. Yet the available evidence, both circumstantial and experimental, implicates learning. Learning has the advantage that the image (template) acquired in this way is a more exact replica of the characteristics to be recognized that would be expected if it were based on the degree of relationship (Beecher, 1982). Although the possibility cannot be ruled out, there is as yet no direct evidence that birds can recognize kin to which they have not been previously exposed.

Throughout this section we have assumed that parent–young recognition was really individual recognition. Yet no evidence has been presented that parents distinguish among their young or that young discriminate between their parents.

V. RECOGNITION OF NEIGHBORS

A. Introduction

Neighbor recognition has been documented mostly in dispersed territorial species and almost exclusively with respect to song or similar long range signals. Marler (1960, p. 359) suggested that song might enable males "to distinguish between new intruders and old rivals" and Thorpe (1961, p. 43) suggested that a "rival might recognize a given male not merely by the position of his territory but actually by his individual voice, independently of his territory." By the time these statements appeared some experimental evidence was already accumulating.

B. Experimental Evidence

1. *Neighbor–Stranger Discrimination*

a. Species with a Single Song. Neighbor–stranger (N–S) discrimination was first demonstrated in the Ovenbird *(Seiurus aurocapillus)*, a species in which an individual male has a distinctive, stereotyped song (Weeden and Falls, 1959).

During the breeding season male Ovenbirds sing exclusively in discrete territories which they defend against other singing males, but they forage in more extended areas overlapping each other's territories (Zach and Falls, 1979). Weeden and Falls played recorded songs from a loudspeaker located within song territories. Males responded with calls and songs, and approached and flew about the speaker as they did in normal territorial encounters. The time taken to react and the number of calls, songs, and movements observed during the experiment were used as criteria of response. While there was considerable individual variation in response, birds reacted more strongly to songs of strangers, recorded some distance away, than to those of immediate neighbors. Discrimination occurred rapidly, birds usually reacting after a single rendition of a strange song but only after a neighbor's song was played a few times. Responses to a bird's own song were intermediate.

Comparable results have been obtained for the White-throated Sparrow *(Zonotrichia albicollis)* (Falls, 1969; Lemon and Harris, 1974; Brooks and Falls, 1975a), Indigo Bunting *(Passerina cyanea)* (Emlen, 1971), Field Sparrow *(Spizella pusilla)* (Goldman, 1973), and Common Yellowthroat *(Geothlypis trichas)* (Wunderle, 1978). Negative results, obtained by Belcher and Thompson (1969) for the Indigo Bunting, may have resulted from their use of a cotton model which introduced a visual stimulus. All these species show noticeable individual variation in their songs.

Falls and McNicholl (1979) conducted similar experiments with the "hoot" of the Blue Grouse *(Dendragapus obscurus)*, which is used in the same manner as territorial song (Stirling and Bendell, 1970). Results were exceptionally clear. Males responded to recordings of neighbors simply by facing and countersinging in the direction of the loudspeaker, while they approached the speaker and performed a variety of displays when a stranger's voice was broadcast. In this study discrimination was more marked (responses stronger to S and weaker to N) when recordings were presented in the sequence S–N–S, as compared to N–S–N. There may have been enhancement of response to the first song type played.

b. Species with Song Repertoires. Several species in this category have been studied using methods comparable to those described above. Although results have been in the same direction (weaker responses to N songs), evidence of discrimination has not usually been as strong as in species with single songs. In most cases, fewer measures of response have shown significant N–S differences and differences have been small.

Krebs (1971), using a simple response–no response criterion, obtained clear evidence of N–S discrimination in Great Tits (*Parus major*), which have three to four individually distinctive songs (Gompertz, 1961). Järvi *et al.* (1977), experimenting with the same species, found significant N–S differences in only one of several measures of response. Chaffinches (*Fringilla coelebs*) with repertoires of two to three song types were investigated by Pickstock and Krebs (1980). In three series of experiments involving several criteria of response, 21 of 26 N–S comparisons were in the expected direction but only six were statistically significant. Repertoires of Song Sparrows *(Zonotrichia melodia)* are larger than those of Great Tits and Chaffinches and neighbors rarely share identical patterns (Harris and Lemon, 1972). Comparatively weak discrimination has been found in this species (Harris and Lemon, 1976; Kroodsma, 1976). The case of the Song Sparrow is also complicated by the occurrence of local dialects (Harris and Lemon, 1972) and the finding that responses are reduced to songs from other dialect areas (Harris and Lemon, 1974). To minimize this problem, both studies of N–S discrimination used songs of strangers (nonneighbors) separated from the individuals tested by only two or more territories. Thus, the strangers used here may have been more "familiar" than those used in other studies. Using a method in which N and S songs were presented on different days, Searcy *et al.* (1981) compared Song Sparrows with Swamp Sparrows (*Z. georgiana)*. Differences in responses to N and S songs were much greater in Swamp Sparrows with repertoires of four to five songs than in Song Sparrows with eight to ten songs. These same authors also used a different experimental design with Swamp Sparrows, playing N and S songs alternately from two loudspeakers 16 m apart near the territory boundary. The birds responded equally to both songs. Using a similar design with speakers 30 m apart, I found that Western Meadowlarks (*Sturnella neglecta*) showed a significant tendency to spend more time at the speaker playing the S song. In a further series of playback experiments, the same birds showed a greater tendency to match S songs than N songs (J. B. Falls, unpublished data).

In all the studies reviewed so far in this section, only a single song from the neighbor's repertoire was used in each experiment. Thus, recognition based on more than one song type was excluded. In recent experiments comparing Eastern Meadowlarks (*S. magna*) with Western Meadowlarks, we played sequences of two songs of both strangers and neighbors from a single loudspeaker. However, discrimination was weak, especially by Eastern Meadowlarks whose repertoires are an order of magnitude larger than those of the western species (Falls and d'Agincourt, 1981).

Brémond (1968) used a different experimental approach with the Robin (*Erithacus rubecula*), a species in which an individual's repertoire may include several hundred motifs. Neighbors do not ordinarily respond to one another's singing but, if a stranger's song is broadcast within one bird's territory, the territory holder will attack the source while the neighbor exhibits agonistic behavior at the common boundary. Since the neighbor responds in a similar way if the territory holder is artifically prevented from responding, it must be responding to the unfamiliar song. If the strange song is broadcast from the common boundary, both birds attack the source in turn but do not attack each other. These results cannot be compared directly with those of the other studies.

Australian Magpies *(Gymnorhina tibicen)* provide an example of chorus singing. Groups of from two to ten birds of both sexes engage in joint caroling, which is a territorial proclamation (Carrick, 1963). Using playback at mutual boundaries, I found that groups responded more strongly to their own group carols and to a single song of a group member than to comparable singing of a neighboring group (J. B. Falls, unpublished data). Less complete evidence suggested that they also distinguished carols of neighboring and strange groups. Though I made no detailed analysis of the structure of these group songs, some were clearly recognizable by ear. Sonagrams indicate that they are not rigidly sterotyped performances so I have classed them here with song repertoires.

2. *Individual Recognition*

Neighbor–stranger discrimination has often been interpreted as individual recognition but all the results presented so far are consistent with a simpler dichotomy between familiar and unfamiliar songs (Beer, 1970). The evidence that more may be involved is provided by experiments in which the speaker location has been varied.

In the research already reviewed, speaker locations were not always reported but in most cases were just inside the boundary adjacent to the neighbor whose song was played. In this location N–S discrimination has nearly always been found. It might be suspected that this result was influenced by the neighbor's response, as described in Brémond's experiments with the Robin. However, as in that case, such a result would still reflect discrimination on the part of the neighbor. In our experiments, the speaker was usually placed about 10 m inside the territorial boundary and directed toward the center of the territory with the result that neighbors seldom interfered. In some cases responses to neighbors and strangers at the center of the territory are equally strong (Wunderle, 1978). In experiments with White-throated Sparrows we were able to demonstrate weak but significant discrimination when the speaker was carefully placed in the territory center (Falls and Brooks, 1975). Both Emlen (1971) and Krebs (1971) also obtained significant discrimination at the center.

The most revealing result was obtained by placing the loudspeaker near the boundary on the opposite side of the territory to the normal location of the neighbor whose song was used. When this was done, Falls and Brooks (1975) found that the responses of White-throated Sparrows to a neighbor's song were no different from responses to a stranger's song at that location. If neighbors' songs were only recognized as a class of familiar sounds, then it should not matter at what part of the boundary the song is played, since neighbors occur on both sides of the territory. This result shows that a bird can discriminate among songs of different neighbors and recognize that the displaced song is not the one that belongs in that location.

Similar results were obtained with the Stripe-backed Wren (*Campylorhynchus nuchalis*) using a different experimental design (Wiley and Wiley, 1977). In this tropical species a principal pair occupies a year-round territory and sings a repertoire of about five duets in territorial encounters. Neighboring pairs have mainly different repertoires. Clear N–S discrimination was shown at the appropriate territory boundary. Wiley and Wiley also used playback at a particular boundary to compare responses to duets of nearby neighbors (N) and neighbors from across the territory (XN). Responses were significantly stronger to XN than to N in trials when XN preceded N. This result recalls the finding of Falls and McNicholl (1979) with Blue Grouse that discrimination was more pronounced when the stronger stimulus was presented first. A further demonstration of this type has recently been carried out with the primary (breeding) calls of Red-headed Woodpeckers (D. Crusoe, unpublished data). Individuals are easily recognizable by voice and calls of color-marked birds remain substantially similar from year to year. Stronger responses to strangers than to neighbors and to neighbors from the opposite as opposed to the usual direction were significant by all of five criteria of response.

3. *Recognition of Other Species' Songs*

Richards (1979) studied neighbor recognition of heterospecific song in a population of Rufous-sided Towhees *(Pipilo erythrophthalmus),* some of whose repertoires included imitations of songs of Carolina Wrens *(Thyrothorus ludovicianus).* He found that neighbors of such individuals responded similarly to playback of their wren imitations and normal towhee songs. However, these same neighbors did not interact with Carolina Wrens, whose territories overlapped their own. More distant towhees distinguished between towhee songs and wren imitations, responding only very weakly to the latter. Thus, the neighbors of the towhees in question must have learned particular wren imitations.

Somewhat comparable findings have emerged from studies of birds' responses to neighbors of different species where interspecific territoriality is involved. For example, Falls and Szijj (1959) studied interactions between Eastern and Western meadowlarks in southern Ontario where the former species is common and widespread and the latter occurs rarely either singly (one male) or in small groups of adjacent territories. All Western Meadowlarks studied responded to playback of Eastern as well as Western song, while only those Eastern Meadowlarks with Western neighbors responded to Western songs. Similar results have recently been obtained by Catchpole (1978), using songs of *Acrocephalus* warblers in areas of sympatry and allopatry. In such cases, birds must learn songs of their neighbors. However, it remains to be shown whether they are able to differentiate individual members of the other species.

4. *Features of Song Used for Neighbor Recognition*

The only attempt to put this question to the birds in the form of playback experiments was carried out on the White-throated Sparrow in which neighbor recognition had already been demonstrated (Brooks and Falls, 1975b) (see Sections V,B,1a and 2). Territory holders treated a stranger's song as such, whether the major song type was the same as or different from that of a neighbor. Neighbor–stranger discrimination was demonstrated using as little as the first three notes of a song. These findings focus attention on details of the first part of the song, which consists of a series of pure whistles that differ in frequency and duration. Songs of neighbors were altered in one or other of these features and it was shown that changes of more than 5 to 10% in frequency of the songs, frequency of the first note, and possibly in frequency change between the first and second notes interfered with recognition of a neighbor's song. Songs altered in these ways were treated as strangers' songs, eliciting stronger responses. In contrast, changes as great as 15% in the length of the song or of the first note had no effect. Reliance on frequency as opposed to timing was consistent with the results of analyses by Borror and Gunn (1965) and Lemon and Harris (1974) who found that frequency was less variable in songs of an individual than was timing. Moreover, the long pure whistles which comprise the song of this species lend themselves to accurate determination of frequency by a listener. Using only the features identified as important in this study, a male White-throated Sparrow should be capable of identifying many more individuals than just his immediate neighbors.

C. Discussion

Neighbor–stranger discrimination by song is evidently widespread among species that occupy large territories. However, it does not appear to have been found in the colonial species in which parent–young and mate recognition by calls have been demonstrated, even when it has been looked for (Beer, 1970, for young and adult Laughing Gulls; White, 1971, for adult Gannets; Wooller, 1978, for adult Black-legged Kittiwakes). In the crowded conditions of a breeding colony visual contact may eliminate the need for vocal recognition between neighbors.

The few studies that have gone beyond neighbor–stranger experiments have shown that birds can recognize their neighbors individually and not simply as a familiar class. Therefore, it seems likely that neighbor recognition usually indicates individual recognition. A number of issues remains to be considered.

1. *The Role of Location*

First consider the significance of location in a typical N–S experiment conducted near a territory boundary, with playback of the two songs separated in time. The territory owner responds weakly to the N song, usually not approaching closely. It has no difficulty determining the direction of the signal since it moves directly toward the speaker when the S song is played. It often overshoots the S song, however, suggesting that the distance is more difficult to estimate. Perhaps a bird fails to approach an N song played from the appropriate direction because it misjudges the distance, expecting the song to originate in the neighbor's territory. Another type of experiment, in which N and S songs are played alternately from two speakers, allows us to evaluate this possibility. Using this design with Swamp Sparrows, which show N–S discrimination when the two songs are presented separately, Searcy *et al.* (1981) obtained equally strong responses to both songs. Apparently, some type of interaction occurred between effects of the N and S playbacks. They suggested that a bird approaching the S song discovered that the N song also originated within its territory and attacked it as well. Thus, when a bird can determine the location accurately, it does not tolerate an N song in its territory. This explanation can also account for the strong response obtained by playing an N song from the opposite side of the territory, since it obviously originates from the wrong direction and, like an S song, might be in the bird's territory. Several studies have obtained N–S discrimination at the territory center. This is more difficult to explain but it could be argued that some errors of location are possible at the center, depending on the position of the bird in relation to the speaker. Using a two-speaker design similar to that of Searcy *et al.* (1981), I recently found N–S discrimination in Western Meadowlarks at the territory boundary. This evidence is directly contradictory to the above explanation. However, boundaries may be less definite in the large territories of meadowlarks than in the smaller territories of Swamp Sparrows.

An alternative explanation would be that birds are more tolerant of N than S song inside their territories, with the degree of tolerance of N song diminishing as the intrusion moves across the territory from the usual location of the N. This explanation can account for all the results of N–S experiments except those of Searcy *et al.* However, it does not rule out mistaken location as a contributing factor. It may well be that species or populations differ in their tolerance of neighbors. Further experiments with two speakers would help to resolve these questions.

What can we conclude about the roles of song and location in neighbor recognition? Clearly, characteristics of the song are necessary, since only a particular N song will elicit an appropriate response at a given location. Locational information may also be necessary, since a given N song in the wrong location is apparently not distinguished from an S song. We could only claim that song alone was sufficient for recognition if birds showed N–S discrimination (i.e., tolerated N song) when the N song was clearly out of place. Since evidence at the territory center is somewhat equivocal, we must be cautious about claiming that individual songs are recognized independently of location. An alternative is to regard both song and location as necessary properties of the identifying signal.

2. *Learning Mechanisms*

At least two kinds of learning may be involved in neighbor–stranger recognition: habituation and associative learning. Typically, responses to familiar neighbors decline over the season, while responses to strange songs remain strong (Brooks and Falls, 1975a). Thus, a stimulus-specific decrement in response occurs with repeated exposure to particular songs. This conforms to the classical definition of habituation (Thorpe, 1956).

Habituation to birdsong has been demonstrated experimentally. Using playback to map territories, I have found that if response wanes it can be restored by changing songs (Falls, 1981). Petrinovich and Peeke (1973) found habituation in responses to playback of particular songs of White-crowned Sparrows. More detailed studies showed that a long-lasting stimulus-specific response decrement occurred with respect to a number of behavioral variables (Petrinovich and Patterson, 1979, 1980; see also Patterson and Petrinovich, 1979).

The reader will recall that discimination between different neighbors was demonstrated by displacing a song to a new location, where response increased. This could result if location of a song was one of the stimulus variables to which habituation occurred. Then, moving the song would constitute a novel stimulus restoring the response. Another experiment supports this explanation. I played conspecific songs to Australian Magpies from a fixed loudspeaker. When, after several hours of intermittent playback, the initial response had ceased, I moved the speaker, restoring the response, which subsequently waned more rapidly than before.

While we could account for most cases of neighbor–stranger discrimination by a process of habituation, associative learning may also be involved. In some unusual cases it would seem to be required. Richards' (1979) finding that Rufous-sided Towhees responded to their neighbors' imitations of Carolina Wren songs (but not to wrens) can be explained if towhees associate the imitations with the normal songs or appearance of their neighbors. A similar explanation would account for the response of neighboring Eastern and Western meadowlarks to each other's songs. Since the songs are rather different, it seems likely that appearance and visual displays, which are similar in these species, are involved as unconditioned stimuli. I believe associative learning rather than convergence of songs accounts for a number of similar cases (see also Catchpole, 1978).

3. Song Repertoires

Evidence reviewed earlier suggests that the degree of N–S discrimination is inversely related to repertoire size, being clearest in species that have single songs. Why should this be so? Neighbor recognition requires both individual differences in the vocalizations of different birds and the ability of listeners to detect and learn these differences. Since there can be individual differences in each song type, a repertoire increases the potential for individuality. Besides some or all of the song types that comprise it, the organization of a repertoire may be individually distinctive (Wildenthal, 1965; Hansen, 1981). Given sufficient time, the repertoire as a whole might be used for neighbor recognition but this has not really been tested. What has been investigated experimentally is rapid identification which must be based on one or a few song types.

In contrast to the repertoire as a whole, single songs seem likely to lose their individuality as repertoire size increases. To serve any of the proposed functions of repertoires (Krebs and Kroodsma, 1980), the song types of which they are composed must be recognizably different from one another. Since each individual should diversify its repertoire, a particular song type of one bird is likely to resemble songs of other birds rather than other songs in the same repertoire. As repertoire size increases so should the similarity between songs of different individuals. In many species precise copying ensures that song types will be shared and this enables birds to match each other's songs. Thus, in achieving functional repertoires the scope for individually distinctive songs may be sacrificed. Not only are songs of different birds likely to be similar but features common to all the songs of each individual seem likely to be reduced since they would run counter to the tendency for contrast with repertoires. I know of no comparative evidence on this point. However, in Great Tits, which have small repertoires, songs of the same bird tend to be similar in frequency (J. B. Falls, unpublished data).

Compared with single songs, repertoires present listeners with more (and perhaps more similar) sounds to be learned and provide less exposure to each. Thus, as repertoires increase in size the task of learning them becomes more difficult. This may explain why clear N–S discrimination was found in Stripe-backed Wrens, which sing year-round affording more opportunity for listeners to learn repertoires of particular pairs.

In the preceding section I suggested that N–S discrimination arises by habituation to neighbors. Yet one of the proposed functions of repertoires is to hold the attention of listeners by reducing habituation (Hartshorne, 1956; Krebs and Kroodsma, 1980). In the following section, I shall argue that N–S discrimination is adaptive in reducing unnecessary strife as the season advances. Other things being equal, we should expect to find more strife between neighbors in species with large repertoires. While no strictly comparative work has been done, Morton's observations on Carolina Wrens (see Chapter 6, Volume 1) may represent such a case.* Whatever the benefits of repertoires, they have to be weighed against possible costs, one of which may be a reduction in N–S discrimination. This is a possible constraint on repertoire size.

If the typical N–S experiment shows the result of previous habituation to N songs, it may be particularly susceptible to effects of repertoire size. We should look for other ways to put the question to the birds; matching experiments (Section IV,A,1,b) represent one such possibility. Although on present evidence we must conclude that repertoires are a hindrance to neighbor recognition, the possibility of recognizing the repertoire as a whole should be investigated. Another possibility is that birds with large repertoires repeat individually distinctive sounds frequently. From casual listening I believe this might apply to the Song Thrush *(Turdus philomelos),* which sings some phrases more often than others (J. Hall-Craggs, unpublished data). In any case birds in stable populations can still use the locations from which their neighbors sing to reduce the chances of mistaken identification.

4. *Adaptive Value*

The main finding of the research reviewed in this section is that, in many species, males respond much less to songs of neighbors than to songs of strangers, often not even approaching when they hear their neighbors. This accords with the general observation that, although chases and fights occur frequently in the establishment of territories, later interactions between neighbors are usually limited to countersinging from well within their respective territories. Judging from results of playback, this temporal deescalation between neighbors is not extended to strangers, to which responses continue to be rapid and strong (Brooks and Falls, 1975a).

Is this difference in response to neighbors and strangers adaptive? If White-throated Sparrows are experimentally removed, their territories are rapidly taken over, not by their immediate neighbors but by birds that did not sing in the area before (J. B. Falls and D. J. Loncke, unpublished data). In terms of the usual playback experiment, these birds are strangers. While they are becoming established, strife with the surrounding birds is temporarily intense. Thus, by responding strongly to strangers, White-throated Sparrows direct more of their aggression toward intruders that represent a serious threat to their territories. By responding weakly to neighbors, with which they have already established a common boundary, they run little risk of losing ground (but see references in preceding section to cases where neighbors may pose a threat). Thus, to some

*An alternative explanation for strong response to neighbors is that, in such cases, they may pose a serious threat to a bird's territory or reproductive potential (Falls and d'Agincourt, 1981; Grove, 1981).

extent, they can conserve time and energy and reduce their exposure to the risks of injury and predation that accompany fighting. Since neighbors recognize one another by song and location, their countersinging may normally be sufficient for mutual deterrence (Schleidt, 1973) while providing a familiar background against which they can recognize any change in the *status quo,* including the relocation of a neighbor, that calls for a more vigorous response.

Neighbor recognition may facilitate reestablishment of territories from year to year in migratory species whose songs remain stable. Although evidence that birds retain the individual characteristics of their songs from 1 year to the next is rather limited (e.g., Borror, 1960; J. B. Falls, personal observations of Ovenbirds and White-throated Sparrows), some observations suggest that birds recognize their previous neighbors. Nolan (1978) observed 21 fights between newly arrived Prairie Warblers, 19 of which involved new neighbors. Previous neighbors seldom fought, although they trespassed on each other's territories as much as did new birds. Knapton (1979) reported similar observations for the Clay-colored Sparrow *(Spizella pallida).* It seems likely that recognition of familiar boundaries as well as familiar neighbors could be responsible for these results.

Ovenbirds and White-throated Sparrows often trespass quietly in neighboring territories (Zach and Falls, 1979; J. G. Jones, unpublished data) and casual observation suggests that this may also occur in other species. It may be important to these trespassers to know what to expect from each of their neighbors. Learning each neighbor's song would enable a bird to know at least the extent of its territory and its whereabouts at any given time.

The ability to recognize the voices of particular territory holders while remaining silent may be important to birds seeking territories. These include individuals with inferior territories as well as "floaters" without territories. An example involving the former group was noted by Krebs (1971) in the Great Tit. When territory holders were removed in a woodland, replacements came within a few hours from birds occupying hedgerows, whose territories did not immediately adjoin the woods. Later, Krebs (1976) found that playing a Great Tit's song in its territory after it has been removed significantly retarded reoccupation of the territory. He argued that birds recognize and monitor all territory holders within earshot, and thus immediately know when a territory is vacated. This should apply not only to birds that already have inferior territories, but also to "floaters" that are present long enough to learn the songs of resident territory holders. For two Species, the Rufous-collared Sparrow *(Zonotrichia capensis)* (Smith, 1978) and the Song Sparrow (Knapton and Krebs, 1976), there is direct evidence that hierarchies of floaters are associated continuously with one or a few territories. For White-throated Sparrows, recent evidence obtained by telemetry indicates that "floaters" may remain silently in an area for at least a few days, overlapping several song territories (J. G. Jones, unpublished data).

5. Dialects and Kin Recognition

The finding, in several species that have song dialects, that birds respond more strongly to local than to different dialects (Lemon, 1967, with Cardinals; Milligan and Verner, 1971, with White-crowned Sparrows, *Zonotrichia leucophrys;* Harris and Lemon, 1974, with Song Sparrows) may seem to contradict the results for neighbor–stranger discrimination, which show weaker responses to the more familiar song. Actually, the difference appears to be one of scale for, as we have seen, Song Sparrows discriminate in the usual way between immediate neighbors and more distant birds in the same dialect area. Baker *et al.* (1981) have demonstrated similar N–S discrimination within a dialect area in the race *Z. l. nuttalli* of the White-crowned Sparrow previously studied by Milligan and Verner. Thus, dialect and neighbor recognition are different phenomena. The remarkable conformity found among songs of different individuals within a dialect area suggests that individual differences may be slight in such cases; these results indicate that N–S discrimination may be correspondingly weak.

In proposing a model to account for dialects, Treisman (1978) suggested that they might permit birds to settle next to kin and avoid using the more destructive types of aggression. He cited cases of neighbor–stranger discrimination, leaving the impression that reduced strife between neighbors was a result of kin recognition. While this point has since beeen clarified (Trainer, 1980; Treisman, 1980), the question of whether there is any relationship between neighbor and kin recognition deserves comment. In relation to Treisman's original hypothesis, many of the species in which N–S discrimination has been studied do not have dialects. As discussed above, the mechanisms and advantages of neighbor recognition seem independent of whether or not the birds are related. Finally, while N–S discrimination seems to be widespread, there is no corresponding evidence that neighbors are usually closely related.

6. Possibilities for Cheating

Are there possibilities for cheating on neighbor recognition? Trespassers in territories are typically silent but this may be an attempt to remain inconspicuous and not stimulate aggressive responses rather than a way of withholding identity. Is there a possibility of cheating by mimicry? Mimicry plays a part in the development of song (see Chapter 3, this volume) but the result is not usually so precise as to preclude individual recognition. If a "floater" taking over a territory could mimic the previous owner it might benefit from the behavior of the mate and neighbors of the previous resident. Alternatively, a new arrival could mimic a successful neighbor and benefit from the responses of other birds. In this case the "model" could be expected to respond strongly to the newcomer. Payne (1981) has observed what seems to be an example of this in the Indigo Bunting.

Some new males mimic the songs of established neighbors and are more successful than others in holding territories and breeding. He argues convincingly (Payne, 1982) that this is a case of competitive mimicry and may represent a more widespread phenomenon among species with song dialects. Rohwer (1982 and personal communication) suggests that, while such mimicry may not deceive other neighbors, it may repel potential intruders that have had interactions with the more experienced model in the course of probing a large number of territories.

VI. GENERAL DISCUSSION

A. Individual Recognition

In different ways, the recognition of other individuals serves both to simplify and complicate the lives of birds. On the one hand, behavior can be channeled efficiently to certain individuals by recognizing such simple dichotomies as between one's mate and others or between one's offspring and others. This is obviously important in the confusion of a large colony. On the other hand, much more subtle interactions can occur with those individuals that are recognized. Recognition opens up the possibility of progressively learning about an individual and building upon that experience. It is not necessary, for example, to continually recycle through territory establishment with a known neighbor. The idiosyncrasies of behavior of familiar individuals become more predictable and the social interactions of a group can be more finely tuned than would otherwise be the case (Simpson, 1973). Conflicts between known individuals may be fundamentally different from those involving strangers, because deception may be replaced by known asymmetries (van Rhijn and Vodegel, 1980). Individual recognition permits the development of various selfish subgroups within what would otherwise be the same class of individuals (Wilson, 1975). Reciprocal altruism, for example, presupposes some form of individual recognition. Social hierarchies select for some sort of status signaling of which individual recognition is a special case (Barnard and Burk, 1979). As Marler (1976) points out, strangeness is threatening and much social behavior is devoted to increasing familiarity. This may help to explain the repetitive nature of much of the vocal signaling of birds.

Before an individual can behave altruistically toward its kin, there must be some mechanism which allows it to single them out. In the case of parents and offspring, individual recognition of the type already described provides one possibility. Where parent–young recognition continues after maturity of the young, more extended groups can develop. Especially if siblings recognize one another, further complexities are possible. Emlen (1978) speculates about such possibilities in the White-fronted Bee-eater (*Merops bullockoides*). There is little evidence to suggest that territorial neighbors are generally kin or that, if they are, they gain any special advantage (Greenwood *et al.*, 1979). With mates the problem would be to avoid inbreeding with close kin. While song recognition could conceivably be used for this purpose, there is little evidence that this is the case (Krebs and Kroodsma, 1980). Much depends on whether a bird acquires its definitive song before or after dispersal from its natal area (see Chapters 1 and 7, this volume). Thus, parent–young recognition may provide a basis for more extended kin recognition but individual recognition in other contexts probably has little to do with genetic relationship.

In general the benefits of individual recognition are mutual. The sender identifies itself and benefits from the way in which it is treated by others. However, under some circumstances, it may be advantageous to remain anonymous. This might be true for chicks approaching strangers or males trying to mate with females other than their own. However, parents and mates appear to be on guard against these offenses. A more extreme type of cheating would involve mimicking a bird of higher status. Although the model is not likely to acquiesce, advantages to mimics might result in the formation of local dialects (Rohwer, 1982). This could limit recognition to individuals familiar enough with each other's songs to identify subtle differences.

Individual recognition is to be expected between birds that are in continuous or repeated contact, and where other mechanisms such as fixed sites or barriers to movement are not sufficient to prevent confusion. In general, available evidence shows this to be the case, although some intriguing problems remain to be resolved. Thus, recognition of mates appears to be virtually universal but has not been studied in lekking species where contact between the sexes is brief. Parent–young recognition is often absent during early development before young become mobile. In altricial species mechanisms of recognition may be absent at an early age. Parent–young recognition has been amply documented in semi-precocial and fledged altricial species but little is known concerning its continuation into adulthood. For truly precocial species there is little available evidence, much of the research focusing on recognition of species characteristics. Observations that young Blue Grouse may transfer between broods and that females are indiscriminately attracted to the calls of young (J. Bendell, unpublished data) suggest that precocial species would repay more detailed investigation. Neighbor recognition is widespread in species with stable territories and relatively simple, stereotyped songs. However, species with complex repertoires pose problems that call for new approaches. Recognition among unrelated members of flocks has not been investigated. Where contacts are transitory simpler forms of status signaling may replace individual recognition.

B. Identification by Sound

Like humans, but unlike many other animals, birds are audiovisual. Thus, where vision is obscured (by other individuals, topography, vegetation, darkness) or ineffective (monomorphic adults, cryptic young), they are likely to use acoustic signals as identifying signatures. Sounds may be heard at a distance and accurately located, serving as beacons. They may be learned in the egg and used after hatching. Identifying information may be encoded in many features of acoustic signals. Unlike some visual signals, sounds may be broadcast or withheld, as the need arises.

Identification by sound is based on individual variants of a species' sound signals. How do these arise? Development of vocalizations involves a complex mix of inherited tendencies and imitation (see Chapter 1, this volume). Some variation is bound to arise if only as a result of copying errors, whether it is adaptive or not. In most species, once vocalizations are acquired they are repeated in a stereotyped manner. Thus, we expect less variation in the repeated performance of one individual than among different individuals, providing some raw material for individual recognition. If there is an advantage to be gained by individual recognition, more extreme variants may be selected for. For a signal that is genetically determined, variation could be controlled by several multiallelic loci (Beecher, 1982a). Where genetic control is not so direct, variation could be enhanced through improvisation or invention.

For evidence that selection enhances individuality, I turn to comparative studies which make it possible to relate the extent of variation in signals of different species to the needs inherent in their life histories and to the occurrence of individual recognition. The results of Beer and Evans for young gulls of different species show relationships between the individuality of adult calls, the age at which chicks begin to mingle, and the ability of chicks to recognize calls of their parents. Comparative studies call for the use of some common currency in which to express the degree of individuality. Jouventin's use of the coefficient of variation for temporal features of penguin calls provides an example. In this case, the amount of variation in calls of both adults and chicks depends on whether or not the species has a fixed nest site. To measure the information capacity of the Bank Swallow chick call, Beecher (1982a) calculated the value of H based on the number of features of the call showing independent individual variation and the extent of variation in each. They also estimated the information capacity needed in a signature system to identify members of the population under study with a small probability of duplication (they assumed that signatures were drawn randomly from a pool). The two values of H for the information available and the information needed agreed reasonably well. While this result suggests that the degree of individual variation is adaptive, the actual values are somewhat arbitrary, depending on selection and measurement of call characteristics in the one case and assumptions of the model in the other. However, if these absolute values are difficult to interpret, comparative aspects of the same study make a more convincing case. Chick calls of the solitary Rough-winged Swallow showed much less individuality than those of the colonial Bank Swallow and playback experiments demonstrated a corresponding difference in the abilities of adults of these two species to recognize their own chicks. Taken together, the results of all these studies strongly suggest that selection has enhanced the individuality of calls in those cases where the advantages of individual recognition are great.

If signature systems are adapted for individual recognition, we might expect to find cases of divergence in vocalizations where more than two individuals are associated. Suggestive evidence was found by Thompson (1969a,b) in a study of individual differences in "caws" in a flock of Common Crows *(Corvus brachyrhynchos)*. Caws of individuals changed over time but no direct evidence of divergence was found. The possibility of divergence should be explored in neighborhoods of territorial birds. However, other factors (learning mechanisms, local adaptation, species recognition) may favor convergence.

Can we predict whether or not a signal will be used for individual recognition from analysis of its variation? Again, absolute measures of variation are hard to interpret in the absence of information concerning which features of a signal would be detected and used by a bird. Field trials of the abilities of birds to detect differences in sounds would help in this regard. However, the comparative studies discussed above suggest that, at a gross level, individuality in structure of vocalizations has predictive value within related groups of birds.

Many studies of individual variation have been carried out after individual recognition has already been demonstrated in an effort to determine which features of a signal are most likely to be used. In the few cases where further experimentation has followed, manipulation of features showing the greatest individual variation has interfered most with the birds' ability to recognize the signal (Tschantz, 1968; Brooks and Falls, 1975b; Jouventin *et al.*, 1979). These results again suggest that structural variation has predictive value, in this case for the features used in individual recognition.

Can any general predictions be made concerning which features of sound are likely to be used for individual recognition? Studies of individual variation have turned up a remarkable array of possibilities. Individuality occurs not only in time–frequency characteristics of vocalizations but also in such properties as syntactical structure in songs of Mockingbirds *(Mimus polyglottos)* (Wildenthal, 1965) and wing-beat intervals in Flappet Larks *(Mirafra rufocinnamomea)* (Bertram, 1977). In tackling this question, it may help to consider a number of physical and biological constraints (see also Chapter 5, Volume 1).

Broadcast distance is an important factor. At very close range or near the nest, a variety of fine details and subtle qualities of a sound may be appreciated and we should expect the sort of multifactorial differences found in young swallows and jays to be useful. However, such fine structure will deteriorate over comparatively short distances and will be hard to perceive as signal-to-noise ratio decreases. At moderate distances, direction becomes important and repetitive patterning of wide-band sounds is appropriate. We commonly find sounds of this type in seabird colonies (e.g., penguins, murres, Gannets, and some gulls and terns) and they have also been described in colonies of Common Grackles *(Quiscalus quiscula)* (Wiley, 1976). At greater distances, especially in forests, narrow-band sounds carry further and information is likely to be encoded in the main time–frequency pattern. White-throated Sparrows provide an example where individuality is found in frequency and frequency changes in a song composed of pure tones (Brooks and Falls, 1975b). Background noise varies independently of distance. Stein (1956) suggests that thrushes *(Catharus),* which sing in the quiet of the evening, express individuality in fine structure. At the other extreme, the background noise of a seabird colony may encourage the use of gross temporal features.

Individuality is usually encoded in signals that have other functions as well, including the communication of behavioral messages and species identity (see Chapter 7, Volume 1). Individual variation cannot be so great as to seriously impair these other functions. This conflict can be resolved by using different features or different ranges of the same feature to convey different information. In White-throated Sparrows, many different individuals can be accommodated within the species range of frequency. Alternatively, individuality may be encoded in some vocalizations of the species' repertoire, leaving others to serve different functions. Beer's studies of the Laughing Gull suggest this. We might expect the demands of species recognition to be paramount in territorial song, especially in communities of high species diversity. Thus, major features of the song are preempted to convey species identity, leaving variations on the main theme to encode individuality. Dialects and repertoires may further reduce the scope for individuality in particular songs. In colonies, especially where there is a stable social structure, most vocalizations will be directed to close associates and the needs of individual variation may be paramount. Clearly, more comparative work is needed before we can make detailed predictions in this area.

ACKNOWLEDGMENTS

I am grateful to many colleagues and students for helpful discussions. Peter Fetterolf in particular assisted with the literature on parent–young recognition. Peter Marler commented on the manuscript. Myron Baker, Michael Beecher, Dale Crusoe, Patrick McArthur, and William Searcy kindly sent me their results before publication. The editors were both helpful and patient.

REFERENCES

Bailey, K. (1978). The structure and variation of the separation call of the Bobwhite Quail (*Colinus virginianus*, Odontophorinae). *Anim. Behav.* **26,** 296–303.

Baker, M. C., Thompson, D. B., and Sherman, G. L. (1981). Neighbor/stranger song discrimination in White-crowned Sparrows. Condor **83,** 265–267.

Balda, R. P., and Balda, J. H. (1978). The care of young Piñon Jays (*Gymnorhinus cyanocephalus*) and their integration into the flock. *J. Ornithol.* **119,** 146–171.

Barnard, C. J., and Burk, T. (1979). Dominance hierarchies and the evolution of "individual recognition." *J. Theor. Biol.* **81,** 65–73.

Bateson, P. (1979). How do sensitive periods arise and what are they for? *Anim. Behav.* **27,** 470–486.

Beecher, M. D. (1982a). Signature systems and kin recognition. *Am. Zool.* **22,** 477–490.

Beecher, M. D. (1982b). Development of parent–offspring recognition in birds. *In* "Genetic and Experiential Factors in Perceptual Development" (R. K. Aslin, J. R. Alberts, and M. R. Petusa, eds.). Academic Press, New York. In press.

Beecher, M. D., Beecher, I. M., and Lumpkin, S. (1981a). Parent–offspring recognition in Bank Swallows (*Riparia riparia*): I. Natural history. *Anim. Behav.* **29,** 86–94.

Beecher, M. D., Beecher, I. M., and Hahn, S. (1981b). Parent–offspring recognition in Bank Swallows *(Riparia riparia):* II. Development and acoustic basis. *Anim. Behav.* **29,** 95–101.

Beer, C. G. (1966). Adaptations to nesting habitat in the reproductive behaviour of the Black-billed Gull, *Larus bulleri. Ibis* **108,** 394–410.

Beer, C. G. (1969). Laughing Gull chicks: Recognition of their parents' voices. *Science* **166,** 1030–1032.

Beer, C. G. (1970). Individual recognition of voice in the social behavior of birds. *In* "Advances in the Study of Behavior" (D. S. Lehrman, R. A. Hinde, and E. Shaw, eds.), Vol. 3, pp. 27–74. Academic Press, New York.

Beer, C. G. (1975). Multiple functions and gull displays. *In* "Function and Evolution of Behaviour—Essays in Honour of Professor Niko Tinbergen, F.R.S." (G. Baerends, C. Beer, and A. Manning, eds.), pp. 16–54. Oxford Univ. Press (Clarendon), London and New York.

Beer, C. G. (1979). Vocal communication between Laughing Gull parents and chicks. *Behaviour* **70,** 118–146.

Belcher, J. W., and Thompson, W. L. (1969). Territorial defense and individual song recognition in the Indigo Bunting. *Jack-Pine Warbler* **47,** 76–83.

Bertram, B. (1970). The vocal behaviour of the Indian Hill Mynah, *Gracula religiosa. Anim. Behav. Monogr.* **3,** 79–192.

Bertram, B. C. R. (1977). Variation in the wing-song of the Flappet Lark. *Anim. Behav.* **25,** 165–170.

Borror, D. J. (1960). The analysis of animal sounds. *In* "Animal Sounds and Communication" (W. E.Lanyon and W. N. Tavolga, eds.), Publ. No. 7, pp. 26–37. Am. Inst. Biol. Sci., Washington, D.C.

Borror, D. J., and Gunn, W. W. H. (1965). Variation in White-throated Sparrow songs. *Auk* **82,** 26–47.

Brémond, J. C. (1968). Recherches sur la sémantique et les éléments vecteurs d'information dans les signaux acoustiques du rouge-gorge (*Erithacus rubecula* L.). *Terre Vie* **2,** 109–220.

Brooke, M. de L. (1978). Sexual differences in the voice and individual vocal recognition in the Manx Shearwater *(Puffinus puffinus). Anim. Behav.* **26,** 622–629.

Brooks, R. J., and Falls, J. B. (1975a). Individual recognition by song in White-throated Sparrows. I. Discrimination of songs of neighbors and strangers. *Can. J. Zool.* **53,** 879–888.

Brooks, R. J., and Falls, J. B. (1975b). Individual recognition by song in White-throated Sparrows. III. Song features used in individual recognition. *Can. J. Zool.* **53,** 1749–1761.

Buckley, P. A., and Buckley, F. G. (1972). Individual egg and chick recognition by adult Royal Terns *(Sterna maxima maxima). Anim. Behav.* **20,** 457–462.

Burger, J. (1974). Breeding adaptations of Franklin's Gull *(Larus pipixcan)* to a marsh habitat. *Anim. Behav.* **22,** 521–567.

Burtt, E. H., Jr. (1977). Some factors in the timing of parent–chick recognition in swallows. *Anim. Behav.* **25,** 231–239.

Busse, K., and Busse, K. (1977). Prägungsbedingte Bindung von Küstenseeschwalbenküken *(Sterna paradisaea* Pont.*)* an die Eltern und ihre Fähigkeit, sie an der Stimme zu erkennen. *Z. Tierpsychol.* **43,** 225–238.

Carrick, R. (1963). Ecological significance of territory in the Australian Magpie, *Gymnorhina tibicen. Proc. Int. Ornithol. Congr., Am. Ornithol. Union* (C. G. Sibley, ed.), pp. 740–753. Allen Press, Lawrence, Kansas.

Catchpole, C.K. (1978). Interspecific territorialism and competition in *Acrocephalus* warblers as revealed by playback experiments in areas of sympatry and allopatry. *Anim. Behav.* **26,** 1072–1080.

Conover, M. R., Klopfer, F. D., and Miller, D. E.(1980). Stimulus features of chicks and other factors evoking parental protective behaviour in Ring-billed Gulls. *Anim. Behav.* **28,** 29–41.

Coulson, J. C. (1966). The influence of the pair bond and age on the breeding biology of the Kittiwake Gull, *Rissa tridactyla. J. Anim. Ecol.* **35,** 269–279.

Craig, W. (1908). The voice of pigeons regarded as a means of vocal control. *Am. J. Sociol.* **14,** 66–100.

Cullen, E. (1957). Adaptations in the kittiwake to cliff nesting. *Ibis* **99,** 275–302.

Davies, S. J. J. F., and Carrick, R. (1962). On the ability of Crested Terns, *Sterna bergii,* to recognize their own chicks. *Aust. J. Zool.* **10,** 171–177.

Emlen, S. T. (1971). The role of song in individual recognition in the Indigo Bunting. *Z. Tierpsychol.* **28,** 241–246.

Emlen, S. T. (1978). Cooperative breeding in birds. *In* "Behavioural Ecology—an Evolutionary Approach" (J. R. Krebs and N. B. Davies, eds.), pp. 245–281. Sinauer, Sunderland, Massachusetts.

Evans, R. M. (1970a). Parental recognition and the "mew call" in Black-billed Gulls *(Larus bulleri). Auk* **87,** 503–513.

Evans, R. M. (1970b). Imprinting and the control of mobility in young Ring-billed Gulls *(Larus delawarensis). Anim. Behav. Monogr.* **3,** 193–248.

Evans, R. M. (1980). Development of individual call recognition in young Ring-billed Gulls *(Larus delawarensis):* an effect of feeding. *Anim. Behav.* **28,** 60–67.

Falls, J. B. (1969). Functions of territorial song in the White-throated Sparrow. *In* "Bird Vocalizations" (R. A. Hinde, ed.), pp. 207–232. Cambridge Univ. Press, London and New York.

Falls, J. B. (1981). Mapping territories with playback: an accurate census method for songbirds. *In* "Estimating the Numbers of Terrestrial Birds" (C. J. Ralph and J. M. Scott, eds.), Studies in Avian Biology, Vol. 6, pp. 86–91. Cooper Ornithol. Soc., Allen Press, Lawrence, Kansas.

Falls, J. B., and Brooks, R. J. (1975). Individual recognition by song in White-throated Sparrows. II. Effects of location. *Can. J. Zool.* **53,** 1412–1420.

Falls, J. B., and d'Agincourt, L. G. (1981). A comparison of neighbor–stranger discrimination in Eastern and Western meadowlarks. *Can. J. Zool.* **59,** 2380–2385.

Falls, J. B., and McNicholl, M. K. (1979). Neighbor–stranger discrimination by song in male Blue Grouse. *Can. J. Zool.* **57,** 457–462.

Falls, J. B., and Szijj, L. J. (1959). Reactions of Eastern and Western meadowlarks in Ontario to each others vocalizations. *Anat. Rec.* **134,** 560.

Goldman, P. (1973). Song recognition by Field Sparrows. *Auk* **90,** 106–113.

Gompertz, T. (1961). The vocabulary of the Great Tit. *Br. Birds* **54,** 369–394, 409–418.

Greenwood, P. J., Harvey, P. H., and Perrins, C. M. (1979). Kin selection and territoriality in birds: A test. *Anim. Behav.* **27,** 645–651.

Grove, P. A. (1981). The effect of location and stage of nesting on neighbor/stranger discrimination in the House Wren. Ph.D. Thesis, City University of New York.

Hansen, P. (1981). Coordinated singing in neighbouring Yellowhammers *(Emberiza citrinella). Natura Jutlandica* **19,** 121–138.

Harris, M. A., and Lemon, R. E. (1972). Songs of Song Sparrows *(Melospiza melodia):* individual variation and dialects. *Can. J. Zool.* **50,** 301–309.

Harris, M. A., and Lemon, R. E. (1974). Songs of Song Sparrows: reactions of males to songs of different localities. *Condor* **76,** 33–44.

Harris, M. A., and Lemon, R. E. (1976). Responses of male Song Sparrows, *Melospiza melodia,* to neighbouring and non-neighbouring individuals. *Ibis* **118,** 421–424.

Hartshorne, C. (1956). The monotony-threshold in singing birds. *Auk* **73,** 176–192.

Hoogland, J. L., and Sherman, P. W. (1976). Advantages and disadvantages of Bank Swallow *(Riparia riparia)* coloniality. *Ecol. Monogr.* **46,** 33–58.

Hutchison, R. E., Stevenson, J. G., and Thorpe, W. H. (1968). The basis for individual recognition by voice in the Sandwich Tern *(Sterna sandvicensis). Behaviour* **32,** 150–157.

Immelmann, K. (1969). Song development in the Zebra Finch and other estrildid finches. *In* "Bird Vocalizations" (R. A. Hinde, ed.), pp. 61–74. Cambridge Univ. Press, London and New York.

Ingold, P. (1973). Zur lautlichen Beziehung des Elters zur seinem Kueken bei Tordalken *(Alca torda). Behaviour* **45,** 154–190.

Järvi, T., Radesäter, T., and Jakobsson, S. (1977). Individual recognition and variation in the Great Tit *(Parus major)* two-syllable song. *Biophon* **5**(1), 4–9.

Jouventin, P. (1972). Un nouveau système de reconnaissance acoustique chez les oiseaux. *Behaviour* **43,** 176–185.

Jouventin, P., and Roux, P. (1979). Le chant du Manchot Adélie *(Pygoscelis adeliae).* Rôle dans la reconnaissance individuelle et comparaison avec le Manchot empereur non territorial. *Oiseau Rev. Franc. Ornithol.* **49,** 31–37.

Jouventin, P., Guillotin, M., and Cornet, A. (1979). Le chant du Manchot empereur et sa signification adaptative. *Behaviour* **70,** 231–250.

Knapton, R. W. (1979). Breeding ecology of the Clay-colored Sparrow. *Living Bird* **17,** 137–158.

Knapton, R. W., and Krebs, J. R. (1976). Dominance hierarchies in winter Song Sparrows. *Condor* **78,** 567–569.

Krebs, J. R. (1971). Territory and breeding density in the Great Tit, *Parus major* L. *Ecology* **52,** 2–22.

Krebs, J. R. (1976). Bird song and territory defence. *New Sci.* **70,** 534–536.

Krebs, J. R., and Kroodsma, D.E. (1980). Repertoires and geographical variation in bird song. *In* "Advances in the Study of Behavior" (J. S. Rosenblatt, R. A. Hinde, C. Beer, and M.-C. Busnell, eds.), Vol. 11, pp. 143–177. Academic Press, New York.

Kroodsma, D. E. (1976). The effect of large song repertoires on neighbor "recognition" in male Song Sparrows. *Condor* **78,** 97–99.

Lashley, K. S. (1913). Notes on the nesting activities of the Noddy and Sooty terns. *Carnegie Inst. Washington Publ.* No. 211, 61–83.

Lemon, R. E. (1967). The response of Cardinals to songs of different dialects. *Anim. Behav.* **15,** 538–545.

Lemon, R. E., and Harris, M. (1974). The question of dialects in the songs of White-throated Sparrows. *Can. J. Zool.* **52,** 83–98.

McArthur, P. D. (1982). Mechanisms and development of parent–young recognition in the Piñon Jays *(Gymnorhinus cyanocephalus). Anim. Behav.* **30,** 62–74.

Marler, P. (1956). The voice of the Chaffinch and its function as a language. *Ibis* **98,** 231–261.

Marler, P. (1960). Bird songs and mate selection. *In* "Animal Sounds and Communication" (W. E. Lanyon and W. N. Tavolga, eds.), Publ. No. 7, pp. 348–367. Am. Inst. Biol. Sci., Washington, D.C.

Marler, P. (1976). On animal aggression. The roles of strangeness and familiarity. *Am. Psychol.* **31,** 239–246.

Marler, P., and Mundinger, P. C. (1975). Vocalizations, social organization and breeding biology of the Twite *Acanthis flavirostris. Ibis* **117,** 1–17.

Miller, D. B. (1978). Species-typical and individually distinctive acoustic features of crow calls in Red Jungle Fowl. *Z. Tierpsychol.* **47,** 182–193.

Miller, D. B. (1979a). The acoustic basis of mate recognition by female Zebra Finches *(Taeniopygia guttata). Anim. Behav.* **27,** 376–380.

Miller, D. B. (1979b). Long-term recognition of father's song by female Zebra Finches. *Nature (London)* **280,** 389–391.

Miller, D. E., and Emlen, J. T. (1975). Individual chick recognition and family integrity in the Ring-billed Gull. *Behaviour* **52,** 124–144.

Miller, E. H. (1979). An approach to the analysis of graded vocalizations of birds. *Behav. Neural Biol.* **27,** 25–38.

Milligan, M. M., and Verner, J. (1971). Inter-populational song dialect discrimination in the White-crowned Sparrow. *Condor* **73,** 208–213.

Mosely, L. J. (1979). Individual auditory recognition in the Least Tern *(Sterna albifrons). Auk* **96,** 31–39.

Mundinger, P. C. (1970). Vocal imitation and individual recognition in finch calls. *Science* **168,** 480–482.

Nice, M. M. (1943). Studies in the life history of the Song Sparrow. II. The behavior of the Song Sparrow and other passerines. *Trans. Linn. Soc. N.Y.* **6.**

Nolan, V., Jr. (1978). "The Ecology and Behavior of the Prairie Warbler *Dendroica discolor,*" Ornithological Monograph, No. 26. Am. Ornithol. Union, Allen Press, Lawrence, Kansas.

Patterson, T. L., and Petrinovich, L. (1979). Field studies of habituation: II. Effect of massed stimulus presentation. *J. Comp. Physiol. Psychol.* **93,** 351–359.

Payne, R. B. (1981). Song learning and social interaction in Indigo Buntings. *Anim. Behav.* **29,** 688–697.

Payne, R. B. (1982). Ecological consequences of song matching: breeding success and intraspecific song mimicry in Indigo Buntings. *Ecology* **63,** 401–411.

Penney, R. L. (1968). Territorial and social behaviour in the Adélie Penguin. *In* "Antarctic Bird Studies" (O. L. Austin, Jr., ed.), Antarctic Research Series, No. 12, pp. 83–131. Am. Geophys. Union, Washington, D.C.

Petrinovich, L., and Patterson, T. L. (1979). Field studies of habituation. I. Effect of reproductive condition, number of trials, and different delay intervals on responses of the White-crowned Sparrow. *J. Comp. Physiol. Psychol.* **93,** 337–350.

Petrinovich, L., and Patterson, T. L. (1980). Field studies of habituation. III. Playback contingent on the response of the White-crowned Sparrow. *Anim. Behav.* **28,** 742–751.

Petrinovich, L., and Peeke, H. V. S. (1973). Habituation to territorial song in the White-crowned Sparrow. *Behav. Biol.* **9,** 719–729.

Pickstock, J. C., and Krebs, J. R. (1980). Neighbour–stranger discrimination in the Chaffinch *(Fringilla coelebs). J. Ornithol.* **121,** 105–108.

Richards, D. G. (1979). Recognition of neighbors by associative learning in Rufous-sided Towhees. *Auk* **96,** 688–693.

Robinson, A. (1949). The biological function of bird song in Australia. *Emu* **49,** 291–315.

Rohwer, S. (1982). The evolution of reliable and unreliable badges of fighting ability. *Am. Zool.* **22,** 531–546.

Schleidt, W. M. (1973). Tonic communication: Continual effects of discrete signs in animal communication systems. *J. Theor. Biol.* **42,** 359–386.

Searcy, W. A., McArthur, P. D., Peters, S. S., and Marler, P. (1981). Response of male Song and Swamp sparrows to neighbor, stranger, and self songs. *Behaviour* **77,** 152–163.

Simpson, M. J. A. (1973). Social displays and the recognition of individuals. *In* "Perspectives in Ethology" (P. P. G. Bateson and P. H. Klopfer, eds.), pp. 225–279. Plenum, New York.

Smith, S. M. (1978). The "underworld" in a territorial sparrow: adaptive strategy for floaters. *Am. Nat.* **112,** 571–582.

Smith, W. J. (1977). "The Behavior of Communicating. An Ethological Approach." Harvard Univ. Press, Cambridge, Massachusetts.

Stein, R. C. (1956). A comparative study of "advertising song" in the *Hylocichla* thrushes. *Auk* **73,** 503–512.

Stevenson, J., Hutchison, R. E., Hutchison, J., Bertram, B. C. R., and Thorpe, W. H. (1970). Individual recognition by auditory cues in the Common Tern *(Sterna hirundo). Nature (London)* **226,** 562–563.

Stirling, I., and Bendell, J. F. (1970). The reproductive behaviour of Blue Grouse. *Syesis* **3,** 161–171.

Stokes, A. W. (1967). Behavior of the Bobwhite, *Colinus virginianus. Auk* **84,** 1–33.

Thompson, D. H., and Emlen, J. T. (1968). Parent–chick individual recognition in the Adélie Penguin. *Antarct. J.* **3,** 132.

Thompson, N. S. (1969a). Individual identification and temporal patterning in the cawing of Common Crows. *Commun. Behav. Biol.* **4,** 29–33.

Thompson, N. S. (1969b). Physical properties of cawing in the Common Crow. *Commun. Behav. Biol.* **4,** 269–271.

Thorpe, W. H. (1956). "Learning and Instinct in Animals." Methuen, London.

Thorpe, W. H. (1961). "Bird-song. The Biology of Vocal Communication and Expression in Birds." Cambridge Univ. Press, London and New York.

Tinbergen, N. (1953). "The Herring Gull's World." Collins, London.

Trainer, J. M.(1980). Comments on a kin association model of bird song dialects. *Anim. Behav.* **28,** 310–311.

Treisman, M. (1978). Bird song dialects, repertoire size, and kin association. *Anim. Behav.* **26,** 814–817.

Treisman, M. (1980). Some difficulties in testing explanations for the occurrence of bird song dialects. *Anim. Behav.* **28,** 311–312.

Tschantz, B. (1968). Trottellummen. *Z. Tierpsychol., Suppl.* No. 4.

Van Rhijn, J. G., and Vodegel, R. (1980). Being honest about one's intentions: An evolutionary stable strategy for animal conflicts. *J. Theor. Biol.* **85,** 623–641.

von Rautenfeld, D. B. (1978). Bemerkungen zur Aurstauschbarkeit von Kueken der Silbermoewe *(Larus argentatus)* nach der ersten Lebelswoche. *Z. Tierpsychol.* **47,** 180–181.

Watson, M. (1969). Significance of antiphonal song in the Eastern Whipbird, *Psophodes olivaceus. Behaviour* **35,** 157–178.

Weeden, J. S., and Falls, J. B. (1959). Differential response of male Ovenbirds to recorded songs of neighboring and more distant individuals. *Auk* **76,** 343–351.

White, S. J. (1971). Selective responsiveness by the Gannet *(Sula bassana)* to played-back calls. *Anim. Behav.* **19,** 125–131.

White, S. J., and White, R. E. C. (1970). Individual voice production in Gannets. *Behaviour* **37,** 40–54.

White, S. J., White, R. E. C., and Thorpe, W. H. (1970). Acoustic basis for individual recognition by voice in the Gannet. *Nature (London)* **225,** 1156–1158.

Wildenthal, J. L. (1965). Structure in primary song of the Mockingbird *(Mimus polyglottos). Auk* **82,** 161–189.

Wiley, R. H. (1976). Communication and spatial relationships in a colony of Common Grackles. *Anim. Behav.* **24,** 570–584.

Wiley, R. H., and Wiley, M. S. (1977). Recognition of neighbors' duets by Stripe-backed Wrens; *Campylorhynchus nuchalis. Behaviour* **62,** 10–34.

Williams, H. W. (1969). Vocal behavior of the adult California Quail. *Auk* **86,** 631–659.

Wilson, E. O. (1975). "Sociobiology: The New Synthesis." Harvard Univ. Press, Cambridge, Massachusetts.

Wooller, R. D. (1978). Individual vocal recognition in the Kittiwake Gull, *Rissa tridactyla* (L.). *Z. Tierpsychol.* **48,** 68–86.

Wunderle, J. M., Jr. (1978). Differential response of territorial Yellowthroats to the songs of neighbors and non-neighbors. *Auk* **95,** 389–395.

Zach, R., and Falls, J. B. (1979). Foraging and territoriality of male Ovenbirds (Aves: Parulidae) in a heterogeneous habitat. *J. Anim. Ecol.* **48,** 33–52.

9

Conceptual Issues in the Study of Communication

C. G. BEER

I. INTRODUCTION

A. Communication

Few of the conceptual issues about acoustic communication in birds are exclusive to it. Most apply also to animal communication in general. It might be argued that all the issues proceed from or come down to the question: what is communication?

We use the word "communication," and its adjective and verb forms, in so many ways, and in connection with such a diversity of matters, that a single

ACOUSTIC COMMUNICATION IN BIRDS
VOLUME 2

ISBN 0-12-426802-1

answer to this question must appear unlikely. We speak, for example, of communicating by telephone, of a professor's failure to communicate his ideas, of a garbled communication, of a communicable disease, of a communicating door between two rooms. Words that would do as synonyms in these cases might be "conferring," "convey," "message," "infectious," and "connecting," respectively. These synonyms will not substitute for one another (for example, a professor's failure to connect his ideas would be a thing apart from his failure to convey them), but they do have some affinity. There is, to use Wittgenstein's (1958) analogy, a "family resemblance" between their several meanings. For instance, they all share a notion of passage, the passing of something from here to there, be it talk, understanding, information, bacteria, or nightly visitors.

The passing of something from here to there is a pervasive feature of biological process at all levels, from the cellular to the social—so much so, indeed, as to be virtually a capsule description of life itself. Think of the transmission of genes in reproduction; the function of messenger RNA; the active and passive transport of ions across cell membranes; the passage of transmitter substance at synapses; sensory coding, hormonal regulation, feeding, copulation, vocalization. Only a few of these come within the customary connotation of "communication"; but they all come within the scope of the science of cybernetics, which comprises the very comprehensive abstractions of control theory and communication theory (Wiener, 1948; Ashby, 1956).

Cybernetics was christened just over 30 years ago, full of promise of unifying developments to come. An important part of its communication theory was Shannon's (1948; Shannon and Weaver, 1949) mathematical expression for "amount of information," which Shannon also showed to be isomorphic to the expression of entropy in statistical mechanics. The quantity so defined is variously referred to as information, uncertainty, and entropy (see, e.g., Attneave, 1959; George, 1961; Johnston, 1976; Hailman, 1977). If the theory really thus identified order with chaos it would be unifying indeed; but, of course, it does nothing of the sort. Nevertheless, the essentially statistical concept and measure of information offered by cybernetic communication theory have found a wide range of application, including many and various studies of animal social behavior. Control theory has also proved to be very comprehensive in its coverage. For example, McFarland (1970, 1971) has drawn parallels between equations describing mechanical and electrical control systems and motivational systems, and thus made a mathematically rigorous case for the notion of motivational energy, which critics of energy models of motivation had dismissed as based on misleading analogy (see, e.g., Hinde, 1960). This unifying argument even seems to imply that motivational energy conforms to the first and second laws of thermodynamics, if information is defined in the way that makes it correspond to entropy.

Although the parallel between information and entropy has what one writer described as "the ring of fundamental importance" (George, 1961, p. 35), its significance is still unclear, and some scientists have advised that it be treated with great care, since the matter is one in which "a very slight change in the conditions or assumptions may make a statement change from rigorously true to ridiculously false" (Ashby, 1956, p. 177). Such fallacy could result from confusion between the mathematically defined concept of information and information in the sense in which it might be demanded of a secret agent during interrogation.

Both senses apply in the study of animal communication. By scoring the numbers and orders of occurrence of the different syllables or phrases in a sample of birdsong, one can work out the quantity of information associated with each. If one does this for a sample of human speech, a word like "the," being common, will turn out to have a low value, and a word like "widershins," being rare, will get a high value. But these values for amount of information will not tell you how "the" is used or what "widershins" means. Likewise knowing how many "bits" of information each of the distinguishable parts of a bird's utterances conveys will not tell you whether these parts have distinguishable meanings or what such meanings might be. Much of the study of animal communication is concerned with this latter kind of deciphering. Consequently, the question of what is to count as communication for an ethologist has not been fully answered by the study of heat engines, or even the principles governing transmission in telephone lines. At least for the time being, the answer must be sought in behavioral terms, i.e., in study of interactions between organisms and the mediation of influence in such interaction.

However, not all that happens with effect in behavioral interactions customarily counts as communication. I might be described as communicating when I give someone a piece of my mind, but not, as a rule, if I give him a black eye, even though he may know more about my feelings after I hit him than before (see Cherry, 1957, who distinguished "jump in the lake" from pushing someone over the side). Although there may often be doubt about where to draw the line, communication is usually distinguished from the use of main force in social interaction as involving the sending and receiving of signs or signals. But talk of signs and signals merely shifts the question of what constitutes communication. We see signs in many circumstances where we should not be inclined to speak of communication: a threatening sky means a storm in the offing; a yellow skin is a symptom of jaundice; a footprint in the sand told Crusoe of other visitors to the island. Virtually anything that an animal does can be informative about it, but requires more than that to be regarded as communication behavior.

What makes the difference, according to the evidence of usage, is whether the sign is a signal; that is to say, whether the sign occurs for the sake of what is conveyed by it, inferred from it, or associated with it. Although we can tell the coming weather from the cloud formation, the cloud formation does not occur to tell us about the weather. Traffic lights, on the other hand, exist for no other reason than to signal when we must stop and when we may go. Communication behavior is behavior the *raison d'être* of which is communication.

This way of delineating communication behavior is still very vague by the standards of definition usually demanded in science. For instance, it does not specify whether the function served by communication behavior is accountable solely in evolutionary terms or as intended by the individual animal. These are neither exclusive nor exhaustive possibilities, but what applies in one context may well differ from what applies in another. This can be a source of conceptual issue. Nevertheless, to dispose of such an issue by more rigorous definition would be to give a false picture of the field. There are also the inevitable borderline cases. Clenched fists and bared teeth may signal threat, but they are also the initial components of the actions they represent, and can lead to hitting and biting. However, vocal behavior seldom presents this sort of problem. Apart from such special cases as the echolocation cries of whales, bats, and the Oilbird *(Steatornis caripensis)*, the sounds that animals make appear to have no other point than the sending of signals. At least on this point then, the vocal behavior of birds does not raise a conceptual issue.

B. Conceptual Issues

To speak of conceptual issues suggests that there are also issues of other kinds. Several possibilities propose themselves, of which two might be regarded as paramount in a science: methodological issues and factual issues. But considerations of method and fact can be intimately entangled with considerations of concept and idea, about which there may be differences of opinion and preference.

Take, for example, the ethologists' use of measurement. Before a quantitative method can be applied in behavioral analysis, decisions have to be made about what is to be counted or measured, what units the counting and measuring are to employ, what can be classed together and what apart. These decisions will depend upon how the behavior under study is conceptualized and the consequent questions it poses. For someone who views the signals of a species as fixed in form and function, and hence the same for all individuals and all occasions, variability in performance or response, if recognized at all, will be regarded as noise in the system, something to be ignored or dealt with by the statistical procedures that filter it out. That same variability will be a focus of interest for someone who conceives communication as conditioned by context and individual relationships. In general, the categories into which the behavior is sorted, the dimensions along which it is ordered, and the scales of division applied within each dimension will depend upon prior assumptions about what is essence and what is accident, prior expectation about where significance will prove to be found, and prior acceptance of a system of descriptive terms and explanatory concepts governing what makes sense as a question and as an answer.

But the conceptual basis of a branch of study is constrained in turn by the methods available or in vogue. Certain ideas or questions are inconceivable prior to the discovery or invention of the technology giving rise to them. Think of the history of the notion of the nerve impulse; of how the distinction between haploid and diploid depended upon microscopy; of how the connotations and denotations of "gene" and "hormone" track progress in the analytic techniques of biochemistry and neuroendocrinology; of how what we now mean by "syllable," "phrase," or "song" in connection with bird vocalization is imbued with effects of the means by which we can now see sounds as shapes and measure durations as distances.

Facts lead lives no freer than those of concepts or methods. What counts as a fact depends upon the conceptual framework within which the matter in question is viewed, and upon what it is within the scope of the methods in use to furnish. Take, for example, the fact that hypothalamic stimulation of testosterone-treated Domestic Fowl (*Gallus domesticus*) will make them call in a way describable as "superimposition of crowing on a series of peeps" (Andrew, 1969, p. 102). This would be inconceivable without the prior conceptions of hormones, the functional anatomy of the brain, and the vocal repertoire; and it could not have been discovered without the means of obtaining the hormone, locating the brain site, applying the stimulation, and so forth. We can know only what we have the means to know and what we turn those means toward.

Not that we are wrong to talk of "brute" facts. Facts are what they are often in spite of what we would have them be or expect them to be. Consequently they can raise problems. Facts are not problematical in themselves, but only as they fail to fit into a prior set of beliefs, expectations, descriptions, and conceptions (see Alexander, 1963). Therefore facts can force conceptual revision or innovation, and lead to modification or invention of methods to explore or exploit new possibilities brought into apprehension. In ethology, for example, our concept of communication has had to accommodate to the evidence that variability in the use of and response to a signal can exceed what would be consistent with earlier ideas (see Beer, 1971, 1976). The surprising discovery of lateralization of neural control of the syrinx in songbirds (Nottebohm, 1970), and subsequent developmental study showing functional plasticity in this vocal system before a certain age, helped to turn attention toward other hitherto unsuspected possibilities of parallels between birdsong and human speech (Marler, 1970b).

The conceptual, the methodological, and the factual thus join in a dance, to the figures of which each is as essential as are the three points to a triangle. However, concepts take the lead, more often than not. I give them pride of place in what follows. Conceptual issues affect all aspects of the study of communication. I divide discussion between questions of description, motivation, function, development, and phylogeny. These are neither mutually exclusive nor collectively exhaustive divisions; indeed, I argue their interdependence and limitations. Few of the issues are of very recent origin; many date back at least to Darwin. In the 10 years between the publications of Sebeok's two encyclopedic collections on animal communication (Sebeok, 1968, 1977), accumulations of new material necessitated enlargement of the second book to twice the size of the first. But this new material is mostly factual information; the ideas informing the two books are much the same. Indeed, understanding of acoustic communication is represented as having stood quite still, the chapter for 1977 simply being a reprint of that for 1968. There has been more movement than this observation suggests; but the impression I have, and so will convey, is that we have to do mainly with variations on old themes.

II. DESCRIPTION

A. Semiotics and Ethology

Science has been accused of describing what everybody knows in words nobody understands. However much everybody may know about animal communication, most of what has been written about the subject is anything but easy reading. "Why is it . . . that those who write about communication have such difficulty communicating?" (Hailman, 1978, p. 771).

Part of the reason, I suspect, is that this literature got off to a bad start. The credit for founding the analysis of communication on a formal and logical footing is generally given to the American philosopher C. S. Peirce. His "theory of signs" or "semiotic" is still invoked by some of the leading authorities in the field (see, e.g., Johnston, 1976; Hailman, 1977). He is so difficult to follow that reference to him runs the risk of being regarded as more a display of erudition than a help to understanding. Here is Peirce's definition of a sign:

> "A sign, or *representamen*, is something which stands to somebody for something in some respect or capacity. It addresses somebody, that is it creates in the mind of that person an equivalent sign, or perhaps a more developed sign. That sign which it creates I call the *interpretant* of the first sign. The sign stands for something, its *object*. It stands for that object, not in all respects, but in reference to a sort of idea, which I have sometimes called the *ground* of the representamen" (Peirce, 1932, Vol. 2, p. 228).

Typically Peircean are the neologisms and the tripartite division. Peirce appears to have been infatuated with the number 3. In his classification of signs he initially used three trichotomies to generate 10 classes of sign (Peirce, 1932, Vol. 2), but later tried increasing the number of trichotomies to ten, which gave him 66 classes (Lieb, 1953). Many of the distinctions seem to be for the sake of the symmetry of the system rather than a reflection of differences between natural kinds; and one all too easily loses one's bearings in the thickets of terminology with their strange-sounding names, such as qualisign, sinsign, legisign, decisign, seme, and rheme. To make matters worse Peirce was no fanatic about consistency. For instance, what he distinguished as "sinsign" and "legisign" in the initial classification he elsewhere referred to as "token" and type."

A token is an individual occurrence of a sign, and so describable in terms of its physical properties, as for example, a sequence of sounds having such and such details of frequency and amplitude modulation. A type is a class of such occurrences as specified by signification—what the sign stands for. The relationship between "signifier" and "signified" (de Saussure, 1916) has been a source of problem which still supplies debate in such fields as linguistics, anthropology, and literary criticism.* In philosophy, for example, Ayer (1968) raised the question of whether a type could be defined in terms of the characteristics of its tokens, or the tokens require the prior notion of the type to distinguish what gives them sign status from what is irrelevant. In ethology also we have to do with the question of whether signals or displays should be differentiated on the basis of form or function (see Beer, 1977). Stimulation serving as a sign will vary in a number of respects from occasion to occasion; consequently some basis has to be found for deciding which respects convey meaning and within what limits of variation.

This point about essence and accident with regard to the characteristics of signs was taken up by Charles Morris (1938; see also Stephenson, 1973). In his terms, "signals" are observable or recordable physical disturbances, and "sign vehicles" are whatever properties of signals convey meaning, meaning being defined as a predictably occurring response to a signal. Thus long calls in gulls are, without doubt, signals; but which of the many aspects of them—the many respects in which they show variation—are sign vehicles is still being sorted out by study to find which of them affect the behavior of recipients and how. What a sign vehicle signifies is its "referent," which amounts to the observable concomitants in the situation of transmission or the behavior of the signal sender. This rewriting of Peirce's semiotic triad in operational terms is reflective of a tough-minded puritanism to which much of behavioral science was converted between the wars, and which persists as a powerful, though perhaps less assuredly held, faith today.

Morris (1938) also reformulated the distinctions on the basis of which Peirce

*For example, this matter is central to what now goes by the name of "structuralism" (and its offshoots, "post-structuralism" and "deconstructionism")—a mainly French movement which embraces all of these fields (see, e.g., Macksey and Donato, 1970; Strickland, 1980). This movement should not be confused with the old brand of psychology to which Titchener gave the same name (and to which some psychologists still refer; see, e.g., Hochberg, 1970), or with the "structuralist" linguistics such as those of Bloomfield, which was a behavioristic kind of approach opposed by the "structuralist" (new style) linguistics of Chomsky.

had argued that "the science of semiotic has three branches" (Peirce, 1932, Vol. 2, p. 228). In place of Peirce's "pure grammar," "logic," and "pure rhetoric," Morris divided the subject into: "syntactics," the study of the relations between signs; "semantics," the study of the relations between signs and their referents; and "pragmatics," the study of the relations between signs and the responses they evoke. Carnap (1942) similarly distinguished syntax, semantics, and pragmatics, in discussing language; and George (1961) rescored the trio for full orchestra by suggesting that "syntax is mathematical logic, semantics is philosophy or philosophy of science, and pragmatics is psychology . . ." (p. 41)—communication theory *über alles!*

In a more sober vein, Marler (1961) drew on the writings of Morris (1938) and Cherry (1957) to introduce semiotics to ethology. Like Morris and Cherry he positioned the analyst of communication outside the process, observing it as a nonparticipant. We can observe signal production and the effects of signal reception, and so apply syntactics and pragmatics to animal communication. However, Marler was doubtful about the usefulness of semantics, since its definition as "the study of the "meaning" of signs" (Marler, 1961) carried a subjective connotation, putting it on the far side of the limits of scientific objectivity. The information content of a signal must consequently be read from its effects.

Smith (1963, 1965) took a slightly different view. He was bothered by the observation that the same signal can have different effects on different occasions. His solution to the problem was that the effect of a signal can depend upon and so vary with the situation in which it arrives—response can be jointly determined by signal and context. The referent of a signal *per se* therefore has to be sought by referring back to the signaler to find the denominator common to all instances of use of the signal. One has to distinguish what is encoded in a signal at the source—its "message"—from what is decoded from the combination of signal and context at the destination—its "meaning." In this scheme semantics is the deciphering of messages, and so retains equal footing with pragmatics (the deciphering of meanings) and syntactics (analysis of signals as they are in themselves). In the remainder of this section I discuss syntactics, returning to semantics in the section on motivation, and to pragmatics in the section on function.

To sum up the contribution of semiotics to the study of animal communication, I think it consists mainly of terminology, classificatory schemes, a way of ordering the subject, rather than explanatory theory. But semiotics is a meeting place of many disciplines; it connects to engineering in one direction, to linguistics in another, to literary criticism in yet another. It thus makes for the seeing of common patterns or parallels of sorts that could have theoretical significance.

B. Ethology and Objectivity

In the paper to which I have referred, Marler (1961) acknowledged that ethology had turned to the study of animal communication independently of philosophy, linguistics, and engineering, and with a different set of interests and biases, concepts and preconceptions. The zoological tradition within which ethology arose was dominated by concern with anatomical forms, their description and comparison as bases for taxonomic classification and functional interpretation. In addition to drawing on the evidence of comparative anatomy to support his theory of evolution, and so giving explanation for that evidence, Darwin showed that the approach of comparative morphology could be applied to behavior. He did so most notably in "The Expression of Emotion in Man and Animals" (Darwin, 1872), which was a pioneer work in the study of communication behavior. Communication behavior has held the greatest fascination for ethologists ever since. From the observations of such people as Howard and Selous, Whitman and Heinroth, Huxley and Craig, Lorenz and Tinbergen, a biology of behavior emerged that concentrated on detailed description of form in the light of comparative, evolutionary, and taxonomic considerations. Marler contrasted this approach with that of comparative psychology: "Instead of approaching animal communication with anthropomorphic preconceptions, they set out to describe the natural behavior in objective terms, seeking to derive evolutionary conclusions about the evolutionary basis of behavior" (Marler, 1961, p. 297).

The implication about comparative psychology can be challenged, but more to my point is the question of whether the ethologists succeeded in their putative quest for objectivity. There is no doubt about the aspiration; the difficulty is with the notion of objectivity. To describe something in objective terms presumably means to describe the thing as it is in itself, free of preconceptions or judgments about it. But the implied "doctrine of immaculate perception" is now generally regarded as mistaken; instead the prevailing view is that "nothing is observed without being in some way interpreted" (Ayer, 1968, p. 113). Whether or not judgment is always involved in observation, it certainly is in description. When we describe we class things, events, relations, or properties together or apart on the basis of similarities and differences. But anything can be like anything else in some respects, and unlike it in others. Hence there is no end to the range of possible descriptions, and no one description that includes all possibilities. What one possible description classes together will be classed apart by another, making what counts as the same kind of thing in the one case what counts as several kinds of things in the other (see Hampshire, 1960). I might, for example, try to categorize the vocalizations of some bird solely in terms of their physical characteristics, and even then I should have to decide what features to go by and what degrees of variation to register. Accordingly, the discussion will be of twitters, trills, warbles, whistles, cackles, and croons. Or I might sort the sounds on the basis of how the bird appears to use them, and so talk of luring calls, alarm calls, food-begging calls, attack calls. Of course, each instance of the luring call or the alarm call can also be described in physical terms as, say, a twitter or a trill; but that may be insufficient to identify it as an instance of its functional class, just as a description of the sounds constituting an utterance of "fall" will not tell you whether you have a token of the word for come down or the word for autumn. How we choose to describe the vocalizations can be influenced by whether our interests are in the physiology of their production, the motivation governing their use, their adaptive significance, or their evolutionary provenance. The general point is this: "A description . . . can no more be absolutely, finally, and ultimately *precise* than it can be absolutely *full* or *complete*" (Austin, 1962a, p. 128, italics in original). Complete description is possible only within a context where convention has established a single set of mutually exclusive, collectively exhaustive possibilities, such as exist for the notation used to describe games of chess. Otherwise, asking for a complete description is like asking for the extent of infinity. Description is necessarily limited by the comparisons necessarily chosen in its making. Even at the descriptive level, a *tabula rasa* conception of science cannot apply, and the notion of objectivity must be qualified accordingly.

The description that ethologists initially made of communication behavior had roots in zoological and Darwinian tradition, and reflected ideas about instinct partially derived from the same source. The communication behavior of a species was viewed as a repertoire of discrete motor patterns—the displays—each having fixed form and articulation for all members, at least of the same sex or age class, analogous to the set of bones and their arrangement in the skeleton. This anatomical analogy also encouraged an atomistic view of the repertoire and how it functions. Lorenz (1935, 1950) explicitly construed the social behavior of animals as a mosiac composed of independent releaser–response couplings, the integration of which is due to their peripheral location and order of occurrence, rather than to a unified central representation of the "social companion." He appears to have shared with the American behaviorist psychology he often ridiculed, a belief that true scientific explanation must be in terms of a billiard ball model of causality: irreducible self-contained entities which knock one another blindly about in accordance with laws of motion (see Taylor, 1964).

As with any viewpoint, the mosaic conception of communication picked out some features and ignored others. For example, it took as typical what Hailman (1977) calls the CB ("citizens' band") model: communicants take it in turn to send and receive signals, like people using a CB radio. The classic example is Tinbergen's (1951; ter Pelkwijk and Tinbergen, 1937) description of the courtship of Three-spined Sticklebacks *(Gasterosteus aculeatus)*. Perhaps the closest thing in bird vocalization is the antiphonal singing of such birds as the Boubou Shrike *(Laniarius ferrugineus;* Hooker and Hooker, 1969; Thorpe, 1972*)*. But antiphonal singing is an oddity which has called for special explanation. Far more common are cases where birds vociferate simultaneously in a manner that can be likened to the shouting contest between player and umpire over a disputed call in a baseball game, or the singing of team supporters at a rugby match. There are also many examples of joint performances of visual displays, such as the choking of gulls (see, e.g., Moynihan, 1955), mutual displays of penguins (see, e.g., Stonehouse, 1953), synchronized antics of courting grebes (Huxley, 1914). The emergence of vocalization as a rival to visual displays for ethological interest probably did much to bring home the limitations of the CB model.

For example, bird vocalizations provided evidence of "tonic communication" (Schleidt, 1973), in which information is encoded in rate of repetition and can have effects persisting beyond immediate reply. My studies of Laughing Gull *(Larus atricilla)* calls have suggested that the gulls might be following syntactic rules enabling them to use the same signal in different combinations with others to convey different messages or compose compound messages (Beer, 1976). The CB model did not exclude these possibilities, but neither did it give any incentive to look in their direction. The gull case also brought in question the classical ethological emphasis on species specificity by showing that to comprehend what is going on in animal communication the observer may need to take into account the individual identities of the participants, their social relationships, and other contextual features. The general trend in animal communication studies has been toward greater and greater complexity both in what is perceived to be happening and in conceptions of what is happening, as each accommodates to the influence of the other.

Evidence of such complexity began to intrude on the simplicity of classical ethological conceptions of communication as soon as quantitative methods were brought seriously to bear on the matter. As Lorenz (1952) feared, when ethologists began measuring and counting and scoring they accumulated data of such variability that they had to resort to statistical analyses implying probabilistic conceptions of order in social interactions. Tables of transition frequencies and correlation coefficients, stochastic analyses and factor analyses, Markov chains, and computations of amounts of information engulfed the once clean lines of ethological communication theory in thickets of numbers so tangled as to baffle belief that communication behavior ever communicates.

However, these quantitative methods have prerequisites. For example, stochastic analysis and information measurement assume what is called a "statistically stationary" or "ergodic" situation—the transition probabilities must not change, at least within ranges of time or place that would make the calculations biologically significant. But in fact the probabilities may well vary because of learning during social development or adjustment to position in a social hierarchy; they may differ according to the social relationships of participants and other contextual features of the circumstances. Prior decision on how to sort the data can therefore profoundly affect the results of analysis. Other problems of this sort have been discussed by Slater (1973; see also Colgan, 1978). But even assuming that all such problems were solved, there could still be doubt about what a sample of statistical data told about communication. Several writers have pointed out, for example, that measures of information apply to the extent to which an observer may be enabled to judge the current behavioral odds given knowledge of preceding performance, rather than to what transpires between the interacting animals (MacKay, 1972; Baylis, 1976; Hailman, 1977, 1978). MacKay (1972) succinctly expressed the point: "relatively little of the puzzlement actually felt by observers of animal communication has been of the right kind to be relieved by numerical answers to numerical questions" (p. 7).

The first and most crucial question is that of what the signals or "sign vehicles" are. To repeat an earlier point, we have to decide what should be counted and measured before we can count and measure, and the meaning of the counts and measurements will depend on the soundness of where we draw the distinctions. The problem has two aspects: the perceptual and the conceptual. On the perceptual side we have to do with the phenomenal or ostensive basis for what we recognize as signal categories; on the conceptual side we have to do with the rationale in terms of which the categories are defined (see Körner, 1959). The distinction can be compared to those drawn in logic between extension and intension, denotation and connotation, reference and sense.

The evidence of the senses gives us what we experience as the long call of a gull, the song of a songbird, the coo of a dove. Such performances strike us self-evidently as signals. What alternative can be imagined for most of the vocal productions of birds and other animals? But this, of course, is only the starting point. To understand the communication system we have to go beyond what our own senses and sensory adjuncts tell us to what the animals register and differentiate as signals. Many years ago von Uexküll (1934; see also transl. in Schiller, 1957) vividly portrayed how the phenomenal worlds of other creatures are, in many cases, different from ours. Marler's (1969) discussion of tonal quality of bird sounds illustrates how our experience may be a poor guide to the animal's experience, and hence fail to direct attention to where the lines between signals actually lie. Even when we share the relevant sensory domain with the animal, what strikes us as salient may not include what is significant to the animal. For example, I did an experiment in which Laughing Gull chicks responded differentially to playback of long calls of their parents, and hence made me aware of an amplitude modulation that specifies address at the start of such calls—a feature which I had not noticed before and probably would never have noticed otherwise (Beer, 1975, 1976).

The last example also illustrates the point that the animal holds the answer to the question of what counts as a signal. A necessary condition for some feature of behavior to be considered as communication is that it elicits response or affects behavior in some way in animals either naturally or artificially presented with it. This condition is often written into definitions of communication; for example: "The ultimate criterion for recognition of the occurrence of communication is that of a resultant change, sometimes delayed, sometimes scarcely perceptible, in the probability of subsequent behavior of other communicants" (Klopfer and

Hatch, 1968, p. 32). However, not everyone agrees that this is a necessary condition (see Lyon's comment on Cullen's chapter in Hinde, 1972, p. 122) and among those who do, some deny that it is a sufficient condition. MacKay (1972), for example, holds that the behavior in question must also be manifestly directed by the performer toward whatever change it promotes in the recipient.

This matter of whether effect or directedness is essential to the notion of communication and the definition of a signal carries us from perceptual to conceptual considerations; for they are candidates for criteria of class membership rather than given characteristics of the particulars to be ordered. There are many patterns of behavior that we perceive as communication, yet about which we can say little with certainty as to what they express, or convey, or are used to effect. Indeed, a methodological precept of ethology is that one should categorize behavior patterns purely on the basis of form, at least until the results of functional or motivational analysis are in and able to establish alternative criteria. Thus the patterns should be given names like "long call" or "crooning," which beg no questions about meaning or message, rather than names like "alarm call" and "luring call," which do.

Nevertheless, one does not perceive these patterns in a vacuum of expectation about how they are used and to what effect. On the contrary, even the ostensive categories are to some extent projections of conceptual suppositions involving notions of source and consequence. The old problem of the relationship between token and type persists for the ethologist of communication no less than for the philosopher of language.*

III. MOTIVATION AND REFERENCE

Motivational concepts can be divided roughly into motive concepts and motor concepts. With motive concepts, actions are accounted for in terms of the goals toward which they are directed; with motor concepts movements are accounted for in terms of the causes that produce them. These cognate words derive from the verb "to move," which itself has active and passive senses—we can move someone and be moved by them, both physically and emotionally.

"Emotion," of course, comes from the same root, and it too can be either cause or effect. According to William James (1890), emotional feeling is consequent on behavioral response to a charged situation—"we see the bear, and run, and feel afraid." For Darwin (1872) on the other hand, emotion was the cause, at least of that which is its expression. Expression of emotion is not something we do so much as something which happens to us. Like sneezing or breaking out in

*For a different and more technical treatment of many of the points I try to make here, see Green and Marler (1979).

spots it is symptom rather than action. We do not blush in order to convey that we are lying; we blush because we are lying. A Cartesian conception of animals as machines generalizes the paradigm of the blush to animal communication as a whole. According to this conception, communication behavior should be thought of as caused by underlying physiological conditions. This is the motor conception of motivation applied to communication behavior.

However, expression is also something that we can engage in actively, or do deliberately, to try to make ourselves understood in some way. If I exhort a student to express himself more clearly I assume that he is trying to get something across, that he is acting with intent to inform. According to this model, communication behavior is action directed by the performer toward a desired outcome. This is the motive conception of motivation applied to communication behavior.

Complex goal-directed machines such as computers are often canvassed as ''intentional systems'' (see, e.g., Dennett, 1978) and so are within the scope of the motive model. Hence the motive model need not imply anything beyond the reach of scientific observation—''a ghost in the machine'' (Ryle, 1949)—which would prohibit its application to animal communication. Taylor (1964; see also Dennett, 1978) argued that such a model is implied when we describe animal communication in terms that imply intention when we use them of ourselves. There is a great deal of such description. Even so, ethological theory has tended to assume that only a motor model will do to account for animal communication behavior.

Lorenz, however, wrote of the appetitive phases of social interaction as goal directed toward consummatory acts, the animals thus ''striving'' to satisfy a ''craving'' (Lorenz, 1937; see also transl. in Schiller, 1957, p. 171). Nevertheless, the hypothetical physiological machinery he proposed made the behavior solely dependent on ''action-specific energy'' and stimuli acting on ''innate releasing mechanisms'' (Lorenz, 1950), which typifies a motor mechanism. Tinbergen's (1951) theory of instinctive motivation assembled Lorenzian units into a hierarchical scheme in which ''motivational impulses'' drove behavior, the successive phases of which were determined by the kinds and order of releasing stimuli consequently encountered. This too was essentially a motor conception of motivation, in spite of its inclusion of ''appetitive behavior.'' According to this theory, much communication behavior is caused by motivational conflict—simultaneous arousal of incompatible drives, such as those for attacking and fleeing. Moynihan (1955), for example, applied this approach to gulls, interpreting the various displays as outward manifestations of internal motivational states consisting of different levels of hostility, fear, and sexuality.

Drive theories of motivation got much adverse criticism during the late 1950s and 1960s, both in psychology and ethology. Hinde, for example, argued against the looseness of the analogies to physical systems and physical energy, against the plausibility of supposing a single variable could deal with all that drive had been called in to explain, and against the implication that alternative measures of behavior supposedly controlled by the same drive correlate closely (see, e.g., Hinde, 1960). Such doubts about drive were corroborated by much of the physiological research and quantitative analysis coming into fashion at this time. But there were also some physiological studies (see, e.g., von Holst and von Saint Paul, 1963) and some studies using quantitative methods such as factor analysis (see, e.g., Wiepkema, 1961), the results of which were interpreted as supporting an interaction-of-drives conception. In spite of such efforts the general picture was, and still is, one of many facts in search of a theory.

A promising alternative to the view that drives determine the categories of communication behavior, is Smith's notion of "behavioral message." In his earlier formulations he said that the message—that which is encoded in a signal by its sender—is ultimately some central neurophysiological state (Smith, 1968), which sounded little different from a drive conception. However, Smith realized that for some patterns of communication behavior, what the signaler does in sequence with them, and the situations in which they occur, are so varied as to confound the belief that the same motivational state governs all performances. For example, Laughing Gulls perform long calls in a range of settings so broad that nearly all the motivational states in the ethological inventory can be inferred (Beer, 1975). Smith sought to maintain belief in a single common message in a case like this by distinguishing reference from motivation (Smith, 1977): a display is principally about the signaler's impending behavior, not underlying motivation. Thus, for example, the "kit-ter" call of the Eastern Kingbird *(Tyrannus tyrannus)* expresses hesitancy about locomotion, whether the hesitancy is caused by conditions associated with rivalry, pair formation, parental care, or antipredator behavior. In addition to such a basic message, or "message set," Smith allows that supplementary information, such as specification of species, sex, and individual identity, can be carried by variation in some features of the signal. But the main burden of a display is behavioral information, which is the same for all occasions.

Must the message always be about behavior? At least equally conceivable is the possibility that a display could refer to such things as predators to avoid, food to eat, places to nest, or directions to take, without including anything about the behavior of the performer. One can also question the necessity for the message of a display always to be the same. Could an animal not use a display to convey different messages on different occasions by varying the other signals performed in concert or in sequence with it, as we do with words and phrases? I suspect that such reference to outside objects, and use of something like grammatical syntax, smacks too much of anthropomorphism and mentalism to appeal to the tough-minded thinking that prevails in animal communication studies. Smith's notion of display as involuntary expression of behavioral tendency can be viewed as yet another motor conception of communication. His statement that "Displays are acts specialized to make information available" (Smith, 1977, p. 69) might suggest a motive concept, but it does not imply intent. Consider his example of the tongue showing of people: "Tongue showing provides the information that the communicator will show a depressed probability of interacting, seeking to interact, or being readily receptive to attempts to interact" (Smith, 1977, p. 409; see also Smith *et al.*, 1974)—but, as a rule, we do not do this deliberately or intentionally any more than we do when we blush.

However, some communication behavior gives the impression of being more than an involuntary symptom of how the signaler is disposed to act. For example, some species of monkeys (Struhsaker, 1967) and birds (Marler, 1959) have different alarm calls for different kinds of predators. In use these calls are more simply interpreted as referring to the predator than to the behavior of the signaler, or even to the best action to be taken, for this could be different for animals in different situations. In gulls there are combinations and sequences that invite syntactic interpretation (Beer, 1973, 1975, 1976; Galusha and Stout, 1977; Amlaner and Stout, 1978). One of the simpler cases is presented by "facing away," a movement consisting of a gull's turning its head to hide its face from the view of the other bird. In the Laughing Gull this can be superimposed on most of the other postures. For one of the forms it takes, its message in these combinations can be read as negation of what the accompanying posture signifies when performed alone. An especially telling configuration often occurs when the partners of a pair are in a fight with a third bird: should one of the paired birds launch an attack on the outsider, it will face away from its mate as it does so, as though to indicate that its aggressiveness is aimed elsewhere. The manifest directedness of this action is more in accord with the motive conception than with the motor conception of the production of communication behavior.

Facing away has also been interpreted as a means by which an animal can cut off visual stimulation, tending to make the animal behave otherwise than it would choose (Chance, 1962). A vocal parallel might be a bird's calling at the same time as others, if by raising its own voice the bird were trying to drown out the other voices so as to avoid reacting to them. The possibility of such reflexive directedness in the performance of displays tends to be left out of consideration by the prevailing emphasis, which represents displays as encoding information.

However, Dawkins and Krebs (1978) have argued that asking about the information content of a display may be less informative than asking about how the display is used.* For them communication is best conceived as instrumental to manipulation: "a means by which one animal makes use of another animal's

*Compare the shift of emphasis from the "picture theory of meaning" of Russell and early Wittgenstein (1922) to the preoccupation with varieties of linguistic use ("Don't ask for the meaning, ask for the use," Urmson, 1956, p. 179) of the later Wittgenstein (1958) and the "ordinary language" philosophers who followed in his wake, such as Ryle and Austin (see, e.g., Austin, 1962b). The ensuing philosophy of "speech acts" (Searle, 1969; Holdcroft, 1978) might be worth the attention of some students of animal communication.

muscle power'' (Dawkins and Krebs, 1978, p. 283). This sounds like a motive conception of communication, for to manipulate, in the ordinary sense of the word, is to *do* something, not to undergo something. However, Dawkins and Krebs say that they imply nothing about ''conscious prediction,'' and hence, presumably, intention, on the part of the animals. By thus disengaging manipulation from its entailment with intentionality they can only be using the word in a figurative sense, which is consistent with their appearing to be concerned with accounting for animal signals in terms of natural selection rather than motivation. In that context communication is no more literally manipulative than the genes are literally selfish or the animal literally lying. But when thoughts get tangled in metaphors they often become hard to keep on course. For much of their discussion Dawkins and Krebs use what might be called the motivational voice, even though the argument is about ultimate function; or else a relationship between the two matters is assumed but not made explicit. When they conclude that ''it is probably better to abandon the concept of information altogether'' (Dawkins and Krebs, 1978, p. 309) they may have a case as far as considerations of natural selection are concerned, but hardly with regard to motivation and proximate function. Deprived of a notion of information we should lack the means to say much that there is to be said about what animals are doing when they communicate with one another. For example, there are times when animals appear to be using signals for manipulation, even in the literal sense of the word, by transmitting information. When a gull induces its chicks to approach from out of sight by calling to them, the call informs the chicks of the identity of the caller and its location (Beer, 1970).

There may even be cases for which it makes sense to distinguish what a signal means from what an animal might mean by it. Thus a gull's ''upright'' posture signifies hostility (Tinbergen, 1959), but when, as in the ''greeting ceremony'' of courtship, the gull faces away after adopting the upright, the message can be interpreted as denial of hostility (Beer, 1975). A related idea is that use of a display might involve ''lying.'' Zahavi (1975) has speculated ingeniously on this possibility. He has argued that certain highly conspicuous displays and bodily features are advertisements of reproductive quality affecting mate selection. By putting the displayer at risk of being preyed on they say something like: ''Listen! I am so brave I can court calamity''; or by hampering action they say something like: ''Look! I am so tough I can win with weights on my wings.'' But for this ''handicap principle'' to work, such display would have to be backed up by the quality it claims, for selection would ultimately call the bluff of fakers by favoring ability to distinguish the true from the false. In fact, there is some experimental evidence to show that when animals are made to misrepresent themselves they are penalized immediately and fail the test of social interaction (Rohwer and Rohwer, 1978). However, some features of Zahavi's theory make it highly controversial and difficult to evaluate (see Dawkins, 1976). In any case the theory is about the evolution and ultimate function of advertisement (including art; see Zahavi, 1978), not motivation. Hence we are again dealing in figures of speech when "lying," "cheating," and "deception" occur in this context. Yet one can at least imagine the possibility that an animal might be capable of deliberately trying to signal deceptively. The metaphoric usage should not be allowed to obscure the more literal sense which could apply to the question of motivation.

In what they say about aesthetics Dawkins and Krebs (1978) do implicitly recognize a distinction between ultimate and proximate considerations, particularly with regard to the old question of why birdsong is so much more elaborate and musically rich than its supposed function of information transmission necessitates. A frequent suggestion is that such singing is done for its own sake, and is hence self-rewarding, the singer being its own audience and indulging its own aesthetic taste (Hall-Craggs, 1969; Hartshorne, 1973; Nelson, 1973; Thorpe and Hall-Craggs, 1976). Dawkins and Krebs admit this to be a plausible possibility, and even cite some developmental evidence supporting it. But they rightly point out that the self-indulgence thesis will not answer the question of how such behavior evolved. The answer they suggest is that the singing has been selected for its effectiveness as "persuasion"; like the rhetoric of oratory and the seduction of advertisement its success is in what it can get others to do rather than in what it enables them to learn. This success ultimately translates into differential genetic replication, but, like motivation and reference, the proximate effects may require study on their own terms for a full understanding of function and consequence.

IV. FUNCTION AND CONSEQUENCE

Parallel to the distinction he drew between message and motivation, Smith (1977) maintained a distinction between meaning and function. In his terms meaning is a joint product of signal and context and is to be read from the change of behavior of recipients. Function is the way in which such change ultimately contributes survival value—promotes reproductive success—to the signaler, or, rather, to certain of the signaler's brands of gene. So, for example, the singing of a male songbird on its territory means threat to a rival male, as manifested by the latter's departure or responding with counterthreat. The song thus serves to establish and maintain territory, and territory serves to secure resources necessary to providing for a family. As I have already exemplified, the contrast can be expressed as being between immediate function or consequence and ultimate function or consequence.

Hinde (1975) and others have drawn a distinction between functional and nonfunctional consequences, the difference being that functional consequences are those that reflect the selection pressures to which the behavior answers, whereas nonfunctional consequences are incidental to or in spite of these pressures. Thus, the intimidating effect of the male bird's song on the rival is a functional consequence; the risk incurred of attracting the attention of a predator is not, since selection will not have favored those features of the song effective in attracting predators (putting aside Zahavi and his handicap principle for the moment). On the contrary, selection from predator pressure will tend to minimize these features. This makes the point that behavioral characteristics, like other products of evolution, are often shaped by multiple selection pressures. Consequently, in many cases, one can be mistaken to ask for *the* function of a pattern of behavior. If several selection pressures have contributed to its evolution the pattern might well serve several proximate and even ultimate functions. For example, birdsong appears to serve for territorial defense, mate attraction, reproductive synchronization, and reproductive isolation.

As far as immediate effects are concerned, differences of context can account for the different "meanings" (Smith, 1963, 1965).* But different effects can also be distributed between different features of a display. For example, species-identifying characteristics may be distinct from individually identifying characteristics; features specifying the "address" of the signal may be distinguishable from features that are vehicles of command, exhortation, persuasion, information, as in the case of the long-call of the Laughing Gull (Beer, 1975, 1976). The possibility of multiple function has obvious methodological implications: it enjoins analysis to attend to details of signal variation to a degree that assumption of unitary function would be unlikely to encourage.

Another methodological point is that although proximate and ultimate function will obviously be closely related, the kind of study that elucidates the one may be incapable of throwing much light on the other. For this as well as other reasons Smith's distinction between meaning and function is important. Hence, I do not agree with Wilson's (1975, p. 218) opinion that "any attempt to separate meaning from function in animal communication seems to create more ambiguity than it removes." Ambiguity usually arises from a failure to draw distinctions. By conflating meaning and function, and also by treating Smith's message categories as though they were functions, Wilson garbles Smith's account and encourages the kind of category confusion to which ethology in general and animal communication study in particular continue to be prone.

The point raises a more general issue. The new synthesis to which sociobiology, as represented by Wilson (1975), aspires is, doubtless, a consummation devoutly to be wished. But it will not be realized without due recognition of the

*Smith was anticipated in his discovery of the importance of context by Haldane (1954), Collias (1960), and Marler (1961), but he has the credit for getting ethologists to pay attention to it. Manley (1960) made at least as compelling a case, but has left his work unpublished.

differences between the disciplines that it would bring together. Colonial powers need to understand the native cultures of the lands they conquer if they are to govern wisely, and at least a measure of self-determination has frequently proved to be in the interest of the continued course of empire. The facts with which the direct study of animal communication deals are of an order that is, for the most part, discontinuous with that dealt with by either population genetics or molecular genetics. The conceptual geography of communication studies does not map onto that of evolutionary biology, or that of cellular biology.

However, the drift of this rhetorical digression is not a denigration of attempts to account for features of animal communication in terms of natural selection. On the contrary, I side with those who plead for more experimental and comparative work to test the implications of the newly thriving theory on this matter. For example, application of games theory to the question of optimal strategy in aggressive interaction has shown that threat display, used in a certain proportion, should be more successful in the long run than either consistent attack or fleeing (see, e.g., Maynard Smith and Price, 1973). But the argument dealt with only imaginary data, leaving research to find cases to test the predictions against the real thing. Another example of how speculation about function can give assignment to research is Krebs's idea of the "Beau Geste effect" (Krebs, 1977). This would explain the large song repertoires of many passerine species as means by which a single individual can seem to be several, and so maintain a territory larger than if it had only one or two songs to sing. The idea is testable by experiment and adaptive correlation study. Indeed, such work is already underway (Smith and Reid, 1979).

However, neither of these conceptions leads directly to consideration of precisely how the signals are used and comprehended, whether there are rules of syntactic combination governing production and reception, or what criteria to choose to define categories for description and classification. The point bears repeating: in general, the study of short-term effects and the study of long-term effects of signal use, although obviously interrelated, may require different methods and involve different concepts. Consequently, their integration should not be forced at the expense of their complementary relationship. The same is true for the genetic and developmental approaches, for which confusion of questions has perhaps caused more muddle than anywhere else in biology.

V. ENDOWMENT AND DEVELOPMENT

The nature/nurture issue recurrently erupts in biology and psychology. Not long ago it got linguists arguing about the nature of grammar. Most recently it helped make sociobiology a household word. A great deal of unnecessary contention might have been avoided had due consideration been given to the extent to which genetic facts and ontogenetic facts convey distinct realms of discourse. Much of the mischief has been due to the ambiguity of words like "innate" and "instinctive," which can mean both "genetically inherited" and "development free of any dependence on experience or other formative interaction with the environment." The common assumption is that these two senses entail one another, and hence that evidence for the one is evidence for the other. The fallacy in such inference has been pointed out many times (see, e.g., Lehrman, 1953, 1970)—so often, indeed, that its repetition has become a tiresome litany which I should prefer not to continue. Suffice it to say then, that the ample evidence of genetic inheritance of vocal characteristics in birds, such as the success with which canaries have been selectively bred for singing, is no substitute for study of how vocal performance develops.

Such study has underlined the point that inheritance and learning are not mutually exclusive. For example, the work of Marler and associates on the acquisition of dialect-differentiated song in White-crowned Sparrows *(Zonotrichia leucophrys)* showed learning to be involved but limited to a period and pattern set independently of experience (Marler, 1970a; Marler and Mundinger, 1971; Konishi, 1965; Konishi and Nottebohm, 1969). Other work on other species has revealed selective learning of syllable types (Marler and Peters, 1977) during sensitive periods different from those for acquisition of song pattern, for which there may or may not be predisposition favoring imitation of the species type (Immelmann, 1969). Auditory deprivation may have different effects at different times, which may differ according to whether the deprivation includes a bird's hearing of its own voice.

Perhaps because most of such study has been on songbirds, it has been preoccupied with syntactics, to the neglect of semantic and pragmatic considerations—focused on how a bird acquires the form of its utterances rather than the competence in their use and comprehension. Songbird song has such complexity and variety of form that the uneven distribution of interest is hardly surprising. In contrast, gull vocalizations are much less complex and variable in syllabic structure, but more puzzling as far as use is concerned, for most of them occur in a wide range of social situations (Beer, 1975). There is syntactic complexity and variability in the stringing of notes into call patterns, combination with postural display and in signal sequences, and some indication that this syntactic flexibility has semantic and pragmatic correlates (Beer, 1975, 1976). Hence, developmental interest in gull communication has been caught as much by considerations of how proficiency in working the system is attained, as by how the forms of the basic motor patterns are acquired.

Choice of species is not the only thing affecting choice of what to look for, however. After all, gulls have long been a favorite subject for ethological study of communication behavior, yet only recently has attention focused on individuality in social relationships, as opposed to what is general to the species. The shift of interest reflects a changed conception of communication. I expect further refocusing of attention when the idea of communication as manipulation (Dawkins and Krebs, 1978) is turned to developmental account. Perhaps even more profound would be application of the thinking of Piaget to animal social development, for his epigenetic concepts of assimilation and accommodation seem to me to offer hope for more penetrating interpretation of developmental transition than the old dichotomy of maturation and learning. Already experimental study of vocal interaction between gull parents and chicks has shown that later phases of social development depend upon adequate negotiation of earlier phases, and that use of and response to different kinds of call vary with age of the family (Beer, 1970, 1979). The general point is that what developmental study looks for depends on what is thought to be there for it to find, which depends on how the social communication system is viewed in its syntactic, semantic, and pragmatic aspects.

VI. EVOLUTIONARY DERIVATION

The concept most at issue in connection with the evolutionary provenance of behavior is homology. Again, confusion arises through failure to mark distinctions, and there is also a problem about analogy. The distinction most often ignored is between description and explanation, or between denotation and connotation. The concept of homology originated in comparative anatomy, where it referred to correspondence between parts with respect to their position in structures sharing the same pattern of organization, such as the segments within a body and skeleton design common to different kinds of bodies (Owen, 1848; Russell, 1916). Tracking down such correspondence was a main preoccupation of comparative anatomists in the eighteenth and nineteenth centuries prior to the coming of evolutionary theory. Darwin used the evidence of homology to support his case that present forms are descended from past forms. But descent with modification also explained the existence of homologies. The circularity became vicious when people started defining homology in terms of evolutionary origin, which is the situation we have today (see, e.g., Wilson, 1975, p. 586). Thus morphological description became confused with phylogenetic explanation.

In anatomical practice, however, homology is still sometimes recognized in the old way. For example, Karten (1969) used patterns of neural connection to argue his judgements of homology between parts of bird and mammal brains, and so speculate about cerebral phylogeny. If anatomy is to be used validly as evidence of phylogeny, there have to be independent criteria for telling what corresponds to what. I have suggested elsewhere (Beer, 1977) that distinguishing the perceptual or ostensive grounds for homology—the denotation of the term—from the intensive concept or connotation—the meaning of the term as used in prevailing theory—might be helpful. This distinction works reasonably well for anatomy, where there is the old "Principle of Connections" (Russell, 1916) for the ostensive reference. Behavior is another matter.

In applying the notion of homology to behavior, Whitman, Heinroth, Lorenz, and others evidently assumed that it meant the same as in anatomy: "Instincts and organs are to be studied from the common viewpoint of phyletic descent" (Whitman, 1899). The assumption makes sense for the connotation of the term—correspondence of form due to descent from a common evolutionary origin—but is problematical when applied to cases. If the parallel with anatomy were complete, there would be a behavioral equivalent of the Principle of Connections. There is not. Each of the candidates—position in sequences, in repertoires, in functional cycles—has limited applicability and numerous exceptions (Beer, 1980). Similarity, although the most frequently used criterion, is insufficient by itself to separate homology from analogy; and there are many instances of dissimilar patterns judged to be homologous on other grounds, such as evidence of intermediate links. Evidence of common phyletic origin is another resort, but this is unavailable when inference is from behavioral comparison to phylogenetic conclusion.

Failure to find consistent criteria for behavioral homology has led some writers to question whether the concept is scientifically useful (Blest, 1961; Atz, 1970; Klopfer, 1973, 1976; Hailman, 1976). They view it as a loose analogy to the anatomical concept, which itself is poorly anchored operationally. One suggestion has been to collapse behavioral homology into structural homology by shifting comparison to the neural substrate of the behavior patterns (Hodos, 1970).

Whether or not this reductive approach could be carried out, it would still require behavioral comparison to pick out the patterns likely to prove homologous by neural criteria. In any case behavioral comparison has convincing accomplishments to its credit. As used by Lorenz (1941) in his work on duck displays, for example, the concepts of behavioral homology and ritualization have a "logic in use" (Kaplan, 1964) producing order which even the critics must concede to be without reasonable alternative.

However, such success as that attained by the comparative morphological approach to behavioral phylogeny has been limited to microevolutionary levels, rarely extending beyond the taxonomic distance of families. One reason is the lack of a fossil record of behavior patterns; another is that behavior is so adaptively labile that the likelihood of divergence and convergence increases with taxonomic distance at a rate too steep for comparative judgment to keep a sure footing for more than a few steps. Another limitation is that the approach has little application to behavior other than the visually perceived. Nearly all the examples interpreted as "derived activities" (Tinbergen, 1952) are postures or movements; their supposed origins in "intention movements," "displacement activities," "ambivalent movements," and "redirected action" have no clear counterparts in other modalities. Some authors write as though the visual were the type for all signals in animal communication: "In animal discourse, redundancy is introduced into this universe of signals which are iconic parts of the signaler's probable response" (Bateson, 1968, p. 624). But few acoustic signals are at all iconic. Perhaps partly for this reason few attempts have been made to apply comparative morphology to the phylogeny of acoustic communication in birds.

In spite of its historical importance in ethology, and Lorenz's (1974) recent reaffirmation of his faith in it, the comparative morphological approach to behavior is at its lowest ebb. In Sebeok's (1977) latest survey Barlow could find no reference more recent than 1966 to exemplify work on ritualization, and Marler could contribute a chapter on the evolution of communication in which there is no mention of homology.

Nevertheless, communication behavior has an evolutionary history, which still presents ethology with questions. For someone whose ear catches the grating quality common to the calls of terns, a metallic note distinctive of most of the songs of icterids, or a flute-like timbre identifying a bird as a shorebird, such family resemblances must at least suggest ancestral origin and the possibility of genealogical study. Again the work would be influenced by how the communication was viewed in its syntactic, semantic, and pragmatic aspects. It might even take inspiration from study of the history of language, for comparative philology gives the best idea of what the comparative morphological approach to communication behavior would be like if its facts were firmer.

VII. ANIMAL COMMUNICATION AND HUMAN LANGUAGE

To what extent might animal communication be like human speech? This old question has recently received new notice as a result of the studies of sign learning by chimpanzees (Gardner and Gardner, 1969; Premack, 1976; Rumbaugh, 1977). But even before this work got people arguing about what language is, attempts had been made to list the "design features" of human speech and score their occurrences in animal communication (Hockett, 1959, 1960a,b, 1963; Altmann, 1962, 1967; Hockett and Altmann, 1968). The field is full of controversy (see, e.g., Harnad, *et al.*, 1976; Ristau and Robbins, 1979; Terrace, 1979; Seidenberg and Petitto, 1979); but the debate has encouraged more liberal speculation about animal and human communication than used to be allowed by the precepts of behaviorism. This freer thinking has included the communication systems of birds.

For example, Marler and Peters (1977, p. 521) have argued that there are "many parallels between avian song learning and the development of the perception and production of speech in human infants" (see also Marler, 1970b, 1975), such as perceptual predispositions affecting what the young organism registers and learns in the course of vocal development. An avian analog of the phoneme/morpheme distinction (minimum units of form distinguished from minimum units of sense) has been claimed, along with other possible parallels to linguistic grammar (Beer, 1975, 1976).

Perhaps the most controversial issue attending such attribution of language-like capacities to animals is what they imply about cognition in the animals concerned. For example, if we accept the claim that an animal's communication is in some respect linguistic, are we thereby obliged to think that the animal acts with intention? Recent developments in animal communication studies, and their comparison with the human case, have led Griffin (1976) to raise "the question of animal awareness" and sponsor what he calls "cognitive ethology."

However, concepts like intention and awareness are anything but simple in application to ourselves. As Anscombe (1956), Hampshire (1960), Rundle (1972), and other philosophers have demonstrated, we talk of intention in so many different contexts, and with so many different shades of meaning, that trying to pin it down is like trying to spear moonlight in a pond. Similarly with the notion of awareness, we might have no hesitation about describing an animal as aware in one sense but not in another.

This kind of worrying about the meanings of words is often dismissed by scientists as idle chatter, the serious business being to get the facts. "A merely semantic" issue is supposed to be of no importance. Sometimes the judgment is right. Nevertheless, we take words for granted at our peril, for they can lead us into muddle and prejudgment if we do not use them with care. Many of the conceptual issues in the study of communication have their sources in confusions of linguistic idioms (see Ryle, 1951). We do not even have a clear answer to the root question: what is communication?

VIII. SUMMARY

Two themes run through this essay: many of the conceptual issues in the study of communication arise from confusion of questions of different kinds; the view taken of one kind of question can influence the formulation of others. The main specific points are as follows:

1. In description of communication behavior, problems are posed by relationships between criteria of form (syntactic), use (semantic), and effect (pragmatic).
2. What an animal's behavior communicates to a person may be different from what it communicates to another animal.

3. Motivational concepts of communication include causal and intentional models, and there is an issue about whether reference must be to bodily state or can be to outside objects.

4. Emphasis on information transmission may obstruct other not necessarily incompatible views, such as the possibility that signals can be used for manipulation.

5. Functional interpretation of communication may need to distinguish between proximate and ultimate effects, and include the possibility of multiple function.

6. The nature/nurture issue is a confusion distracting from the need for *both* genetic and ontogenetic study and hence the complex interdependence that can be revealed by their integration.

7. Understanding of the evolutionary provenance of communication behavior, especially the vocal signals, is hampered by lack of a precise concept of behavioral homology, the current notion being a confusion between ostensive and connotative criteria.

8. Comparison of animal communication and human speech raises numerous issues, some of which involve the logic of the language used in communication study.

REFERENCES

Alexander, P. (1963). "Sensationism and Scientific Explanation." Routledge & Kegan Paul, London.

Altmann, S. A. (1962). Social behavior of anthropoid primates: Analysis of recent concepts. *In* "Roots of Behavior" (E. L. Bliss, ed.), pp. 277–285. Harper, New York.

Altmann, S. A. (1967). The structure of primate social communication. *In* "Social Communication among Primates" (S. A. Altmann, ed.), pp. 325–362. Chicago Univ. Press, Chicago, Illinois.

Amlaner, C. J., and Stout, J. F. (1978). Aggressive communication by *Larus glaucescens*. Part VI: Interactions of territory residents with a remotely controlled, locomotory model. *Behaviour* **66,** 223–251.

Andrew, R. J. (1969). The effects of testosterone on avian vocalizations. *In* "Bird Vocalizations" (R. A.Hinde, ed.), pp. 97–130. Cambridge Univ. Press, London and New York.

Anscombe, G. E. M. (1957). "Intention." Blackwell, Oxford.

Ashby, W. R. (1956). "An Introduction to Cybernetics." Chapman & Hall, London.

Attneave, F. (1959). "Applications of Information Theory to Psychology." Holt, New York.

Atz, J. W. (1970). The application of the idea of homology to behavior. *In* "Development and Evolution of Behavior" (L. R. Aronson, E. Tobach, D. S. Lehrman, and J. S. Rosenblatt, eds.), pp. 53–74. Freeman, San Francisco, California.

Austin, J. L. (1962a). "Sense and Sensibilia." Oxford Univ. Press (Clarendon), London and New York.

Austin, J. L. (1962b). "How to Do Things with Words." Oxford Univ. Press (Clarendon), London and New York.

Ayer, A. J. (1968). "The Origins of Pragmatism." Macmillan, New York.

Barlow, G. W. (1977). Modal action patterns. *In* "How Animals Communicate" (T. A. Sebeok, ed.), pp. 98–134. Indiana Univ. Press, Bloomington.

Bateson, G. (1968). Redundancy and coding. *In* "Animal Communication" (T. A. Sebeok, ed.), pp. 614–626. Indiana Univ. Press, Bloomington.

Baylis, J. R. (1976). A quantitative study of long-term courtship. I. Ethological isolation between sympatric populations of the midas cichlid, *Cichlosoma citrinellum,* and the arrow cichlid, *C. zaliosum. Behaviour* **59,** 59–69.

Beer, C. G. (1970). Individual recognition of voice in the social behavior of birds. *In* "Advances in the Study of Behavior" (D. S. Lehrman, R. A. Hinde, and E. Shaw, eds.), Vol. 3, pp. 27–74. Academic Press, New York.

Beer, C. G. (1971). Diversity in the study of the development of social behavior. *In* "The Biopsychology of Development" (E. Tobach, L. R. Aronson, and E. Shaw, eds.), pp. 433–455. Academic Press, New York.

Beer, C. G. (1973). A view of birds. *In* "Minnesota Symposia on Child Psychology" (A. Pick, ed.), Vol. 7, pp. 47–86. Minnesota Univ. Press, Minneapolis.

Beer, C. G. (1975). Multiple functions and gull displays. *In* "Function and Evolution in Behaviour—Essays in Honour of Professor Niko Tinbergen, F.R.S." (G. P. Baerends, C. G. Beer, and A. Manning, eds.), pp. 16–54. Oxford Univ. Press (Clarendon), London and New York.

Beer, C. G. (1976). Some complexities in the communication behavior of gulls. *Ann. N.Y. Acad. Sci.* **280,** 413–432.

Beer, C. G. (1977). What is a display? *Am. Zool.* **17,** 155–165.

Beer, C. G. (1979). Vocal communication between Laughing Gull parents and chicks. *Behaviour* **70,** 118–146.

Beer, C. G. (1980). Perspectives on animal behavior comparisons. *In* "Comparative Methods in Psychology" (M. H. Bornstein, ed.), pp. 17–64. Erlbaum, Hillsdale, New Jersey.

Blest, A. D. (1961). The concept of ritualisation. *In* "Current Problems in Animal Behaviour" (W. H. Thorpe and O. L. Zangwill, eds.), pp. 102–124. Cambridge Univ. Press, London and New York.

Carnap, R. (1942). "Introduction to Semantics." Harvard Univ. Press, Cambridge, Massachusetts.

Chance, M. R. A. (1962). An interpretation of some agonistic postures: The role of "cut-off" acts and postures. *Symp. Zool. Soc. London* **8,** 71–89.

Cherry, C. (1957). "On Human Communication." Wiley, New York.

Colgan, P. W., ed. (1978). "Quantitative Ethology." Wiley, New York.

Collias, N. E. (1960). An ecological and functional classification of animal sounds. *In* "Animal Sounds and Communication" (W. E. Lanyon and W. N. Tavolga, eds.), Publ. No. 7, pp. 368–391. Am. Inst. Biol. Sci., Washington, D.C.

Cullen, J. M. (1972). Some principles of animal communication. *In* "Non-Verbal Communication" (R. A. Hinde, ed.), pp. 101–122. Cambridge Univ. Press,

Darwin, C. (1872). "The Expression of Emotion in Man and Animals." Murray, London.

Dawkins, R. (1976). "The Selfish Gene." Oxford Univ. Press, London and New York.

Dawkins, R., and Krebs, J. R. (1978). Animal signals: information and manipulation? *In* "Behavioural Ecology—an Evolutionary Approach" (J. R. Krebs and N. B. Davies, eds.), pp. 282–309. Blackwell, Oxford.

Dennett, D. C.(1978). "Brainstorms." Bradford Books, Montgomery, Vermont.

de Saussure, F. (1916). "Cours de Linguistique Générale." Payot, Paris.

Galusha, J. G., and Stout, J. F. (1977). Aggressive communication by *Larus glaucescens.* Part IV. Experiments on visual communication. *Behaviour* **62,** 222–235.

Gardner, R. A., and Gardner, B. T. (1969). Teaching sign language to a chimpanzee. *Science* **165,** 664–672.

George, F. H. (1961). "The Brain as a Computer." Pergamon, Oxford.

Green, S., and Marler, P. (1979). The analysis of animal communication. *In* "Handbook of Behavioral Neurobiology. Vol. 3: Social Behavior and Communication" (P. Marler and J. G. Vandenbergh, eds.), pp. 73–158. Plenum, New York.

Griffin, D.R. (1976). "The Question of Animal Awareness." Rockefeller Univ. Press, New York.

Hailman, J. P. (1976). Homology: Logic, information and efficiency. *In* "Evolution, Brain and Behavior: Persistent Problems" (R. B. Masterson, W. Hodos, and J. Jerison, eds.), pp. 181–198. Erlbaum, Hillsdale, New Jersey.

Hailman, J. P. (1977). "Optical Signals—Animal Communication and Light." Indiana Univ. Press, Bloomington.

Hailman, J. P. (1978). Review of "The behavior of communicating: An ethological approach" by W. J. Smith. *Auk* **95,** 771–774.

Haldane, J. B. S. (1954). La signalisation animale. *Ann. Biol.* **30,** 89–98.

Hall-Craggs, J. (1969). The aesthetic content of bird song. *In* "Bird Vocalizations" (R. A. Hinde, ed.), pp. 367–381. Cambridge Univ. Press, London and New York.

Hampshire, S. (1960). "Thought and Action." Chatto & Windus, London.

Harnad, S. R., Steklis, H. D., and Lancaster, J., eds. (1976). Origin and evolution of language and speech. *Ann. N.Y. Acad. Sci.* **280.**

Hartshorne, C. (1973). "Born to Sing." Indiana Univ. Press, Bloomington.

Hinde, R. A. (1960). Energy models of motivation. *Symp. Soc. Exp. Biol.* **14,** 199–213.

Hinde, R. A., ed. (1972). "Non-Verbal Communication." Cambridge Univ. Press, London and New York.

Hinde, R. A. (1975). The concept of function. *In* "Function and Evolution in Behaviour—Essays in Honour of Professor Niko Tinbergen, F.R.S." (G. P. Baerends, C. G. Beer, and A. Manning, eds.), pp. 3–15. Oxford Univ. Press (Clarendon), London and New York.

Hochberg, J. (1970). The representation of things and people. *In* "Art, Perception and Reality" (E. H. Gombrich, J. Hochberg, and M. Black), pp. 47–94. Johns Hopkins Press, Baltimore, Maryland.

Hockett, C. F. (1959). Animal "languages" and human language. *In* "The Evolution of Man's Capacity for Culture" (J. N. Spuhler, ed.), pp. 32–39. Chicago Univ. Press, Chicago, Illinois.

Hockett, C. F. (1960a). Logical considerations in the study of animal communication. *In* "Animal Sounds and Communication" (W. E. Lanyon and W. N. Tavolga, eds.), Publ. No. 7, pp. 392–430. Am. Inst. Biol. Sci., Washington, D.C.

Hockett, C.F. (1960b). The origin of speech. *Sci. Am.* **203,** 89–96.

Hockett, C. F. (1963). The problem of universals in language. *In* "Universals of Language" (J. H. Greenberg, ed.), pp. 1–29. MIT Press, Cambridge, Massachusetts.

Hockett, C. F., and Altmann, S. A. (1968). A note on design features. *In* "Animal Communication" (T. A. Sebeok, ed.), pp. 61–72. Indiana Univ. Press, Bloomington.

Hodos, W. (1970). Evolutionary interpretation of neural and behavioral studies of living vertebrates. *In* "The Neurosciences: Second Study Program" (F. O. Schmitt, ed.), pp. 26–39. Rockefeller Univ. Press, New York.

Holdcroft, D. (1978). "Words and Deeds." Oxford Univ. Press (Clarendon), London and New York.

Hooker, T., and Hooker, B. I. (1969). Duetting. *In* "Bird Vocalizations" (R. A.Hinde, ed.), pp. 185–205. Cambridge Univ. Press, London and New York.

Huxley, J. (1914). The courtship habits of the Great Crested Grebe *(Podiceps cristatus);* with an addition to the theory of sexual selection. *Proc. Zool. Soc. London* **35,** 491–562.

Immelmann, K. (1969). Song development in the Zebra Finch and other estrildine finches. *In* "Bird Vocalizations" (R. A. Hinde, ed.), pp. 61–77. Cambridge Univ. Press, London and New York.

James, W. (1890). "The Principles of Psychology," 2 vols. Holt, New York.
Johnston, T. D. (1976). Theoretical considerations in the adaptation of animal communication systems. *J. Theor. Biol.* **57,** 43–72.
Kaplan, A. (1964). "The Conduct of Inquiry." Chandler, San Francisco, California.
Karten, H. J. (1969). The organization of the avian telencephalon and some speculations on the phylogeny of the amniote telencephalon. *Ann. N.Y. Acad. Sci.* **167,** 164–179.
Klopfer, P. H. (1973). Does behavior evolve? *Ann. N.Y. Acad. Sci.* **223,** 113–119.
Klopfer, P. H. (1976). Evolution, behavior and language. *In* "Communicative Behavior and Evolution" (M. E. Hahn and E. C. Simmel, eds.), pp. 7–21. Academic Press, New York.
Klopfer, P. H., and Hatch, J. J. (1968). Experimental considerations. *In* "Animal Communication" (T. A. Sebeok, ed.), pp. 31–60. Indiana Univ. Press, Bloomington.
Körner, S. (1959). "Conceptual Thinking—A Logical Inquiry." Dover, New York.
Konishi, M.(1965). The role of auditory feedback in the control of vocalization in the White-crowned Sparrow. *Z. Tierpsychol.* **22,** 770–783.
Konishi, M., and Nottebohm, F. (1969). Experimental studies in the ontogeny of avian vocalizations. *In* "Bird Vocalizations" (R. A. Hinde, ed.), pp. 29–48. Cambridge Univ. Press, London and New York.
Krebs, J. R. (1977). The significance of song repertoires: the Beau Geste hypothesis. *Anim. Behav.* **25,** 475–478.
Lehrman, D. S. (1953). A critique of Konrad Lorenz's theory of instinctive behavior. *Q. Rev. Biol.* **28,** 337–363.
Lehrman, D. S. (1970). Semantic and conceptual issues in the nature-nurture problem. *In* "Development and Evolution of Behavior" (L. R. Aronson, E. Tobach, D. S. Lehrman, and J. S. Rosenblatt, eds.), pp. 17–52. Freeman, San Francisco, California.
Lieb, I. C., ed. (1953). "Charles S. Peirce's Letters to Lady Welby." Whitlock, New Haven, Connecticut.
Lorenz, K. (1935). Der Kumpan in der Umwelt des Vögels. *J. Ornithol.* **83,** 137–312, 289–413.
Lorenz, K. (1937). Über die Bildung des Instinktbegriffes. *Naturwissenschaften* **25,** 289–300, 307–318, 324–331.
Lorenz, K. (1941). Vergleichende Bewegungsstudien an Anatinen. *J. Ornithol.* **89,** 19–29, 194–293.
Lorenz, K. (1950). The comparative method of studying innate behaviour patterns. *Symp. Soc. Exp. Biol.* **4,** 221–268.
Lorenz, K. (1952). Die Entwicklung der vergleichenden Verhaltensforschung in den letzten 12 Jahren. *Verh. Dtsch. Zool. Ges. Freiburg* pp. 36–58.
Lorenz, K. (1974). Analogy as a source of knowledge. *Science* **185,** 229–234.
McFarland, D. J. (1970). Behavioral aspects of homeostasis. *In* "Advances in the Study of Behavior" (D. S. Lehrman, R. A. Hinde, and E. Shaw, eds.), Vol. 3, pp. 1–26. Academic Press, New York.
McFarland, D. J. (1971). "Feedback Mechanisms in Animal Behaviour." Academic Press, New York.
MacKay, D. M. (1972). Formal analysis of communicative processes. *In* "Non-Verbal Communication" (R. A. Hinde, ed.), pp. 3–25. Cambridge Univ. Press, London and New York.
Macksey, R., and Donato, E., eds. (1970). "The Structuralist Controversy." Johns Hopkins Press, Baltimore, Maryland.
Manley, G. H. (1960). The agonistic behaviour of the Black-headed Gull. Ph.D. Thesis, Bodleian Library, Oxford.
Marler, P. (1959). Developments in the study of animal communication. *In* "Darwin's Biological Work" (P. R. Bell, ed.), pp. 150–206. Cambridge Univ. Press, London and New York.
Marler, P. (1961). The logical analysis of animal communication. *J. Theor. Biol.* **1,** 295–317.

Marler, P. (1969). Tonal quality of bird sounds. *In* "Bird Vocalizations" (R. A. Hinde, ed.), pp. 5–18. Cambridge Univ. Press, London and New York.

Marler, P. (1970a). A comparative approach to vocal learning: Song development in White-crowned Sparrows. *J. Comp. Physiol. Psychol.* **71,** 1–25.

Marler, P. (1970b). Birdsong and speech development: Could there be parallels? *Am. Sci.* **6,** 669–673.

Marler, P. (1975). On the origin of speech from animal sounds. *In* "The Role of Speech in Language" (J. F. Kavanagh and J. E. Cutting, eds.), pp. 389–449. MIT Press, Cambridge, Massachusetts.

Marler, P. (1977). The evolution of communication. *In* "How Animals Communicate" (T. A. Sebeok, ed.), pp. 45–70. Indiana Univ. Press, Bloomington.

Marler, P., and Mundinger, P. (1971). Vocal learning in birds. *In* "Ontogeny of Vertebrate Behavior" (H. Moltz, ed.), pp. 389–450. Academic Press, New York.

Marler, P., and Peters, S. (1977). Selective vocal learning in a sparrow. *Science* **198,** 519–522.

Maynard Smith, J., and Price, G. R. (1973). The logic of animal conflict. *Nature (London)* **246,** 15–18.

Morris, C. W. (1938). "Foundations of the Theory of Signs." Chicago Univ. Press, Chicago, Illinois.

Moynihan, M. (1955). Some aspects of reproductive behavior in the Black-headed Gull (*Larus r. ridibundus*) and related species. *Behaviour, Suppl.* No. 4.

Nelson, K. (1973). Does the holistic study of behavior have a future? *In* "Perspectives in Ethology" (P. P. G. Bateson and P. H. Klopfer, eds.), pp. 281–328. Plenum, New York.

Nottebohm, F. (1970). Ontogeny of bird song. *Science* **167,** 950–956.

Owen, R. (1848). "On the Archetypes and Homologies of the Vertebrate Skeleton." Richard & John Taylor, London.

Peirce, C. S. (1931–1935). "The Collected Papers of Charles Sanders Peirce" (C. Hartshorne and P. Weiss, eds.), Vols. 1–6; (A. W. Burke, ed.), Vols. 7 and 8. Harvard Univ. Press, Cambridge, Massachusetts, 1958.

Premack, D. (1976). "Intelligence in Ape and Man." Erlbaum & Wiley, New York.

Ristau, C. A., and Robbins, D. (1979). A threat to man's uniqueness? Language and communication in the chimpanzee. *J. Psycholinguist. Res.* **8,** 267–300.

Rohwer, S., and Rohwer, F. C. (1978). Status signalling in Harris' Sparrows: Experimental deception achieved. *Anim. Behav.* **26,** 1012–1022.

Rumbaugh, D. M., ed. (1977). "Language Learning by a Chimpanzee: The Lana Project." Academic Press, New York.

Rundle, B. (1972). "Perception, Sensation and Verification." Oxford Univ. Press (Clarendon), London and New York.

Russell, E. S. (1916). "Form and Function." Murray, London.

Ryle, G. (1949). "The Concept of Mind." Hutchinson, London.

Ryle, G. (1951). Systematically misleading expression. *In* "Logic and Language, First Series" (A. G. N. Flew, ed.), pp. 11–36. Blackwell, Oxford.

Schiller, C. H., ed. (1957). "Instinctive Behavior." International Univ. Press, New York.

Schleidt, W. M. (1973). Tonic communication: Continual effects of discrete signs in animal communication systems. *J. Theor. Biol.* **42,** 359–386.

Searle, J. R. (1969). "Speech Acts." Cambridge Univ. Press, London and New York.

Sebeok, T. A., ed. (1968). "Animal Communication." Indiana Univ. Press, Bloomington.

Sebeok, T. A., ed. (1977). "How Animals Communicate." Indiana Univ. Press, Bloomington.

Seidenberg, M. S., and Petitto, L. A. (1979). Signing behavior in apes: A critical review. *Cognition* **7,** 177–215.

Shannon, C. E. (1948). Mathematical theory of communication. *Bell Syst. Tech. J.* **27,** 379–423, 623–656.

Shannon, C. E., and Weaver, W. (1949). "The Mathematical Theory of Communication." Univ. of Illinois Press, Urbana.

Slater, P. J. B. (1973). Describing sequences of behavior. *In* "Perspectives in Ethology" (P. P. G. Bateson and P. H. Klopfer, eds.), pp. 131–153. Plenum, New York.

Smith, D. G., and Reid, F. A. (1979). Roles of the song repertoire in Red-winged Blackbirds. *Behav. Ecol. Sociobiol.* **5,** 279–290.

Smith, W. J. (1963). Vocal communication of information in birds. *Am. Nat.* **97,** 117–125.

Smith, W. J. (1965). Message, meaning and context in ethology. *Am. Nat.* **99,** 405–409.

Smith, W. J. (1968). Message-meaning analysis. *In* "Animal Communication" (T. A. Seboek, ed.), pp. 44–60. Indiana Univ. Press, Bloomington.

Smith, W. J. (1977). "The Behavior of Communicating." Harvard Univ. Press, Cambridge, Massachusetts.

Smith, W. J., Chase, J., and Lieblich, A. K. (1974). Tongue showing: A facial display of humans and other primate species. *Semiotica* **11,** 201–246.

Stephenson, G. R. (1973). Testing for group specific communication patterns in Japanese macaques. *Proc. Int. Congr. Primatol., 4th, 1972* **1,** 51–75.

Stonehouse, B. (1953). The Emperor Penguin, *Aptenodytes forsteri* Gray. I—Breeding behavior and development. *Falkland Isl. Depend. Surv. Sci. Rep.* No. 6, 1–33.

Strickland, G. (1980). Structuralism: A retrospective view. *Literary Rev.* **16,** 46–48.

Struhsaker, T. T. (1967). Auditory communication among vervet monkeys *(Cercopithecus aethiops)*. *In* "Social Communication among Primates" (S. A. Altmann, ed.), pp. 281–324. Univ. of Chicago Press, Chicago, Illinois.

Taylor, C. (1964). "The Explanation of Behaviour." Humanities Press, New York.

ter Pelkwijk, J. J., and Tinbergen, N. (1937). Eine reizbiologische Analyse einiger Verhaltensweisen von *Gasterosteus aculeatus* L. *Z. Tierpsychol.* **1,** 201–218.

Terrace, H. S. (1979). Is problem solving language? *J. Exp. Anal. Behav.* **31,** 161–175.

Thorpe, W. H. (1972). Duetting and antiphonal singing in birds. Its extent and significance. *Behaviour, Suppl.* No. 18, 1–197.

Thorpe, W. H., and Hall-Craggs, J. (1976). Sound production and perception in birds as related to the general principles of pattern perception. *In* "Growing Points in Ethology" (P. P. G. Bateson and R. A. Hinde, eds.), pp. 171–189. Cambridge Univ. Press, London and New York.

Tinbergen, N. (1951). "The Study of Instinct." Oxford Univ. Press (Clarendon), London and New York.

Tinbergen, N. (1952). Derived activities: Their causation, biological significance, origin, and emancipation during evolution. *Q. Rev. Biol.* **27,** 1–32.

Tinbergen, N. (1959). Comparative studies of the behaviour of gulls (Laridae): a progress report. *Behaviour* **15,** 1–70.

Urmson, J. O. (1956). "Philosophical Analysis—Its Development between the Two World Wars." Oxford Univ. Press (Clarendon), London and New York.

von Holst, E., and von Saint Paul, U. (1963). On the functional organisation of drives. *Anim. Behav.* **11,** 1–20.

von Uexküll, J. (1934). "Streitzüge durch die Umwelten von Tieren und Menschen." Springer-Verlag, Berlin.

Whitman, C. O. (1899). Animal behavior. *Biol. Lect. Mar. Biol. Lab., Woods Hole, 1898,* pp. 329–331.

Wiener, N. (1948). "Cybernetics." Wiley, New York.

Wiepkema, P. R. (1961). An ethological analysis of the reproductive behaviour of the bitterling. *Arch. Neerl. Zool.* **14,** 103–199.
Wilson, E. O. (1975). "Sociobiology." Harvard Univ. Press, Cambridge, Massachusetts.
Wittgenstein, L. (1922). "Tractatus Logico-philosophicus." Kegan Paul, London.
Wittgenstein, L. (1958). "Philosophical Investigations," 2nd ed. Blackwell, Oxford.
Zahavi, A. (1975). Mate selection—A selection for a handicap. *J. Theor. Biol.* **53,** 205–214.
Zahavi, A. (1978). Decorative patterns and the evolution of art. *New Sci.* **80,** 182–184.

Appendix: A World Survey of Evidence for Vocal Learning in Birds

DONALD E. KROODSMA
JEFFREY R. BAYLIS

The following is a world survey of evidence for vocal learning in birds. Evidence for vocal learning consists of:

1. Vocal imitation, under controlled laboratory conditions, of
   a. conspecific sounds,
   b. heterospecific avian sounds, or
   c. non-avian sounds, including the human voice;
2. Interspecific vocal imitation (mimicry) by free-living birds;
3. Intraspecific vocal imitation among free-living birds, as indicated by microgeographic distributions of vocal behaviors (i.e., dialects); and
4. Abnormal vocal development under acoustic deprivation in the laboratory.

A double asterisk (**) following a species name designates it as a *persistent* mimic; a single asterisk (*) denotes a *consistent* mimic; and lack of an asterisk signifies a *casual* mimic. Species followed by a plus (+) were ranked by Hartshorne (1973) as "superior" singers. (See Chapter 3 of this volume for fuller discussion of these categories of mimicry.)

ACOUSTIC COMMUNICATION IN BIRDS
VOLUME 2

ISBN 0-12-426802-1

Galliformes			
Tetraonidae			
Greater Prairie Chicken	*Tympanuchus cupido*	1a??	Sparling (1979)
Sharp-tailed Grouse	*T. phasianellus*	1a??	Sparling (1979)
Psittaciformes			
Loriidae			
Purple-naped Lory	*Lorius domicellus*	1c	Bechstein, cited in Armstrong (1963)
Psittacidae			
Budgerigar	*Melopsittacus undulatus*	1bc	Gramza (1970)
African Gray Parrot	*Psittacus erithacus*	1bc	Rauch (1978); Nottebohm (1970); Todt (1975)
Turquoise-fronted Parrot	*Amazona aestiva*	1c	Lorenz (1952)
Orange-winged Parrot	*A. amazonica*	3	Nottebohm (1970)
Yellow-headed Parrot	*A. ochrocephala*	1c	L. F. Baptista (unpublished data)
Apodiformes			
Trochilidae			
Little Hermit	*Phaethornis longuemareus*	3	Snow (1968); Wiley (1971)
Piciformes			
Ramphastidae			
Emerald Toucanet	*Aulacorhynchus prasinus*	2?	Wagner (1944)
Passeriformes			
Cotingidae			
Three-wattled Bellbird	*Procnias tricarunculata*	3?	Snow (1977)
Menuridae			
Superb Lyrebird**, +	*Menura novaehollandiae*	1b,2	Robinson (1974, 1975); Chisholm (1932, 1946); Bell (1976)
Albert Lyrebird**, +	*M. alberti*	2	Robinson (1974, 1975); Chisholm (1932, 1946)
Atrichornithidae			
Rufous Scrub-Bird**	*Atrichornis rufescens*	2	Chisholm (1946); Robinson (1973, 1975)
Noisy Scrub-Bird*	*A. clamosus*	2	Robinson (1975); Smith and Robinson (1976)
Alaudidae			
Singing Bush-lark**	*Mirafra javanica*	2	Bourke (1947); Chisholm (1946); Vernon (1973)

Southern Singing Bush-lark*	*M. cheniana*	2	Vernon (1973)
Rufous-naped Lark	*M. africana*	2	Vernon (1973)
Clapper Lark	*M. apiata*	2	Vernon (1973)
Fawn-colored Lark*	*M. africanoides*	2	Vernon (1973)
Sabota Lark*	*M. sabota*	2	Vernon (1973)
Calandra Lark**, +	*Melanocorypha calandra*	2	Armstrong (1963); Hartshorne (1973); Witherby *et al.* (1941); Alexander (1927)
Red-capped Lark	*Calandrella cinerea*	2	Vernon (1973)
Botha's Lark	*C. fringillaris*	2	Vernon (1973)
Crested Lark	*Galerida cristata*	2	Tretzel (1965)
Wood Lark	*Lullula arborea*	1c	Godman (1955)
Skylark	*Alauda arvensis*	1c,2	Hartshorne (1973); Witherby *et al.* (1941); Godman (1955)
Hirundinidae			
Barn Swallow	*Hirundo rustica*	3	M. McVey (unpublished data)
Motacillidae			
Cape Longclaw	*Macronyx capensis*	2	Farkas, cited in Vernon (1973)
Richard's Pipit*	*Anthus novaeseelandiae*	2	Chisholm (1946)
Tree Pipit	*A. trivialis*	2	Armstrong (1963)
Pycnonotidae			
Yellow-spotted Nicator	*Nicator chloris*	2	Bevin and Chapin, cited in Vernon (1973)
Irenidae			
Leafbirds**, +	*Chloropsis* spp.	2	Ali (1941, 1949); Bertram (1970); Hartshorne (1973)
Laniidae			
Boubou Shrike	*Laniarius ferrugineus*	3	Harcus (1977a);. Thorpe and North (1966); Thorpe (1972); Hooker and Hooker (1969)
Bokmakierie	*Telophorus zeylonus*	2	Brookhuysen, cited in Vernon (1973)
Red-backed Shrike**	*Lanius collurio*	2	Blase (1960); Armstrong (1963); Dowsett-Lemaire (1979); Witherby *et al.* (1941)
Lesser Gray Shrike	*L. minor*	2	Witherby *et al.* (1941); Dowsett-Lemaire (1979)
Northern Shrike	*L. excubitor*	2	Witherby *et al.* (1941)

(*continued*)

Fiscal Shrike*	*L. collaris*	2	Moreau, cited in Armstrong (1963); Vernon (1973)
Woodchat Shrike	*L. senator*	2	Witherby *et al.* (1941); Dowsett-Lemaire (1979)
Troglodytidae			
Rock Wren	*Salpinctes obsoletus*	3	Kroodsma (1975)
Short-billed Marsh Wren	*Cistothorus platensis*	1a	Kroodsma and Verner (1978)
Long-billed Marsh Wren	*C. palustris*	1ab,3,4	Verner (1975); Kroodsma (1978, 1979)
Bewick's Wren+	*Thryomanes bewickii*	2,3	Kroodsma (1972, 1974); Thomas (1943)
Happy Wren	*Thryothorus felix*	3	Brown and Lemon (1979)
Carolina Wren+	*T. ludovicianus*	2,3	E. S. Morton and P. Chu (unpublished data); McAtee (1950)
Bar-vented Wren	*T. sinaloa*	3	Brown and Lemon (1979)
Winter Wren	*Troglodytes troglodytes*	3	Kreutzer (1973, 1974); Kroodsma (1980)
House Wren	*T. aedon*	2,3	Kroodsma (1973 and unpublished data); Murray (1944); Nolan (1961)
Mimidae			
Gray Catbird*	*Dumetella carolinensis*	2,3	Harcus (1973); Thompson and Jane (1969); Boughey and Thompson (1976)
Mockingbird**,+	*Mimus polyglottos*	1abc,2,3,4	Wildenthal (1965); Borror and Reese (1956); Hatch (1967); Howard (1974); Laskey (1944)
White-banded Mockingbird**,+	*M. triurus*	2	Hartshorne (1973); Armstrong (1963)
Charles Mockingbird	*Nesomimus trifasciatus*	2	R. I. Bowman (unpublished data)
Brown Thrasher+	*Toxostoma rufum*	2	Bent (1948); D. E. Kroodsma (unpublished data)
California Thrasher	*T. redivivum*	2	Hartshorne (1973); Bent (1948)
Muscicapidae (Turdinae)			
Red-backed Scrub-robin	*Erythropygia leucophrys*	2	Vernon (1973)
Kalahari Scrub-robin	*E. paena*	2	Vernon (1973)
Eastern Bearded Scrub-robin**	*E. quadrivirgata*	2	Vernon (1973)

Bearded Scrub-robin	*Erythropygia barbata*	2	Pakenham, cited in Vernon (1973)
Japanese Robin	*Erithacus akahige*	1(a–c?)	M. Konishi (unpublished data)
Ryukyu Robin	*E. komadori*	1(a–c?)	M. Konishi (unpublished data)
Nightingale	*E. megarhynchos*	1c,2,3	Witherby *et al.* (1941); Todt *et al.* (1979); Godman (1955)
Robin	*E. rubecula*	2	Witherby *et al.* (1941)
Bluethroat	*E. svecicus*	2	Hartshorne (1973); Witherby *et al.* (1941)
Red-capped Robin-chat**,+	*Cossypha natalensis*	2	Oatley (1970); Thorpe (1972); Vernon (1973); Farkas (1969)
Gray-winged Robin-chat	*C. polioptera*	2	Oatley (1970)
Chorister Robin-chat**,+	*C. dichroa*	2	Thorpe (1972); Harcus (1977b); Oatley (1970); Vernon (1973)
Ruppell's Robin-chat**,+	*C. semirufa*	2	Benson (1946, 1948); Thorpe (1972); Hartshorne (1973); Oatley (1970)
White-browed Robin-chat	*C. heuglini*	2	Benson (1946); Oatley (1970)
Blue-shouldered Robin-chat	*C. cyanocampter*	2	Oatley (1970)
Cape Robin-chat	*C. caffra*	2	Oatley (1970)
Snowy-headed Robin-chat	*C. niveicapilla*	2	Thorpe (1972); Oatley (1970)
Spotted Morning Warbler+	*Cichladusa guttata*	2	Van Someren, cited in Vernon (1973)
Morning Warbler	*C. arquata*	2	Vernon (1973)
Red-tailed Morning Warbler+	*C. ruficauda*	2	Niven, cited in Vernon (1973)
White-rumped Shama**,+	*Copsychus malabaricus*	2	Gwinner and Kneutgen (1962); Hartshorne (1973); Armstrong (1963)
Black Redstart	*Phoenicurus ochruros*	2	Armstrong (1963)
Redstart**	*P. phoenicurus*	2	Witherby *et al.* (1941); Armstrong (1963)
Eastern Bluebird	*Sialia sialis*	1a?,4	Hartshorne, cited in Lanyon (1960)
Whinchat	*Saxicola rubetra*	2	Witherby *et al.* (1941); Witchell (1896)
Stonechat*	*S. torquata*	2	Vernon (1973); Witchell (1896); Armstrong (1963)

(continued)

Ant-eating Chat+	*Myrmecocichla formicivora*	2	Vernon (1973)
White-headed Black Chat	*M. arnotti*	2	Vernon (1973)
Cliff-chat+	*Thamnolaea cinnamomeiventris*	2	Vernon (1973)
Buff-streaked Chat	*Oenanthe bifasciata*	2	McLachlan and Liversidge, cited in Vernon (1973)
Isabelline Wheatear	*O. isabellina*	2	Witherby *et al.* (1941)
Wheatear	*O. oenanthe*	2	Witherby *et al.* (1941); Conder, cited in Armstrong (1963)
Pied Wheatear	*O. pleschanka*	2	Witherby *et al.* (1941)
Mountain Chat	*O. monticola*	2	Plowes, cited in Vernon (1973)
Capped Wheatear**,+	*O. pileata*	2	Vernon (1973)
Cape Rock Thrush	*Monticola rupestris*	2	Vernon (1973)
Sentinel Rock Thrush	*M. explorator*	2	Vernon (1973)
Short-toed Rock Thrush	*M. brevipes*	2	Vernon (1973)
Mottled Rock Thrush	*M. angolensis*	2	Vernon (1973)
Rock Thrush	*M. saxatilis*	2	Witherby *et al.* (1941)
Malabar Whistling Thrush	*Myiophoneus horsfieldii*	1c,2	McCann (1931), Bertram (1970)
Wood Thrush	*Hylocichla mustelina*	4	Lanyon (1979)
Olive Thrush	*Turdus olivaceus*	2	Vernon (1973)
Kurrichane Thrush	*T. libonyanus*	2	Vernon (1973)
Blackbird+	*T. merula*	1abc,2,3,4	Tretzel (1967); Witherby *et al.* (1941); Messmer and Messmer (1956); Thielcke-Poltz and Thielcke (1960)
Redwing	*T. iliacus*	3	Bjerke (1974); Bjerke and Bjerke (1981)
Song Thrush+	*T. philomelos*	2	Witherby *et al.* (1941); Barret (1976)
Mistle Thrush	*T. viscivorus*	3	Isaac and Marler (1963)
Glossy-black Thrush**,+	*T. serranus*	2	Skutch (1950); Hartshorne (1973); Armstrong (1963)
American Robin	*T. migratorius*	1b,3,4	Konishi (1965a)
Muscicapidae (Orthonychinae)			
Eastern Whipbird	*Psophodes olivaceus*	2	Chisholm (1946)

Muscicapidae (Timaliinae)			
Hwa-Mei	*Garrulax canorus*	1c	L. F. Baptista (unpublished data)
Muscicapidae (Sylviinae)			
Sedge Warbler*	*Acrocephalus schoenobaenus*	2	Witherby *et al.* (1941); Armstrong (1963)
Blyth's Reed Warbler**, +	*A. dumetorum*	2	Witherby *et al.* (1941); Hartshorne (1973)
African Reed Warbler**	*A. baeticatus*	2	Vernon (1973)
Reed Warbler	*A. scirpaceus*	2	Lemaire (1977); Witherby *et al.* (1941)
Marsh Warbler**, +	*A. palustris*	2,3	Witherby *et al.* (1941); Lemaire (1974, 1975a,b); Dowsett-Lemaire (1979)
Great Reed Warbler	*A. arundinaceus*	2	Witherby *et al.* (1941)
Icterine Warbler*, +	*Hippolais icterina*	2	Witherby *et al.* (1941); Hartshorne (1973)
Melodious Warbler	*H. polyglotta*	2	Witherby *et al.* (1941)
Barred Warbler	*Sylvia nisoria*	2	Hartshorne (1973)
Orphean Warbler	*S. hortensis*	2	Witherby *et al.* (1941)
Garden Warbler +	*S. borin*	2	Witherby *et al.* (1941); Tretzel (1966)
Common Whitethroat	*S. communis*	1a,2	Köpke (1970); Bergmann (1973); Witherby *et al.* (1941); Sauer, cited in Lanyon (1960)
Willow Warbler	*Phylloscopus trochilus*	2	Gwinner and Dorka (1965); M. Schubert (1969); G. Schubert (1971); Ausobsky (1960)
Bonelli's Warbler	*P. bonelli*	3	Brémond (1976)
Goldcrest	*Regulus regulus*	2,3	Becker (1977a,b)
Firecrest	*R. ignicapillus*	2,3	Becker (1977a,b)
Hunter's Warbler	*Cisticola hunteri*	3	Todt (1970)
Green-backed Camaroptera	*Camaroptera brachyura*	2	Bowbrick, cited in Vernon (1973)
Gray-backed Camaroptera	*C. brevicaudata*	2	Harwin, cited in Vernon (1973)
Yellow Eremomela	*Eremomela icteropygialis*	2	Vernon (1973)
Tit-babbler	*Parisoma subcaeruleum*	2	Vernon (1973)

(continued)

Muscicapidae (Malurinae)			
Brown Bristle-bird+	*Dasyornis brachypterus*	2	Chisholm (1932)
Brown Thornbill	*Acanthiza pusilla*	2	Armstrong (1963); Chisholm (1946)
Western Thornbill	*A. inornata*	2	Chisholm (1946)
Broad-tailed Thornbill	*A. apicalis*	2	Chisholm (1946)
Chestnut-tailed Thornbill	*A. uropygialis*	2	Chisholm (1946)
White-browed Scrub Wren*	*Sericornis frontalis*	2	Chisholm (1946)
Large-billed Scrub Wren	*S. magnirostris*	2	Chisholm (1946)
Yellow-throated Scrub Wren*	*S. lathami*	2	Chisholm (1946)
Chestnut-tailed Heath-wren**,+	*Hylacola pyrrhopygia*	2	Hartshorne (1973); Chisholm (1946)
Redthroat*,+	*Pyrrholaemus brunneus*	2	Chisholm (1946)
Speckled Warbler*	*Chthonicola sagittata*	2	Chisholm (1946)
Rock Warbler	*Origma solitaria*	2	Gilbert (1937); Chisholm (1946)
Muscicapidae (Muscicapinae)			
South African Black Flycatcher*	*Melaenornis pammelaina*	2	Vernon (1973)
Fiscal Flycatcher*	*M. silens*	2	Vernon (1973)
Brown Flycatcher	*Microeca leucophaea*	2	Chisholm (1946)
Lemon-breasted Flycatcher	*M. flavigaster*	2	Chisholm (1946)
Muscicapidae (Rhipidurinae)			
Gray Fantail	*Rhipidura fuliginosa*	2	Chisholm (1946)
Muscicapidae (Pachycephalinae)			
Shrike-tit*	*Falcunculus frontatus*	2	Chisholm (1946)
Gray Shrike-thrush	*Colluricincla harmonica*	2	Chisholm (1946)

Paridae			
Marsh Tit	*Parus palustris*	4	Becker (1978)
Carolina Chickadee	*P. carolinensis*	3	Ward (1966)
Coal Tit	*P. ater*	1a,4	Thielcke (1973a)
Great Tit	*P. major*	3	Gompertz (1961); Hunter and Krebs (1979)
Plain Titmouse	*P. inornatus*	3	Dixon (1969)
Tufted Titmouse	*P. bicolor*	3	Lemon (1968a)
Certhiidae			
Brown Creeper	*Certhia familiaris*	1a,2,3,4	Thielcke (1962, 1965b, 1970a,b, 1972, 1973b)
Short-toed Tree Creeper	*C. brachydactyla*	1a,2,3,4	Thielcke (1961, 1965a,b, 1969, 1970a,b, 1973b)
Dicaeidae			
Mistletoe-bird*	*Dicaeum hirundinaceum*	2	Chisholm (1946)
Nectariniidae			
Amethyst Sunbird	*Nectarinia amethystina*	2	Vernon (1973)
Scarlet-chested Sunbird	*N. senegalensis*	2	Vernon (1973)
Variable Sunbird	*N. venusta*	2	Vernon (1973)
White-bellied Sunbird	*N. talatala*	2	Vernon (1973)
Lesser Double-collared Sunbird	*N. chalybea*	2	Vernon (1973)
Orange-breasted Sunbird	*N. violacea*	2	Vernon (1973)
Splendid Sunbird	*N. coccinigastra*	3	Grimes (1974); Payne (1978)
Zosteropidae			
Oriental White-eye	*Zosterops palpebrosa*	1(a–c?)	L. F. Baptista (unpublished data)
Gray-breasted Silvereye*,+	*Z. lateralis*	2	Chisholm (1932)
Cape White-eye	*Z. pallida*	2	Vernon (1973)
Yellow White-eye	*Z. senegalensis*	2	Vernon (1973)
Meliphagidae			
Tui	*Prosthemadera novaeseelandiae*	1c	Gilliard (1958)
Emberizidae			
Corn Bunting	*Emberiza calandra*	3	McGregor (1980)

(*continued*)

Yellowhammer	*E. citrinella*	3,4	Kaiser (1965); Thorpe (1961)
Long-tailed Bunting	*E. cioides*	1(a–c?)	M. Konishi (unpublished data)
Ortolan Bunting	*E. hortulana*	3	Conrads (1976); Conrads and Conrads (1971)
Cirl Bunting	*E. cirlus*	2,3	Thorpe (1961); Kreutzer (1979)
Japanese Yellow Bunting	*E. sulphurata*	1(a–c?)	M. Konishi (unpublished data)
Snow Bunting	*Plectrophenax nivalis*	3	Tinbergen (1939); Chapman, cited in Gatty (1958)
Fox Sparrow	*Zonotrichia iliaca*	3	Martin (1977, 1979)
Song Sparrow	*Z. melodia*	1ab,2,3,4	Mulligan (1966); Harris and Lemon (1972); Townsend (1924); Kroodsma (1977); Eberhardt and Baptista (1977); Borror (1965)
Lincoln's Sparrow	*Z. lincolnii*	2	L. F. Baptista (unpublished data)
Swamp Sparrow	*Z. georgiana*	1ab,3,4	Marler and Peters (1977); D. E. Kroodsma (unpublished data)
Rufous-collared Sparrow	*Z. capensis*	1a,3	Nottebohm (1969, 1975); King (1972); Egli (1971)
White-crowned Sparrow	*Z. leucophrys*	1ab,3,4	Orejuela and Morton (1975); Baptista (1974, 1975a,b, 1976); Marler and Tamura (1962, 1964); Baker (1974, 1975); Marler (1967, 1970); Konishi (1965b); Baptista and Morton (1981); Lein (1979); Baptista and King (1980)
White-throated Sparrow	*Z. albicollis*	1a	Thorneycroft (1967); Lemon and Harris (1974)
Dark-eyed Junco	*Junco hyemalis*	1a,2,3,4	Williams and MacRoberts (1977); Konishi (1964); Marler (1967); Marler *et al.* (1962)
Yellow-eyed Junco	*J. phaeonotus*	1a,4	Konishi (1964); Marler (1960, 1967)
Savannah Sparrow	*Ammodramus sandwichensis*	3	Bradley (1977)
Chipping Sparrow	*Spizella passerina*	2	Borror (1968); Tasker (1955)
Field Sparrow	*S. pusilla*	2	Short (1966)
Vesper Sparrow	*Pooecetes gramineus*	2,3	Kroodsma (1972)
Black-throated Sparrow	*Amphispiza bilineata*	3	Heckenlively (1970)
Sage Sparrow	*A. belli*	3	Rich (1980); J. A. Wiens (unpublished data)
Melodious Grassquit	*Tiaris canora*	1a	Baptista (1978)

Small Tree Finch (Geospizine Finches)	*Camarhynchus parvulus*	3	Bowman (1979)
		1a,3	Bowman (1982)
Green-tailed Towhee	*Pipilo chlorurus*	2	D. Dobkin (unpublished data)
Rufous-sided Towhee	*P. erythrophthalmus*	2,3,4	Borror (1961, 1977); Kroodsma (1971); Ewert (1978, 1979)
Brown Towhee	*P. fuscus*	2	Marshall (1964); Hunt (1922)
Emberizidae (Thraupinae) Rose-breasted Grosbeak	*Pheucticus ludovicianus*	1bc,3	Lemon and Chatfield (1973); Pellett, cited in Scott (1902); Scott (1904b,c)
Black-headed Grosbeak	*P. melanocephalus*	4	Konishi (1965a)
Cardinal	*Cardinalis cardinalis*	1a,3,4	Lemon (1966, 1967, 1968b); Lemon and Scott (1966); Dittus and Lemon (1969, 1970)
Pyrrhuloxia	*C. sinuatus*	3	Lemon and Herzog (1969)
Indigo Bunting	*Passerina cyanea*	1a,2,3,4	Rice and Thompson (1968); Thompson (1970); Emlen *et al.* (1975)
Lazuli Bunting	*P. amoena*	2,3	Emlen *et al.* (1975); D. E. Kroodsma and R. Clover (unpublished data)
Emberizidae (Cardinalinae) Violaceous Euphonia**	*Euphonia violacea*	2	Snow (1974)
Thick-billed Euphonia**	*E. laniirostris*	2	Remsen (1976); Morton (1976)
Parulidae			
Blue-winged Warbler	*Vermivora pinus*	3	Kroodsma (1981)
Chestnut-sided Warbler	*Dendroica pensylvanica*	1b,3,4	Kroodsma (1981); Kroodsma *et al.* (1982); Lein (1978)
Yellowthroat	*Geothlypis trichas*	2,4	Kroodsma *et al.* (1982)
Yellow-breasted Chat*	*Icteria virens*	2	Hartshorne (1973); Scott (1902); (perhaps) Armstrong (1963); Grinnell *et al.* (1930); Cook (1935)
Vireonidae			
White-eyed Vireo**	*Vireo griseus*	2,3?	Adkisson and Conner (1978); Townsend (1924); Bradley (1980)

(*continued*)

Solitary Vireo	*V. solitarius*	3	James (1973, 1976)
Yellow-throated Vireo	*V. flavifrons*	3	James (1973, 1976)
Warbling Vireo	*V. gilvus*	2	James (1976)
Icteridae			
Yellow-rumped Cacique	*Cacicus cela*	2,3	Feekes (1977); Pearson (1974)
Yellow Oriole**	*Icterus nigrogularis*	2	Armstrong (1963)
Yellow Troupial	*I. icterus*	2	Wallace (1853)
Northern Oriole	*I. galbula*	1b	Scott (1901; but see Sanborn, 1932)
Red-winged Blackbird	*Agelaius phoeniceus*	1ab,3,4	Marler *et al.* (1972); Scott (1902, 1904a); E. S. Morton (unpublished data); K. Yasukawa (unpublished data)
Eastern Meadowlark	*Sturnella magna*	1b,2	Lanyon (1957, 1960); H. Kale (unpublished data)
Western Meadowlark	*S. neglecta*	1b,2,3	Lanyon (1957, 1960); D. E. Kroodsma (unpublished data)
Brown-headed Cowbird	*Molothrus ater*	4	King and West (1977); West *et al.* (1979); King *et al.* (1980); West and King (1980)
Bobolink	*Dolichonyx oryzivorus*	3,4	Avery and Oring (1977); Scott (1904a)
Fringillidae			
Chaffinch	*Fringilla coelebs*	1ab,2,3,4	Promptoff (1930); Poulsen (1951, 1958); Hulme (1950); Nottebohm (1967, 1968); Metzmacher and Mairy (1974); Baptista (1976); Thorpe (1958a,b, 1961); Thielcke (1962); Knecht and Scheer (1968); Marler (1952, 1956a,b); Hinde (1958); Conrads (1977); Slater and Ince (1979); Ince *et al.* (1980)
Canary	*Serinus canaria*	1abc,4	Godman (1955); Nottebohm and Nottebohm (1978); Poulsen (1959); Marler *et al.* (1973); Marler and Waser (1977); Waser and Marler (1977); Güttinger (1979)
White-rumped Seedeater	*S. leucopygius*	1b	K. Pereyra (unpublished data)
Yellow-rumped Seedeater⁺	*S. atrogularis*	2	Vernon (1973)
Yellow-fronted Canary	*S. mozambicus*	2	Vernon (1973)
Yellow Canary	*S. flaviventris*	2	Vernon (1973)

Brimstone Canary	*S. sulphuratus*	2	Vernon (1973)
White-throated Seedeater	*S. albogularis*	2	Vernon (1973)
Streaky-headed Seedeater	*S. gularis*	2	Vernon (1973)
White-winged Seedeater	*S. leucopterus*	2	Vernon (1973)
Greenfinch	*Carduelis chloris*	1ab,3,4	Thorpe (1955); Güttinger (1974, 1977, 1979)
Oriental Greenfinch	*C. sinica*	1b	Baptista (1972)
Himalayan Greenfinch	*C. spinoides*	1a	Güttinger (1978)
Siskin	*C. spinus*	1a	Mundinger (1970, 1979)
Pine Siskin	*C. pinus*	1ab	Mundinger (1970, 1979)
American Goldfinch	*C. tristis*	1a	Mundinger (1970, 1979)
Goldfinch	*C. carduelis*	1a	Güttinger (1978)
Common Redpoll	*Acanthis flammea*	1b	Mundinger (1979)
Twite	*A. flavirostris*	1a,3	Marler and Mundinger (1975); Mundinger (1979)
Linnet	*A. cannabina*	1abc	Poulsen (1954); Godman (1955)
Cassin's Finch	*Carpodacus cassinii*	3	Samson (1978)
House Finch	*C. mexicanus*	1ab,2,3,4	Mundinger (1975, 1979, and unpublished data); Bitterbaum and Baptista (1979); Baptista (1972)
Pine Grosbeak	*Pinicola enucleator*	2,3	C. S. Adkisson (unpublished data)
White-winged Crossbill	*Loxia leucoptera*	3	Mundinger (1979)
Bullfinch	*Pyrrhula pyrrhula*	1abc,3,4	Nicolai (1959); Wilkinson and Howse (1975); Schubert (1976); Godman (1955); Thorpe (1955)
Hawfinch	*Coccothraustes coccothraustes*	1bc	Thomson (1964)
Estrildidae			
Avadavat	*Amandava amandava*	1b	Goodwin (1960)
Zebra Finch	*Poephila guttata*	1ab,4	Immelmann (1967, 1969); Price (1979); Arnold (1975a,b)
Blue-faced Parrot Finch	*Erythrura trichroa*	4	Güttinger (1972)
Red-throated Parrot Finch	*E. psittacea*	1b	Güttinger (1972)
Gouldian Finch	*Chloebia gouldiae*	1b	Baptista (1973)

(continued)

Bronze Mannikin	*Lonchura cucullata*	1a,4	Güttinger and Acherman (1972)
White-backed Munia	*L. striata*	1ab	L. F. Baptista (unpublished data); Immelmann (1969); Dietrich (1980)
Spotted Munia	*L. punctulata*	1ab	Güttinger (1973)
Yellow-rumped Finch	*L. flaviprymna*	1a	Güttinger (1973)
Silverbill	*L. malabarica*	1b	Immelmann (1967, 1969)
Cut-throat	*Amadina fasciata*	1b	Ziswiler *et al.* (1972); H. R. Güttinger (unpublished data)
Ploceidae			
Senegal Indigo-bird**	*Vidua chalybeata*	2,3	Payne (1973a); Nicolai (1964, 1974)
Purple Indigo-bird	*V. purpurascens*	2	Payne (1973a, 1980)
Dusky Indigo-bird**	*V. funerea*	2	Payne (1973a)
Pale-winged Indigo-bird	*V. wilsoni*	2	Payne (1973a)
Fischer's Whydah**	*V. fischeri*	2	Nicolai (1964, 1973, 1974)
Shaft-tailed Whydah**	*V. regia*	2	Nicolai (1964, 1974)
Paradise Whydah**	*V. paradisaea*	2	Nicolai (1964); Payne (1973b, 1980)
Broad-tailed Paradise Whydah**	*V. orientalis*	2	Nicolai (1964)
Sturnidae			
Red-winged Starling	*Onychognathus morio*	1b	Vernon (1973)
Black-breasted Glossy Starling	*Lamprotornis corruscus*	2	Parkenham and Ranger, cited in Vernon (1973)
Red-shouldered Glossy Starling	*L. nitens*	2	Vernon (1973)
Glossy Starling	*Lamprotornis* spp.	1b	Vincent (1936)
Wattled Starling	*Creatophora cinerea*	2	Joubert, cited in Vernon (1973)
Starling*,**	*Sturnus vulgaris*	1c,2	Armstrong (1963); Witherby *et al.* (1941); Godman (1955); Vernon (1973); Moss (1977); Townsend (1924); Allard (1939)
Indian Myna	*Acridotheres tristis*	1b	Dean, cited in Vernon (1973)

Crested Myna	*A. cristatellus*	1c,2	L. F. Baptista (unpublished data)
Hill Myna	*Gracula religiosa*	1abc,2,3,4	Thorpe (1959, 1967); Bertram (1970); Tenaza (1976); Godman (1955)
Oriolidae			
White-bellied Oriole*	*Oriolus sagittatus*	2	Gilbert (1937); Chisholm (1946)
Black-headed Forest Oriole	*O. monacha*	2	Vincent (1936)
Black-headed Oriole	*O. larvatus*	2	Vernon (1973)
Dicruridae			
Square-tailed Drongo**	*Dicrurus ludwigii*	2	Vernon (1973); Ali (1941, 1949); Armstrong (1963)
Fork-tailed Drongo**	*D. adsimilis*	2	Vernon (1973); Ali (1941, 1949); Armstrong (1963)
Spangled Drongo	*D. hottentottus*	2	Chisholm (1946)
Large Racket-tailed Drongo**,+	*D. paradiseus*	2	Hartshorne (1973)
Callaeidae			
Saddleback	*Creadion carunculatus*	3	Jenkins (1977)
Artamidae			
White-browed Wood Swallow	*Artamus superciliosus*	2	Chisholm (1946)
Dusky Wood Swallow	*A. cyanopterus*	2	Chisholm (1946)
Cracticidae			
Gray Butcherbird*	*Cracticus torquatus*	2	Chisholm (1946)
Pied Butcherbird	*C. nigrogularis*	2	Chisholm (1946)
Australian Magpie	*Gymnorhina tibicen*	1c,2	Waite (1903); Robinson (1975); Chisholm (1946)
Ptilonorhynchidae			
Tooth-billed Bowerbird*,+	*Scenopoeetes dentirostris*	2	Chisholm (1946); but see Chaffer (1958) and Warham (1962)
Gardener Bowerbird	*Amblyornis inornatus*	2	Gilliard (1969)
Golden Bowerbird*	*Prionodura newtoniana*	2	Chisholm (1946); but see Warham (1962)
Regent Bowerbird	*Sericulus chrysocephalus*	1b,2	Goddard, cited in Marshall (1950); Chisholm (1946)
Satin Bowerbird*	*Ptilonorhynchus violaceus*	2	Robinson (1974, 1975); Chisholm (1946); Gilliard (1969)

(continued)

Spotted Bowerbird*, +	*Chlamydera maculata*	2	Hartshorne (1973); Chisholm (1946)
Great Bowerbird	*C. nuchalis*	2	Chisholm (1946)
Corvidae			
Blue Jay	*Cyanocitta cristata*	1bc,3	Ramsey (1972, 1973); Kramer and Thompson (1979); Scott (1902)
Steller's Jay	*C. stelleri*	3	Brown (1964)
Jay**	*Garrulus glandarius*	1c,2	Armstrong (1963); Scott (1902); Witherby *et al.* (1941)
Jackdaw	*Corvus monedula*	2	Witherby *et al.* (1941)
Common Crow	*C. brachyrhynchos*	1bc	Hartshorne (1973)
Raven	*C. corax*	1bc	Hartshorne (1973); Witherby *et al.* (1941)
African White-necked Raven	*C. albicollis*	2	Vernon (1973)
Crow	*C. corone*	1c	Lorenz (1952)

ACKNOWLEDGMENTS

We thank L. F. Baptista, H. R. Güttinger, and M. Konishi for help with certain species groups. This world survey was completed while DEK was funded by the National Science Foundation (BNS78-02753 and BNS80-40282).

REFERENCES

Adkisson, C. S., and Conner, R. N. (1978). Interspecific vocal imitation by White-eyed Vireos. *Auk* **95,** 602–606.

Alexander, H. G. (1927). The birds of Latium, Italy. *Ibis* **3,** 245–271.

Ali, S. (1941). "The Book of Indian Birds." Bombay Nat. His. Soc., Bombay.

Ali, S. (1949). "Indian Hill Birds." Oxford Univ. Press, London and New York.

Allard, H. A. (1939). Vocal mimicry of the Starling and Mockingbird. *Science* **90,** 370–371.

Armstrong, E. A. (1963). "A Study of Bird Song." Oxford Univ. Press, London and New York.

Arnold, A. P. (1975a). The effects of castration on song development in Zebra Finches (*Poephila guttata*). *J. Exp. Zool.* **191,** 261–278.

Arnold, A. P. (1975b). The effects of castration and androgen replacement on song, courtship, and aggression in Zebra Finches *(Poephila guttata). J. Exp. Zool.* **191,** 309–325.

Ausobsky, A. (1960). Ein weiterer Fitis-Zilpzalp Mischsanger. *Egretta* **3,** 49–52.

Avery, M., and Oring, L. W. (1977). Song dialects in the Bobolink *(Dolichonyx oryzivorus). Condor* **79,** 113–118.

Baker, M. C. (1974). Genetic structure of two populations of White-crowned Sparrows with different song dialects. *Condor* **76,** 351–356.

Baker, M. C. (1975). Song dialects and genetic differences in White-crowned Sparrows *(Zonotrichia leucophrys). Evolution* **29,** 226–241.

Baptista, L. F. (1972). Wild House Finch sings White-crowned Sparrow song. *Z. Tierpsychol.* **30,** 266–270.

Baptista, L. F. (1973). Song mimesis by a captive Gouldian Finch. *Auk* **90,** 891–894.

Baptista, L. F. (1974). The effects of songs of wintering White-crowned Sparrows on song development in sedentary populations of the species. *Z. Tierpsychol.* **34,** 147–171.

Baptista, L. F. (1975a). Song dialects and demes in sedentary populations of the White-crowned Sparrow *(Zonotrichia leucophrys nuttalli). Univ. Calif., Berkeley, Publ. Zool.* **105,** 1–52.

Baptista, L. F. (1975b). Additional evidence of song-misimprinting in the White-crowned Sparrow. *Bird-Banding* **46,** 269–272.

Baptista, L. F.(1976). Geographic variation and "dialects" in the Chaffinch rain-call. *Wilson Ornithol. Soc., Annu. Meet., 57th, Symp. Avian Bioacoust.*

Baptista, L. F. (1978). Territorial, courtship and duet songs of the Cuban Grassquit *(Tiaris canora). J. Ornithol.* **119,** 91–101.

Baptista, L. F., and King, J. R. (1980). Geographical variation in song and song dialects of montane White-crowned Sparrows. *Condor* **82,** 267–284.

Baptista, L. F., and Morton, M. L. (1981). Interspecific song acquisition by a White-crowned Sparrow. *Auk* **98,** 383–385.

Barret, J. H. (1976). Song Thrush imitating Greenshank. *Nat. Wales* **15,** 91.

Becker, P. H. (1977a). Geographische Variation des Gesanges von Winter- und Sommergoldhähnchen *(Regulus regulus, R. ignicapillus). Vogelwarte* **29,** 1–37.

Becker, P. H. (1977b). Verhalten auf Lautäusserungen der Zwillingsart, interspezifische Territorialität und Habitatansprüche von Winter- und Sommergoldhähnchen *(Regulus regulus, R. ignicapillus). J. Ornithol.* **118,** 233–260.

Becker, P. H. (1978). Der Einfluss des Lernens auf einfache und komplexe Gesangsstrophen der Sumpfmeise *(Parus palustris). J. Ornithol.* **119,** 388–411.

Bell, K. (1976). Song of the Superb Lyrebird in southeastern New South Wales, Australia with some observations on habitat. *Emu* **76,** 59–63.

Benson, C. W. (1946). The genera *Turdus,* etc. in Nyasaland. *Ostrich* **17,** 156–164.

Benson, C. W. (1948). Geographical voice-variation in African birds. *Ibis* **90,** 48–71.

Bent, A. C. (1948). Life histories of North American nuthatches, wrens, thrashers, and their allies. *Bull.—U.S. Natl. Mus.* No. 203.

Bergmann, H. H. (1973). Die Imitationsleistung einer Mischsanger-Dorngrasmucke *(Sylvia communis). J. Ornithol.* **114,** 317–338.

Bertram, B. (1970). The vocal behaviour of the Indian Hill Mynah, *Gracula religiosa. Anim. Behav. Monogr.* **3**(2), 79–192.

Bitterbaum, E., and Baptista, L. F. (1979). Geographic variation in songs of California House Finches. *Auk* **96,** 462–474.

Bjerke, T. K. (1974). Geografisk sangvariasjon hos rodvingetrost, *Turdus iliacus. Sterna* **13,** 65–76.

Bjerke, T. K., and Bjerke, T. H. (1981). Song dialects in the Redwing *(Turdus iliacus). Ornis. Scand.* **12,** 40–50.

Blase, B. (1960). Die Lautäusserungen des Neuntöters *(Lanius c. collurio),* Freilanbeobachtungen und Kasper-Hauser-Versuche. *Z. Tierpsychol.* **17,** 293–344.

Borror, D. J. (1961). Intraspecific variation in passerine bird songs. *Wilson Bull.* **73,** 57–78.

Borror, D. J. (1965). Song variation in Maine Song Sparrows. *Wilson Bull.* **77,** 5–37.

Borror, D. J. (1968). Unusual songs in Passerine birds. *Ohio J. Sci.* **68,** 129–138.

Borror, D. J. (1977). Rufous-sided Towhees mimicking Carolina Wren and Field Sparrow. *Wilson Bull.* **89,** 477–480.

Borror, D. J., and Reese, C. R. (1956). Mockingbird imitations of Carolina Wren. *Bull. Mass. Audubon Soc.* **40,** 244–250, 309–318.

Boughey, M. J., and Thompson, N. S. (1976). Species specificity and individual variation in the songs of the Brown Thrasher *(Toxostoma rufum)* and Catbird *(Dumetella carolinensis). Behaviour* **57,** 64–90.

Bourke, P. A. (1947). Notes on the Horsfield Bush-lark. *Emu* **47,** 1–7.

Bowman, R. I. (1979). Adaptive morphology of song dialects in Darwin's finches. *J. Ornithol.* **120,** 353–389.

Bowman, R. I. The evolution of song in Darwin's finches. *In* "Patterns of Evolution in Galapagos Organisms" (R. I. Bowman and A. E. Leviton, eds.). Am. Assoc. Adv. Sci., Pac. Div., San Francisco, California. In press.

Bradley, R. A. (1977). Geographic variation in the song of Belding's Savannah Sparrow *(Passerculus sandwichensis beldingi). Bull. Fla. State Mus., Biol. Sci.* **22**(2), 57–100.

Bradley, R. A. (1980). Vocal and territorial behavior in the White-eyed Vireo. *Wilson Bull.* **92,** 302–311.

Brémond, J. C. (1976). Specific recognition in the song of Bonelli's Warbler *(Phylloscopus bonelli). Behaviour* **58,** 99–116.

Brown, J. L. (1964). The integration of agonistic behavior in the Steller's Jay, *Cyanocitta stelleri* (Gmelin). *Univ. Calif., Berkeley, Publ. Zool.* **60,** 223–328.

Brown, R. N., and Lemon, R. E. (1979). Structure and evolution of song form in the wrens *Thryothorus sinaloa* and *T. felix. Behav. Ecol. Sociobiol.* **5,** 111–131.

Chaffer, R. N. (1958). "Mimicry" of the "Stagemaker." *Emu* **58,** 53–55.

Chisholm, A. G. (1932). Vocal mimicry among Australian birds. *Ibis* **2,** 605–624.

Chisholm, A. G. (1946). Nature's linguists: a study of the problem of vocal mimicry. Brown, Prior, Anderson Pty. Ltd., Melbourne.

Conrads, K. (1976). Studien an Fremddialekt-Sängern und Dialekt-Mischsängern des Ortolans *(Emberiza hortulana)*. *J. Ornithol.* **117,** 438–450.

Conrads, K. (1977). Entwicklung einer Kombinationsstrophe des Buchfinken *(Fringilla c. coelebs* L.) aus einer Grünlings—Imitation und arteigenen elementen in Freiland. *Ber. Naturwiss. Ver. Bielefeld* pp. 91–101.

Conrads, K., and Conrads, W. (1971). Regionaldialekta des Ortolans *(Emberiza hortulana)* in Deutschland. *Vogelwelt* **92,** 81–100.

Cook, H. P. (1935). The song of the Yellow-breasted Chat. *Wilson Bull.* **42,** 297–298.

Dietrich, K. (1980). Vorbidwahl in der Gesangsentwicklung beim Japanischen Movchen *(Lonchura striata* var. *domestica,* Estrildidae). *Z. Tierpsychol.* **52,** 57–76.

Dittus, W. P. J., and Lemon, R. E. (1969). Effects of song tutoring and acoustic isolation on the song repertoires of Cardinals. *Anim. Behav.* **17,** 523–533.

Dittus, W. P. J., and Lemon, R. E. (1970). Auditory feedback in the singing of Cardinals. *Ibis* **112,** 544–548.

Dixon, K. L. (1969). Patterns of singing in a population of the Plain Titmouse. *Condor* **71,** 94–101.

Dowsett-Lemaire, F. (1979). The imitative range of the song of the Marsh Warbler, *Acrocephalus palustris,* with special reference to imitations of African birds. *Ibis* **121,** 453–468.

Eberhardt, D., and Baptista, L. F. (1977). Intraspecific and interspecific song mimesis in California Song Sparrows. *Bird-Banding* **48,** 193–205.

Egli, W. (1971). Investigaciones sobre el canto de *Zonotrichia capensis chilensis* (Meyen) (Aves, Passeriformes). *Bol. Mus. Nac. Hist. Nat. Chile* **32,** 173–190.

Emlen, S. T., Rising, J. D., and Thompson, W. L. (1975). A behavioral and morphological study of sympatry in the Indigo and Lazuli buntings of the Great Plains. *Wilson Bull.* **87,** 145–179.

Ewert, D. N. (1978). Song of the Rufous-sided Towhee *(Pipilo erythrophthalmus)* on Long Island, New York. Ph.D. Thesis, City Univ. of New York, New York.

Ewert, D. N. (1979). Development of song of a Rufous-sided Towhee raised in acoustic isolation. *Condor* **81,** 313–316.

Farkas, T. (1969). Notes on the biology and ethology of the Natal Robin, *Cossypha natalensis. Ibis* **111,** 281–291.

Feekes, F. (1977). Colony-specific song in *Cacicus cela* (Icteridae, Aves): the pass-word hypothesis. *Ardea* **65,** 197–202. Dutton, New York.

Gatty, H. (1958). Nature is your guide. Dutton, New York.

Gilbert, P. A. (1937). Field notes from New South Wales. *Emu* **37,** 28–31.

Gilliard, E. T. (1958). "Living Birds of the World." Doubleday, Garden City, New York.

Gilliard, E. T. (1969). "Birds of Paradise and Bower Birds." Nat. Hist. Press, Garden City, New York.

Godman, S. (1955). "The bird fancyer's delight." *Ibis* **97,** 240–246.

Gompertz, T. (1961). The vocabulary of the Great Tit. *Br. Birds* **54,** 369–418.

Goodwin, D. (1960). Observations on Avadavats and Golden-breasted Waxbills. *Avicult. Mag.* **66,** 174–199.

Gramza, A. F. (1970). Vocal mimicry in captive Budgerigars *(Melopsittacus undulatus). Z. Tierpsychol.* **27,** 971–983.

Grimes, L. G. (1974). Dialects and geographical variation in the song of the Splendid Sunbird, *Nectarinia coccinigaster. Ibis* **116,** 314–329.

Grinnell, J., Dixon, J., and Linsdale, J. M. (1930). Vertebrate natural history of a section of northern California through the Lassen Peak region. *Univ. Calif., Berkeley, Publ. Zool.* **35,** 1–594.

Güttinger, H. R. (1972). Elementwahl und Stophenaufbau in der Gesangsentwicklung einiger Papageiamadinen-Arten (Gattung: *Erythrura,* Familie: Estrildidae). *Z. Tierpsychol.* **31,** 26–38.

Güttinger, H. R. (1973). Kopiervermogen von Rhythmus und Stophenaufbau in der Gesangsentwicklung einiger *Lonchura*-Arten (Estrildidae). *Z. Tierpsychol.* **32,** 374–385.

Güttinger, H. R. (1974). Gesang des Grunlings *(Chloris chloris).* Lokale Unterscheide und Entwicklung bei Schallisolation. *J. Ornithol.* **115,** 321–337.

Güttinger, H. R. (1977). Variable and constant structures in Greenfinch songs *(Chloris chloris)* in different locations. *Behaviour* **60,** 304–318.

Güttinger, H. R. (1978). Verwandtschaftsbeziehungen und Gesangsaufbau bei Stieglitz *(Carduelis carduelis)* und Grünlingsverwandten *(Chloris* spec.). *J. Ornithol.* **119,** 172–190.

Güttinger, H. R. (1979). The integration of learned and genetically programmed behavior: hierarchical organization in songs of Canaries and Greenfinches and their hybrids. *Z. Tierpsychol.* **49,** 285–303.

Güttinger, H. R., and Achermann, J. (1972). Die Gesangsentwicklung des Kleinelsterchens *(Spermestes cucullata). J. Ornithol.* **113,** 37–48.

Gwinner, E., and Dorka, V. (1965). Beobachtungen an Zilpzalp-Fitis-Mischsangern. *Vogelwelt* **86,** 146–151.

Gwinner, E., and Kneutgen, J. (1962). Über die biologische Bedeutung der zweckdienlichen Anwendung erlernter Laute bie Vögeln. *Z. Tierpsychol.* **19,** 692–696.

Harcus, J. L. (1973). Song studies in the breeding biology of the Catbird, *Dumetella carolinensis.* Ph.D. Thesis, Univ. of Toronto, Toronto.

Harcus, J. L. (1977a). The functions of vocal duetting in some African birds. *Z. Tierpsychol.* **43,** 23–45.

Harcus, J. L. (1977b). The functions of mimicry in the vocal behaviour of the Chorister Robin. *Z. Tierpsychol.* **44,** 178–193.

Harris, M., and Lemon, R. E. (1972). Songs of Song Sparrows *(Melospiza melodia):* individual variation and dialects. *Can. J. Zool.* **50,** 301–309.

Hartshorne, C. (1973). "Born to Sing. An Interpretation and World Survey of Bird Song." Indiana Univ. Press, Bloomington.

Hatch, J. J. (1967). Diversity of the song of Mockingbirds *Mimus polyglottos* reared in different auditory environments. Unpubl. Ph.D. thesis, Duke Univ., Durham, North Carolina.

Heckenlively, D. B. (1970). Song in a population of Black-throated Sparrows. *Condor* **72,** 24–36.

Hinde, R. A. (1958). Alternative motor patterns in Chaffinch song. *Anim. Behav.* **6,** 211–218.

Hooker, T., and Hooker, B. I. (1969). Duetting. *In* "Bird Vocalizations" (R. A. Hinde, ed.), pp. 185–205. Cambridge Univ. Press, London and New York.

Howard, R. D. (1974). The influence of sexual selection and interspecific communication on Mockingbird song *(Mimus polyglottos). Evolution* **28,** 428–438.

Hulme, D. C. (1950). Chaffinch *(Fingilla coelebs)* mimicking Hedge Sparrow's *(Prunella modularis)* song. *Br. Birds* **43,** 222.

Hunt, R. (1922). Evidence of musical "tastes" in the Brown Towhee. *Condor* **24,** 193–203.

Hunter, M. L., Jr., and Krebs, J. R. (1979). Geographical variation in the song of the Great Tit *(Parus major)* in relation to ecological factors. *J. Anim. Ecol.* **48,** 759–785.

Immelmann, K. (1967). Zur ontogenetischen Gesangsentwicklung bei Prachtfinken. *Verh. Dtsch. Zool. Ges.* **30,** 320–332.

Immelmann, K. (1969). Song development in the Zebra Finch and other estrildid finches. *In* "Bird Vocalizations" (R. A. Hinde, ed.), pp. 61–74. Cambridge Univ. Press, London and New York.

Ince, S. A., Slater, P. J. B., and Weismann, C. (1980). Changes with time in the songs of a population of Chaffinches. *Condor* **80,** 285–290.

Isaac, D., and Marler, P. (1963). Ordering of sequences of singing behavior of Mistle Thrushes in relationship to timing. *Anim. Behav.* **30,** 344–374.

James, R. D. (1973). Ethological and ecological relationships of the Yellow-throated and Solitary Vireos (Aves: Vireonidae) in Ontario. Ph.D. Thesis, Univ. of Toronto, Toronto.

James, R. D. (1976). Unusual songs with comments on song learning among vireos. *Can. J. Zool.* **54,** 1223–1226.

Jenkins, P. F. (1977). Cultural transmission of song patterns and dialect development in a free-living bird population. *Anim. Behav.* **25,** 50–78.

Kaiser, W. (1965). Der Gesang der Goldammer und die Verbreitung ihrer Dialekte. *Falke* **12,** 40–42, 92–93, 131–135, 169–170, 188–191.

King, A. P., and West, M. J. (1977). Species identification in the North American Cowbird: appropriate responses to abnormal song. *Science* **195,** 1002–1004.

King, A. P., West, M. J., and Eastzer, D. H. (1980). Song structure and song development as potential contributors to reproductive isolation in cowbirds *(Molothrus ater). J. Comp. Physiol. Psychol.* **94,** 1028–1039.

King, J. R. (1972). Variation in the song of the Rufous-collared Sparrows, *Zonotrichia capensis,* in northwestern Argentina. *Z. Tierpsychol.* **30,** 344–373.

Knecht, A., and Scheer, U. (1968). Lautausserung und Verhalten des Azoren Buchfinken *(Fringilla coelebs moreletti* Pucheran). *Z. Tierpsychol.* **25,** 115–169.

Köpke, G. (1970). Beobachtungen an einer Mischsanger-Dorngrasmücke *(Sylvia communis). Ornithol. Mitt.* **22,** 146–149.

Konishi, M. (1964). Effects of deafening on song development in two species of juncos. *Condor* **66,** 85–102.

Konishi, M. (1965a). Effects of deafening on song development in American Robins and Black-headed Grosbeaks. *Z. Tierpsychol.* **22,** 584–599.

Konishi, M. (1965b). The role of auditory feedback in the control of vocalization in the White-crowned Sparrow. *Z. Tierpsychol.* **22,** 770–783.

Kramer, H. G., and Thompson, N. S. (1979). Geographic variation in the bell calls of the Blue Jay *(Cyanocitta cristata). Auk* **96,** 423–425.

Kreutzer, M. (1973). Contribution a l'etude du chant de proclamation territoriale du Troglodyte, *Troglodytes troglodytes.* Ph.D. Thesis, Univ. Paris.

Kreutzer, M. (1974). Réponses comportementales des males *Troglodytes* (Passeriformes) à des chants spécifiques de dialectes différents. *Rev. Comp. Anim.* **8,** 287–295.

Kreutzer, M. (1979). Etude du chant chez Le Bruant Zizi *(Emberiza cirlus).* Le repertoire, caracteristiques et distribution. *Behaviour* **71,** 291–321.

Kroodsma, D. E. (1971). Song variations and singing behavior in the Rufous-sided Towhee, *Pipilo erythrophthalmus oregonus. Condor* **73,** 303–308.

Kroodsma, D. E. (1972). Variation in songs of Vesper Sparrows in Oregon. *Wilson Bull.* **84,** 173–178.

Kroodsma, D. E. (1973). Coexistence of Bewick's Wrens and House Wrens in Oregon. *Auk* **90,** 341–352.

Kroodsma, D. E. (1974). Song learning, dialects, and dispersal in the Bewick's Wren. *Z. Tierpsychol.* **35,** 352–380.

Kroodsma, D. E. (1975). Song patterning in the Rock Wren. *Condor* **77,** 294–303.

Kroodsma, D. E. (1977). A re-evaluation of song development in the Song Sparrow. *Anim. Behav.* **25,** 390–399.

Kroodsma, D. E. (1978). Aspects of learning in the ontogeny of bird song: where, from whom, when, how many, which, and how accurately? *In* "Development of Behavior" (G. Burghardt and M. Bekoff, eds.), pp. 215–230. Garland, New York.

Kroodsma, D. E. (1979). Vocal dueling among male Marsh Wrens: evidence for ritualized expressions of dominance/subordinance. *Auk* **96,** 506–515.

Kroodsma, D. E. (1980). Winter Wren singing behavior: a pinnacle of song complexity. *Condor* **82,** 357–365.

Kroodsma, D. E. (1981). Geographical variation and functions of song types in warblers (Parulidae) *Auk* **98,** 743–751.

Kroodsma, D. E., and Verner, J. (1978). Complex singing behaviors among *Cistothorus* wrens. *Auk* **95,** 703–716.

Kroodsma, D. E., Meservey, W. R., and Pickert, R. Vocal learning in the Parulidae. *Wilson Bull.* (in press).

Lanyon, W. E. (1957). The comparative biology of the Meadowlarks *(Sturnella)* in Wisconsin. *Publ. Nuttall Ornithol. Club* No. 1.

Lanyon, W. E. (1960). The ontogeny of vocalizations in birds. *In* "Animal Sounds and Communications" (W. E. Lanyon and W. N. Tavolga, eds.), Publ. No. 7, pp. 321–347. Am. Inst. Biol. Sci., Washington, D.C.

Lanyon, W. E. (1979). Development of song in the Wood Thrush *(Hylocichla mustelina),* with notes on a technique for hand-rearing passerines from the egg. *Am. Mus. Novit.* No. 2666.

Laskey, A. R. (1944). A Mockingbird acquires his song repertory. *Auk* **61,** 211–219.

Lein, M. R. (1978). Song variation in a population of Chestnut-sided Warblers *(Dendroica pensylvanica):* its nature and suggested significance. *Can. J. Zool.* **56,** 1266–1283.

Lein, M. R. (1979). Song patterns of the Cypress Hills (Alberta, Canada) population of White-crowned Sparrows. *Can. Field-Nat.* **93,** 272–275.

Lemaire, F. (1974). Le chant de la Rousserolle verderolle *(Acrocephalus palustris):* étendue du répertoire imitatif, construction rythmique et musicalité. *Gerfaut* **64,** 3–28.

Lemaire, F. (1975a). Dialectal variations in the imitative song of the Marsh Warbler *(Acrocephalus palustris)* in western and eastern Belgium. *Gerfaut* **65,** 95–106.

Lemaire, F. (1975b). Le chant de la Rousserolle verderolle *(Acrocephalus palustris):* fidélité des imitations et relations avec les espèces imitées et avec les congénères. *Gerfaut* **65,** 3–28.

Lemaire, F. (1977). Mixed song, interspecific competition and hybridisation in the Reed and Marsh warblers *(Acrocephalus scirpaceus* and *palustris). Behaviour* **63,** 215–240.

Lemon, R. E. (1966). Geographic variation in the song of Cardinals. *Can. J. Zool.* **44,** 413–428.

Lemon, R. E. (1967). The response of Cardinals to songs of different dialects. *Anim. Behav.* **15,** 538–545.

Lemon, R. E. (1968a). Coordinated singing by Black-crested Titmice. *Can. J. Zool.* **46,** 1163–1167.

Lemon, R. E. (1968b). The relation between the organisation and function of song in Cardinals. *Behaviour* **32,** 158–178.

Lemon, R. E., and Chatfield, C. (1973). Organization of song of Rose-breasted Grosbeaks. *Anim. Behav.* **21,** 28–44.

Lemon, R. E., and Harris, M. (1974). The question of dialects in the songs of White-throated Sparrows. *Can. J. Zool.* **52,** 83–98.

Lemon, R. E., and Herzog, A. (1969). The vocal behavior of Cardinals and Pyrrhuloxias in Texas. *Condor* **71,** 1–15.

Lemon, R. E., and Scott, D. M. (1966). On the development of song in young cardinals. *Can. J. Zool.* **44,** 191–199.

Lorenz, K. (1952). "King Solomon's Ring." Methuen, London.

McAtee, W. L. (1950). The Carolina Wren, *Thryothorus ludovicianus,* as a mimic. *Wilson Bull.* **62,** 136.

McCann, C. (1931). Notes on the whistling schoolboy or Malabar Whistling Thrush. *Bombay Nat. Hist. Soc.* **35,** 202–204.

McGregor, P. K. (1980). Song dialects in the Corn Bunting *(Emberiza calandra). Z. Tierpsychol.* **54,** 285–297.

Marler, P. (1952). Variation in the song of the Chaffinch, *Fringilla coelebs. Ibis* **94,** 458–472.

Marler, P. (1956a). Behaviour of the Chaffinch *(Fringilla coelebs). Behaviour, Suppl.* No. 5.
Marler, P. (1956b). The voice of the Chaffinch and its function as a language. *Ibis* **98,** 221–261.
Marler, P. (1960). Bird songs and mate selection. *In* ''Animal Sounds and Communication'' (W. E. Lanyon and W. N. Tavolga, eds.), Publ. No. 7, pp. 348–367. Am. Inst. Biol. Sci., Washington, D.C.
Marler, P. (1967). Comparative study of song development in sparrows. *Proc. Int. Ornithol. Congr.* **14,** 231–244.
Marler, P. (1970). A comparative approach to vocal learning: song development in White-crowned Sparrows. *J. Comp. Physiol. Psychol. Monograph* **71**(2), 1–25.
Marler, P., and Mundinger, P. C. (1975). Vocalizations, social organization and breeding biology of the Twite *Acanthis flavirostris. Ibis* **117,** 1–17.
Marler, P., and Peters, S. (1977). Selective vocal learning in a sparrow. *Science* **198,** 519–521.
Marler, P., and Tamura, M. (1962). Song ''dialects'' in three populations of White-crowned Sparrows. *Condor* **64,** 368–377.
Marler, P., and Tamura, M. (1964). Culturally transmitted patterns of vocal behavior in sparrows. *Science* **146,** 1483–1486.
Marler, P., and Waser, M. S. (1977). The role of auditory feedback in Canary song development. *J. Comp. Physiol. Psychol.* **91,** 8–16.
Marler, P., Kreith, M., and Tamura, M. (1962). Song development in hand-raised Oregon Juncos. *Auk* **79,** 12–30.
Marler, P., Mundinger, P., Waser, M. S., and Lutjen, A. (1972). Effects of acoustical stimulation and deprivation on song development in Red-winged Blackbirds *(Agelaius phoeniceus). Anim. Behav.* **20,** 586–606.
Marler, P., Konishi, M., Lutjen, A., and Waser, M. S. (1973). Effects of continous noise on avian hearing and vocal development. *Proc. Natl. Acad. Sci. U.S.A.* **70,** 1393–1396.
Marshall, A. J. (1950). The function of vocal mimicry in birds. *Emu* **50,** 5–16.
Marshall, J. T. (1964). Voice in communication and relationship among Brown Towhees. *Condor* **66,** 345–356.
Martin, D. J. (1977). Songs of the Fox Sparrow. I. Structure of song and its comparison with other Emberizidae. *Condor* **79,** 209–221.
Martin, D. J. (1979). Songs of the Fox Sparrow. II. Intra- and interpopulation variation. *Condor* **81,** 173–184.
Messmer, E., and Messmer, I. (1956). Die Entwicklung der Lautausserungen und einiger Verhaltensweisen der Amsel *(Turdus merula merula)* unter naturlichen Bedingungen und nach Einzelaufzucht in schalldichten Raumen. *Z. Tierpsychol.* **13,** 341–441.
Metzmacher, M., and Mairy, F. (1974). Variations gëographiques de la figure finale du chant du Pinson des arbres, *Fringilla c. coelebs* L. *Gerfaut* **62,** 215–244.
Morton, E. S. (1976). Vocal mimicry in the Thick-billed Euphonia. *Wilson Bull.* **88,** 485–487.
Moss, S. (1977). Starling imitating Cetti's Warbler. *Br. Birds* **70,** 36.
Mulligan, J. A. (1966). Singing behavior and its development in the Song Sparrow, *Melospiza melodia. Univ. Calif., Berkeley, Publ. Zool.* **81,** 1–76.
Mundinger, P. C. (1970). Vocal imitation and individual recognition of finch calls. *Science* **168,** 480–482.
Mundinger, P. C. (1975). Song dialects and colonization in the House Finch, *Carpodacus mexicanus,* on the east coast. *Condor* **77,** 407–422.
Mundinger, P. C. (1979). Call learning in the Carduelinae: ethological and systematic considerations. *Syst. Zool.* **28,** 270–283.
Murray, J. J. (1944). An unusual song from a House Wren. *Wilson Bull.* **56,** 59.
Nicolai, J. (1959). Familientradition in der Gesangsentwicklung des Gimpels *(Pyrrhula pyrrhula). J. Ornithol.* **100,** 39–46.

Nicolai, J. (1964). Der Brutparasitismus der Viduinae als ethologisches Problem. *Z. Tierpsychol.* **21,** 129–204.

Nicolai, J. (1973). Das Lernprogramm in der Gesangsausbildung der Strohwitwe *Tetraenura fischeri* Reichenow. *Z. Tierpsychol.* **32,** 113–138.

Nicolai, J. (1974). Mimicry in parasitic birds. *Sci. Am.* **231,** 92–98.

Nolan, V. (1961). A wren singing combined House and Carolina wren song. *Wilson Bull.* **74,** 83–84.

Nottebohm, F. (1967). The role of sensory feedback in the development of avian vocalizations. *Proc. Int. Ornithol. Cong.* **14,** 265–280.

Nottebohm, F. (1968). Auditory experience and song development in the Chaffinch, *Fringilla coelebs. Ibis* **110,** 549–568.

Nottebohm, F. (1969). The song of the Chingolo, *Zonotrichia capensis,* in Argentina: description and evaluation of a system of dialects. *Condor* **71,** 299–315.

Nottebohm, F. (1970). Ontogeny of bird song. *Science* **167,** 950–956.

Nottebohm, F. (1975). Continental patterns of song variability in *Zonotrichia capensis:* some possible ecological correlates. *Am. Nat.* **109,** 605–624.

Nottebohm, F., and Nottebohm, M. E. (1978). Relationship between song repertoire and age in the Canary, *Serinus canarius. Z. Tierpsychol.* **46,** 298–305.

Oatley, T. B. (1970). The functions of vocal imitations by African *Cossyphas. Ostrich, Suppl.* No. 8, 85–89.

Orejuela, J. E., and Morton, M. (1975). Song dialects in several populations of Mountain White-crowned Sparrows *(Zonotrichia leucophrys oriantha)* in the Sierra Nevada. *Condor* **77,** 145–153.

Payne, R. B. (1973a). Behavior, mimetic songs and song dialects, and relationships of the indigobirds *(Vidua)* of Africa. *Ornithol. Monogr.* **11.**

Payne, R. B. (1973b). Vocal mimicry in the Paradise Whydahs *(Vidua)* and response of female whydahs to the songs of their hosts *(Pytilia)* and their mimics. *Anim. Behav.* **21,** 762–771.

Payne, R. B. (1978). Microgeographic variation in songs of Splendid Sunbirds, *Nectarinia coccinigaster:* population phenetics, habitats, and song dialects. *Behaviour* **65,** 282–308.

Payne, R. B. (1980). Behavior and songs of hybrid parasitic finches. *Auk* **97,** 118–134.

Pearson, D. L. (1974). Use of abandoned Cacique nests by nesting Troupials *(Icterus icterus):* precursor to parasitism? *Wilson Bull.* **86,** 290–291.

Poulsen, H. (1951). Inheritance and learning in the song of the Chaffinch *(Fringilla coelebs* L.). *Behaviour* **3,** 216–228.

Poulsen, H. (1954). On the song of the Linnet *(Carduelis cannabina* (L.)). *Dan. Ornithol.* **48,** 32–37.

Poulsen, H. (1958). The calls of the Chaffinch *(Fringilla coelebs* L.) in Denmark. *Dan. Ornithol. Foren. Tidsskr.* **52,** 89–105.

Poulsen, H. (1959). Song learning in the domestic Canary. *Z. Tierpsychol.* **16,** 173–178.

Price, P. H. (1979). Developmental determinants of structure in Zebra Finch Song. *J. Comp. Physiol. Psychol.* **93,** 260–277.

Promptoff, A. N. (1930). Die geographische Variabilitat des Buchfinkenschlags *(Fringilla coelebs* L.). *Biol. Zentralbl.* **50,** 478–503.

Ramsey, A. O. (1972). Mimesis in hand-reared Blue Jays. *Bird-Banding* **43,** 214–215.

Ramsey, A. O. (1973). Mimesis in Blue Jays. *EBBA News* p. 23.

Rauch, N. (1978). Struktur der Lautausserungen eines sprache imitierenden Graupapageis *(Psittacus erithacus* L.). *Behaviour* **66,** 56–105.

Remsen, J. V., Jr. (1976). Vocal mimicry in the Thick-billed Euphonia, *Euphonia laniirostris. Wilson Bull.* **88,** 487–488.

Rice, J. O, and Thompson, W. L. (1968). Song development in the Indigo Bunting. *Anim. Behav.* **16,** 462–469.
Rich, T. (1980). Territorial behavior of the Sage Sparrow: spatial and random aspects. *Wilson Bull.* **92,** 425–438.
Robinson, F. N. (1973). Vocal mimicry and bird song evolution. *New Sci.* **21,** 742–743.
Robinson, F. N. (1974). The functions of vocal mimicry in some avian displays. *Emu* **75,** 9–10.
Robinson, F. N. (1975). Vocal mimicry and the evolution of bird song. *Emu* **74,** 23–27.
Samson, F.B. (1978). Vocalizations of Cassin's Finch in northern Utah. *Condor* **80,** 203–210.
Sanborn, H. C. (1932). The inheritance of song in birds. *J. Comp. Psychol.* **13,** 345–364.
Schubert, G. (1971). Experimentalle Untersuchengen über die Artkennzeichnenden Parameter im Gesang des Zilpzalps, *Phylloscopus c. collybita* (Vieillot). *Behaviour* **38,** 289–314.
Schubert, M. (1969). Untersuchung über die akustischen Parameter von Zilpzalp-Fitis-Mischgesangen. *Beitr. Vogelkd.* **14,** 354–368.
Schubert, M. (1976). Uber die Variabilitat des Lockrufen des Gimpels *Pyrrhula pyrrhula. Ardea* **64,** 62–71.
Scott, W. E. D. (1901). Data on songbirds: observations on the song of Baltimore Orioles in captivity. *Science* **14,** 522–526.
Scott, W. E. D. (1902). Data on song in birds: the acquisition of new songs. *Science* **15,** 178–181.
Scott, W. E. D. (1904a). The inheritance of song in passerine birds. Remarks and observations on the song of hand-reared Bobolinks and Red-winged Blackbirds *(Dolichonyx oryzivorus* and *Agelaius phoeniceus). Science* **19,** 154–155.
Scott, W. E. D. (1904b). The inheritance of song in passerine birds. Remarks on the development of song in the Rose-breasted Grosbeak, *Zamelodia ludoviciana* (Linnaeus) and the Meadowlark, *Sturnella magna* (Linnaeus). *Science* **19,** 957–959.
Scott, W. E. D. (1904c). The inheritance of song in passerine birds. Further observations on the development of song and nest-building in hand-reared Rose-breasted Grosbeaks, *Zamelodia ludoviciana* (Linnaeus). *Science* **20,** 282–283.
Short, L. L., Jr. (1966). Field Sparrow sings Chipping Sparrow song. *Auk* **83,** 665.
Skutch, A. F. (1950). Life history of the White-breasted Blue Mockingbird. *Condor* **52,** 220–227.
Slater, P. J. B., and Ince, S. A. (1979). Cultural evolution in the Chaffinch *(Fringilla coelebs)* song. *Behaviour* **71,** 146–166.
Smith, G. T., and Robinson, F. N. (1976). The Noisy Scrub Bird: an interim report. *Emu* **76,** 37–42.
Snow, B. K. (1974). Vocal mimicry in the Violaceous Euphonia, *Euphonia violacea. Wilson Bull.* **86,** 179–180.
Snow, B. K. (1977). Territorial behavior and courtship of the male Three-wattled Bellbird. *Auk* **94,** 623–645.
Snow, D. W. (1968). The singing assemblies of Little Hermits. *Living Bird* **7,** 47–55.
Sparling, D. W. (1979). Evidence for vocal learning in Prairie Grouse. *Wilson Bull.* **91,** 618–621.
Tasker, R. R. (1955). Chipping Sparrow with song of Clay-colored Sparrow at Toronto. *Auk* **72,** 303.
Tenaza, R. R. (1976). Wild Mynahs mimic wild primates. *Nature (London)* **259,** 561.
Thielcke, G. (1961). Stammegeschichte und geographische Variation des Gesanges unserer Baumlaufer *(Certhia familiaris* L. und *Certhia brachydactyla* Brehm). *Z. Tierpsychol.* **18,** 188–204.
Thielcke, G. (1962). Die geographische Variation eines erlernten Elements im Gesang des Buchfinken *(Fringilla coelebs)* und des Waldbaumlaufers *(Certhia familiaris). Vogelwarte* **21,** 199–202.
Thielcke, G. (1965a). Gesangsgeographische Variation des Gartenbaumlaufers *(Certhia brachydactyla)* im Hinblick auf das Artbildungsproblem. *Z. Tierpsychol.* **22,** 542–566.

Thielcke, G. (1965b). Die Ontogenese der Bettlelaute von Garten und Waldbaumlaufer *(Certhia brachydactyla* Brehm und *C. familiaris* L.) *Zool. Anz.* **174,** 237–241.

Thielcke, G. (1969). Geographic variation in bird vocalizations. *In* "Bird Vocalizations" (R. A. Hinde, ed.), pp. 311–339. Cambridge Univ. Press, London and New York.

Thielcke, G. (1970a). "Vogelstimmen." Springer-Verlag, Berlin and New York.

Thielcke, G. (1970b). Lernen von Gesang als moglicher Schrittmacher der Evolution. *Z. Zool. Syst. Evolutionsforsch.* **8,** 309–320.

Thielcke, G. (1972). Waldbaumlaufer *(Certhia familiaris)* ahmen artfremdes Signal nach und reagieren darauf. *J. Ornithol.* **113,** 287–296.

Thielcke, G. (1973a). Uniformierung des Gesangs der Tannenmeise *(Parus ater)* durch Lernen. *J. Ornithol.* **114,** 443–454.

Thielcke, G. (1973b). On the origin of divergence of learned signals (songs) in isolated populations. *Ibis* **115,** 511–516.

Thielcke-Poltz, H., and Thielcke, G. (1960). Akustisches Lernen verschieden alter schallisolierter Amseln *(Turdus merula)* und die Entwicklung erlernter Motive ohne und mit kunstlichem Einfluss von Testosteron. *Z. Tierpsychol.* **17,** 211–244.

Thomas, E. S. (1943). A wren singing the songs of both Bewick's and House wren. *Wilson Bull.* **55,** 192–193.

Thompson, W. L. (1970). Song variation in a population of Indigo Buntings. *Auk* **87,** 58–71.

Thompson, W. L., and Jane, P. L. (1969). An analysis of catbird song. *Jack-Pine Warbler* **47,** 115–125.

Thomson, A. L., ed. (1964). "A New Dictionary of Birds." Nelson, London.

Thorneycroft, H. B. (1967). Aspects of the development of vocalizations in the White-throated Sparrow, *Zonotrichia albicollis* (Gmelin). Master's Thesis, Univ. of Toronto, Toronto.

Thorpe, W. H. (1955). Comments on "The Bird Fancyer's Delight": Together with notes on imitation in the subsong of the Chaffinch. *Ibis* **97,** 247–251.

Thorpe, W. H. (1958a). The learning of song patterns by birds, with especial reference to the song of the Chaffinch *Fringilla coelebs. Ibis* **100,** 535–570.

Thorpe, W. H. (1958b). Further studies on the process of song learning in the Chaffinch, *Fringilla coelebs gengleri. Nature (London)* **182,** 554–557.

Thorpe, W. H. (1959). Talking birds and the mode of action of the vocal apparatus of birds. *Proc. Zool. Soc. London* **132,** 441–455.

Thorpe, W. H. (1961). "Bird-Song. The Biology of Vocal Communication and Expression in Birds." Cambridge Univ. Press, London and New York.

Thorpe, W. H. (1967). Vocal imitation and antiphonal song and its implications. *Proc. Int. Ornithol. Congr.* **14,** 245–263.

Thorpe, W. H. (1972). Duetting and antiphonal song in birds. Its extent and significance. *Behaviour, Suppl.* No. 18.

Thorpe, W. H., and North, M. E. W. (1966). Vocal imitation in the tropical Bou-bou Shrike *Laniarius aethiopicus major* as a means of establishing and maintaining social bonds. *Ibis* **108,** 432–435.

Tinbergen, N. (1939). The behaviour of the Snow Bunting in spring. *Trans. Linn. Soc. N.Y.* **5,** 1–95.

Todt, D. (1970). Die antiphonen Paargesänge des ostafrikanischen Grassänger *Cisticola hunteri prinoides* Neumann. *J. Ornithol.* **111,** 332–356.

Todt, D. (1975). Social learning of vocal patterns and modes of their application in Grey Parrots *(Psittacus erithacus). Z. Tierpsychol.* **39,** 178–188.

Todt, D., Hultsch, H., and Heike, D. (1979). Conditions affecting song acquisition in Nightingales *(Luscinia megarhynchos). Z. Tierpsychol.* **51,** 23–35.

Townsend, C. W. (1924). Mimicry of voice in birds. *Auk* **41,** 541–552.

Tretzel, E. (1965). Imitation und Variation von Schaferpfiffen durch Haubenlerchen *(Galerida c. cristata* L.) *Z. Tierpsychol.* **22,** 784–809.

Tretzel, E. (1966). Spottmotivprädisposition und akustische Abstraktion bei Gartengrasmucken *(Sylvia borin borin). Verh. Dtsch. Zool. Ges., Suppl.* No. 30, 333–343.

Tretzel, E. (1967). Imitation und transposition menschlicher Pfiffe durch Amseln *(Turdus m. merula). Z. Tierpsychol.* **24,** 137–161.

Verner, J. (1975). Complex song repertoire of male Long-billed Marsh Wrens in eastern Washington. *Living Bird* **14,** 263–300.

Vernon, C. J. (1973). Vocal imitation by southern African Birds. *Ostrich* **44,** 23–30.

Vincent, A. W. (1936). The birds of north-eastern Africa, etc. *Ibis* **6,** 48–123.

Wagner, H. O. (1944). Notes on the history of the Emerald Toucanet. *Wilson Bull.* **56,** 65–76.

Waite, E. R. (1903). Sympathetic song in birds. *Nature (London)* **68,** 322.

Wallace, A. R. (1853). "A Narrative of Travels on the Amazon and Rio Negro." Reeve & Co., London.

Ward, R. (1966). Regional variation in the song of the Carolina Chickadee. *Living Bird* **5,** 127–150.

Warham, J. (1962). Field notes on Australian Bower-birds and Cat-birds. *Emu* **62,** 1–30.

Waser, M. S., and Marler, P. (1977). Song learning in Canaries. *J. Comp. Physiol. Psychol.* **91,** 1–7.

West, M. J., and King, A. P. (1980). Enriching cowbird song by social deprivation. *J. Comp. Physiol. Psychol.* **94,** 263–270.

West, M. J., King, A. P., Eastzer, D. H., and Staddon, J. E. R. (1979). A bioassay of isolate cowbird song. *J. Comp. Physiol. Psychol.* **93,** 124–133.

Wildenthal, J. L. (1965). Structure in primary song of the Mockingbird *(Mimus polyglottos). Auk* **82,** 161–189.

Wiley, R. H. (1971). Song groups in a singing assembly of Little Hermits. *Condor* **73,** 28–35.

Wilkinson, R., and Howse, P. E. (1975). Variation in the temporal characteristics of the vocalizations of Bullfinches, *Pyrrhula pyrrhula. Z. Tierpsychol.* **38,** 200–211.

Williams, L., and MacRoberts, M. H. (1977). Individual variation in songs of Dark-eyed Juncos. *Condor* **79,** 106–112.

Witchell, C.A. (1896). "The Evolution of Bird-song with Observations on the Influence of Heredity and Imitation." Adam & Charles Black, London.

Witherby, H. F., Jourdain, F. C. R., Ticehurst, N. F., and Tucker, B. W. (1938–1941). "The Handbook of British Birds." H. F. & G. Witherby, London.

Ziswiler, V., Güttinger, H. R., and Bregulla, H. (1972). Monographie der Gattung *Erythrura* Swainson, 1837 (Aves, Passeres, Estrildidae). *Bonner Zool. Monogr.* No. 2.

Taxonomic Index

Roman numerals in boldface refer to volume numbers.

B

H

N

O

T

X

Subject Index

Roman numerals in boldface refer to volume numbers.

J

K

L

M

U

V

W

Contents of Volume 1